AF361962

Mobile Robots Navigation

Mobile Robots Navigation

Special Issue Editors

Oscar Reinoso
Luis Payá

MDPI • Basel • Beijing • Wuhan • Barcelona • Belgrade • Manchester • Tokyo • Cluj • Tianjin

Special Issue Editors
Oscar Reinoso
Universidad Miguel Hernández
Spain

Luis Payá
Miguel Hernandez University
Spain

Editorial Office
MDPI
St. Alban-Anlage 66
4052 Basel, Switzerland

This is a reprint of articles from the Special Issue published online in the open access journal *Applied Sciences* (ISSN 2076-3417) (available at: https://www.mdpi.com/journal/applsci/special_issues/Mobile_Robots_Navigation).

For citation purposes, cite each article independently as indicated on the article page online and as indicated below:

LastName, A.A.; LastName, B.B.; LastName, C.C. Article Title. *Journal Name* **Year**, *Article Number, Page Range.*

ISBN 978-3-03928-670-6 (Pbk)
ISBN 978-3-03928-671-3 (PDF)

Contents

About the Special Issue Editors

Oscar Reinoso is full professor at the Miguel Hernández University (Spain). He received a M.S. in Industrial Engineering from Polytechnic University of Madrid (UPM) in 1991, and a Ph.D. from the Polytechnic University of Madrid in 1996. From 1994 to 1997, he worked in the research & development department of Protos Desarrollo on a visual inspection system. Since 1997, he has been at the Miguel Hernandez University. He has been full professor since 2011 in control, robotics, and computer vision. His research interests include robotics, teleoperated robots, computer vision, parallel robots, visual inspection systems. He has authored over 200 peer-review research articles in international journals, books, and conferences. He has been associate editor of several journals.

Luis Payá holds a M.Eng. in Industrial Engineering (Spain, 2002). He obtained his Ph.D. in Industrial Technologies (Spain, 2014) for his work on omnidirectional imaging, global appearance descriptors, and topological map building and localization of mobile robots. He is currently an Associate Professor of Automatic Control, Electronics, Robotics, and Computer Vision at Miguel Hernández University, Elche, Spain. His current research interests include navigation of mobile robots in social environments, deep learning techniques applied to map building, and localization and deployment of virtual laboratories. He is the author of numerous papers, communications, and books in the cited topics. He has been a visiting researcher at the University of Bristol and at Imperial College London, United Kingdom. He has been an associate editor of the *Journal of Sensors*, he currently belongs to the editorial board of *Mathematical Problems in Engineering*, and he has edited some Special Issues for the journals *Sensors* and *Applied Sciences*.

Preface to "Mobile Robots Navigation"

The presence of mobile robots in diverse scenarios is considerably increasing to solve a variety of issues. Among them, many developments have occurred in the fields of ground, underwater, and flying robotics.

Independent of the environment in which they move, navigation is a fundamental ability of mobile robots so that they can complete autonomously high-level tasks in a specific environment. This problem can be efficiently addressed through the following actions: First, the environment in which the robot has to move must be perceived, and some relevant information must be extracted from it (mapping problem). Second, the robot must be able to estimate its position and orientation within this environment (localization problem). With this information, a trajectory toward the target points must be planned (path planning), and the vehicle has to be reactively guided along this trajectory considering either possible changes or interactions with the environment or with the user (control).

To perceive the environment, some kinds of onboard sensors can be used, such as laser rangefinders, visual systems, or RGB-D platforms. This perception task can be conducted either beforehand or once the navigation task has started while the robot moves through the environment, and the result is a model or map of the environment. The localization task must be designed considering several issues: the available sensors, the structure of the map, and the movement constraints of the robot (i.e., trajectories in 3D or 6D). Integrated exploration systems jointly consider all these issues and they achieve trajectory planning and control while a model of the environment is obtained, and the robot estimates its position and orientation. Finally, the availability of versatile tools to simulate any new development in mobile robots navigation is crucial to quickly testing and comparing navigation algorithms.

Given this information, this book presents current frameworks in these fields and, in general, approaches to any problem related to the navigation of mobile robots. The chapters can be classified into two main groups: (a) map building and localization and (b) path planning and motion control.

The first group of chapters addresses the problems of map building and/or localization. In most applications, mobile robots have to move through a priori unknown environments. Therefore, either a partial knowledge of the surroundings of the robot or building a more complete model or map of this environment that can be used is necessary to subsequently estimate the position and orientation of the robot and to plan optimal trajectories to the target points so that the robot can autonomously move and perform the task for which it has been designed.

Occupancy grids are a traditional and efficient mechanism used to represent the environment. Distance maps (DMs) are a tool widely used to encode the search space in robotics path planning and obstacle avoidance as they can provide the robot with a certain safety radius to efficiently search for collision-free paths and to avoid obstacles in motion. In this field, Chapter 1 proposes an efficient algorithm (canonical ordering dynamic brushfire, CODB) to incrementally update the cell values when cell states are changed due to changes in the environment. CODB requires fewer cell visits and computation costs than previous algorithms. The authors also propose an algorithm to compute DM-based subgoal graphs that are able to provide high-level collision-free roadmaps for the mobile robot. Finally, they verify that optimal trajectories can be obtained with these subgoal graphs and a real-time search algorithm.

Chapter 2 focuses on multilegged walking robots. Their redundant limb structure usually confers good stability and maneuverability, even in complex environments. However, their ability to perceive the surrounding environment impacts their autonomous mobile ability. One of the critical issues is accurate terrain identification. The authors propose an image infilling method for terrain classification based on a bag-of-words (BoW) approach, speeded-up robust features and binary robust invariant scalable keypoints (SURF-BRISK), and support vector machine (SVM). Broadly, the obstacle regions are infilled by surrounding terrain to improve the classification accuracy, a super-pixel image infilling method for mixed terrain is developed, and multiple labels can be given to complex terrains. Their experiments show the validity of the approach in mixed terrain and terrain with obstacles.

The exploration problem is closely related to the mapping problem. It tries to discover unknown areas of the environment to add them to the map so that the robot can more completely know the surroundings. Chapter 3 outlines a method for exploring and modelling unknown indoor environments . The authors approach the problem as a multiple-objective exploration, using the multi-objective grey wolf optimizer (MOGWO) algorithm, which employs static waypoints in the process that promote the efficient exploration of indoor environments. This exploration process works by optimizing both the search of unexplored areas and the accuracy of the map. The simulation results show the ability of the approach to build complete maps in comparison with deterministic and hybrid stochastic exploration algorithms.

In mapping and localization, hierarchical models play an important role. They are used to arrange the information of the environment into several layers with different levels of granularity, which permit solving the efficient solving of the localization problem. Chapter 4 concentrates on this topic, and proposes an algorithm to compress topological models to create the high-level layer of the map by means of a clustering approach. The authors use omnidirectional images and global appearance descriptors. The experiments show the efficiency of the algorithm for creating compact hierarchical models and to solve the problem of visual place recognition with a good balance between computation time and localization accuracy.

Continuing with the localization problem, another relevant issue in mobile robotics is the incremental estimation of egomotion from the exteroceptive sensors with which the robot is equipped, typically visual sensors. This problem is commonly known as visual odometry (VO). Chapter 5 presents a visual-inertial odometry approach based on the cubature information filter and H_∞ filter, namely MCH_∞ IF-VIO, which uses a raw intensity-based measurement model for the camera. The measurements from the inertial measurement unit (IMU) and the camera are fused by means of a hybrid information filter that applies two cubature rules in the time update and the measurement update phases to guarantee numerical stability. The H_∞ filter is used in the measurement update phase to achieve robustness against non-Gaussian noises in the camera measurements. The authors validate their proposal experimentally with a publicly available outdoors dataset, comparing its performance to other previously reported approaches.

Finally, versatile tools that enable researchers to quickly test and compare navigation algorithms in real operation conditions are key in autonomous mobile robotics. In this field, in chapter 6, a framework is developed for fast experimental testing of navigation algorithms in autonomous robotics, which is based on the robot operating system (ROS). The authors provide a basic structure arranged into a number of abstraction levels, which allows researchers to implement and test their algorithms, focusing on any sub-problem of interest such as mapping or localization. The chapter proves the validity of the framework by showing how to implement the localization module of a ground robot that uses global navigation satellite system positioning and Monte Carlo localization with a Kalman filter, and is tested with large outdoor environments.

The second group of chapters address the problems of path planning and motion control. Once a local representation or a complete map of the environment is available, the robot can focus on the planning of optimal trajectories and the motion control to conduct a specific task or series of tasks considering a set of constraints that depend basically on the tasks, the architecture of the robot, and the environment within which the robot moves.

Chapter 7 focuses on the path planning problem, and more concisely, on artificial potential field (APF) approaches. They are an efficient alternative for motion planning in mobile robotics, but they are often limited by the presence of local minima in which the robot may get trapped. For this reason, the authors propose an improved version of this method (dynamic APF, DAPF) that uses a dynamic window approach to avoid local minima regions. They address the problem of dynamic obstacles avoidance by means of a danger index that considers the relative distance between robot and obstacle as well as their relative velocity. The experimental section proves the ability of the algorithm to find optimal paths that avoid both local minima and moving obstacles.

Chapter 8 presents a two-level hierarchical framework for continuous robot navigation in dynamic environments named jump point search improved asynchronous advantage actor-critic (JPS-IA3C). The JPS+ (P) global planner, which is a variant of JPS, efficiently computes a sequence of subgoals for the motion controller, which can eliminate first move lag and avoid local minima. The low-level motion controller IA3C learns the control policies of the robots' local motion to satisfy the kinematic constraints and adapt to changing environments (moving obstacles). IA3C builds a novel reward function framework that avoids learning inefficiencies dues to sparse reward. The authors perform a set of simulation experiments that prove that this hierarchy is able to cope with incomplete and noisy information and navigate robots in unseen and large environments with shorter path lengths and low execution time.

In some applications, the collaboration between the members of a team of robots can be of interest. Chapter 9 proposes a multi-robot path planning algorithm that tries to overcome some of the shortcomings of conventional methods, such as the adaptation to complex and dynamic systems and environments. In multi-robot navigation, depending on the situation of the mission, each robot can be seen either as a moving obstacle that performs independent actions or as a cooperative robot that collaborates with other robots. To address these issues, the authors propose a framework based on the use of deep q learning combined with convolutional neural networks, using visual information from the surrounding of the robots. The simulation results prove the flexible and efficient navigation provided by the method.

Chapter 10 presents a method for path planning for unmanned surface vehicles (USVs), which considers the risk of water depth. This is a crucial factor for safe navigation in shallow waters. With this aim, the authors study the stability of USVs in a variety of situations and calculate the minimum safe water depth. To plan the path, a water depth risk level A* (WDRLA*) algorithm is proposed and its performance is compared with the traditional A* shortest path and safest path. The authors use the depth point of the electronic navigation chart (ENC) and a spline function interpolation algorithm to obtain a grid environment model considering water depth. The numerical simulations prove that the algorithm guarantees navigation safety in different conditions.

Chapter 11 focuses on space robotics that are designed to work in outer space in a variety of tasks, such as assembly and maintenance of space stations. In this kind of robot, the multitask-based trajectory-planning problem (MTTP) is of utmost importance, as it enables the robot to perform two or more tasks in each mission, what would suppose a save of energy. The authors use piecewise continuous sine functions to create the trajectories along the waypoints and transform this problem into parameter optimization, using an improved genetic algorithm to optimize the unknown parameters. Numerical simulations are conducted with a base spacecraft and a seven degrees of freedom manipulator in two simulation cases, and they prove the efficiency of the approach.

Trajectory planning is also a relevant technology for autonomous unmanned aerial vehicles (UAVs). Chapter 12 proposes a flight path planning algorithm to find collision-free, minimum length, and flyable paths for UAVs in three-dimensional urban environments with fixed obstacles for coverage missions. This problem consists of finding a low cost path that covers the free space of an area of interest with minimal overlapping. The authors address this problem based on a novel footprints' sweeps fitting method. They generate a sparse waypoint graph by connecting footprints' sweeps endpoints considering the obstacles, maneuverability constraints, and footprints' sweeps visiting sequence. The simulation results prove the performance of the algorithm in a variety of scenarios.

In the field of movement control of legged mobile robots, chapter 13 addresses the problem of energy consumption, since it can be considered a performance index of quadruped robots. In the chapter, the authors model and analyze the energy consumption of the SCalf robot with a trot gait, and they study the effect of different gait parameters, such as step length, gait cycle, step height, and duty cycle. The experiments show the optimal choice for these relevant parameters, as far as energy consumption is concerned. For this purpose, the authors build a dynamics model of the robot based on an analysis of the foot force distribution and derive a complete energy model that includes mechanical power and heat rate. They use foot trajectory based on cubic spline interpolation to describe the motions of the robot.

Finally, path planning is another important problem in the field of industrial robotics. Chapter 14 concentrates on the field of industrial manipulator robots, more concisely on machining and fabrication applications, among which deburring plays an important role. The authors develop a hybrid manipulator robot with five degrees of freedom. They propose a deburring framework focusing on tool path planning (position and orientation), robotic layered deburring planning, and a process parameter control based on fuzzy logic. A variety of experiments are performed to prove the dexterous manipulation and the orientation reachability of the manipulator and to verify the effectiveness of the two proposed methods in the deburring process.

Oscar Reinoso, Luis Payá
Special Issue Editors

 *applied sciences*

Editorial

Special Issue on Mobile Robots Navigation

Oscar Reinoso * and Luis Payá *

Department of Systems Engineering and Automation, Miguel Hernández University, 03202 Elche (Alicante), Spain
* Correspondence: o.reinoso@umh.es (O.R.); lpaya@umh.es (L.P.)

Received: 4 February 2020; Accepted: 5 February 2020; Published: 15 February 2020

1. Introduction

In recent years, the presence of mobile robots in diverse scenarios has considerably increased, to solve a variety of tasks. Among them, many developments have been carried out over the past few years in the fields of ground, underwater, and flying robotics.

Independently on the environment where they move, navigation is one of the fundamental abilities that mobile robots must be endowed with, so that they can carry out high-level tasks autonomously, in a specific environment. This problem can be addressed efficiently through the following actions. First, it is necessary to perceive the environment in which the robot has to move, and extract some relevant information from it (mapping problem). Second, the robot must be able to solve the localization problem within this environment (localization problem). With this information, a trajectory towards the target points must be planned (path planning), and the vehicle has to be guided along this trajectory, in a reactive way, considering either possible changes or interactions with the environment or with the user (control).

To perceive the environment, some kinds of onboard sensors can be used, such as laser rangefinders, visual systems, or RGB-D platforms. This perception task can be carried out either beforehand or once the navigation task has started, while the robot moves through the environment, and the result is a model or map of the environment. Regarding the localization task, it must be designed considering several issues: the available sensors, the structure of the map, and the movement constraints that the robot presents (i.e., trajectories in 3D or 6D). Furthermore, integrated exploration systems consider all these issues jointly, and they develop trajectory planning and control, while a model of the environment is obtained, and the robot estimates its position and orientation within it.

Finally, the existence of versatile tools to simulate any new development in mobile robots navigation is crucial to quickly test and compare navigation algorithms.

In light of the previous information, this special issue was introduced to present current frameworks in these fields and, in general, approaches to any problem related to the navigation of mobile robots. There were 39 papers submitted to this special issue, from which 14 papers were accepted (i.e., 36% acceptance rate), which addressed a variety of topics, as detailed in the next sections.

2. Map Building and Localization of Mobile Robots

In most applications, mobile robots have to move through a priori unknown environments. Therefore, it is often necessary either to have a partial knowledge of the surroundings of the robot or to build a more complete model or map of this environment that can be used, subsequently, to estimate the position and orientation of the robot and to plan optimal trajectories to the target points, in such a way that the robot can autonomously move and perform the task which it has been designed for.

Occupancy grids constitute a traditional and efficient mechanism to represent the environment. In this respect, Distance Maps (DM) are a widely used tool to encode the search space in robotics path planning and obstacle avoidance, as they can provide the robot with a certain safety radius to efficiently search out collision-free paths and to avoid obstacles in motion. In this field, Qin et al. [1]

propose an efficient algorithm (Canonical Ordering Dynamic Brushfire (CODB)) to incrementally update the cell values when cell states are changed (due to changes in the environment). It requires fewer cell visits and computation costs than previous algorithms. They also propose an algorithm to compute DM-based subgoal graphs which are able to provide high-level collision-free roadmaps for the mobile robot. Finally, they verify that optimal trajectories can be obtained with these subgoal graphs and a real-time search algorithm.

Zhu et al. [2] focus their study on multilegged walking robots. Their redundant limb structure usually confers them good stability and maneuverability even in complex environments. However, their ability to perceive the surrounding environment impacts upon their autonomous mobile ability. One of the critical issues is the accurate terrain identification. The authors propose an image infilling method for terrain classification based on a Bag-of-Words (BoW) approach, Speeded-Up Robust Features and Binary Robust Invariant Scalable Keypoints (SURF-BRISK) and Support Vector Machine (SVM). In broad lines, the obstacle regions are infilled by surrounding terrain to improve the classification accuracy, a super-pixel image infilling method for mixed terrain is developed and multiple labels can be given to complex terrains. Their experiments show the validity of the approach in mixed terrains and terrains with obstacles.

The exploration problem is closely related to the mapping problem. It tries to discover unknown areas of the environment to add them to the map so that the robot can have a more complete knowledge of the surroundings. Kamalova et al. [3] develop a method to explore unknown indoor environments, with the purpose of building a model of them. They approach the problem as a multiple-objective exploration, using the Multi-Objective Grey Wolf Optimizer (MOGWO) algorithm, which employs static waypoints in the process, promoting the efficient exploration of indoor environments. The philosophy of this exploration process is to optimize both the search of unexplored areas and the accuracy of the map. The simulation results show the ability of the approach to build complete maps, comparing to deterministic and hybrid stochastic exploration algorithms.

As far as mapping and localization are concerned, hierarchical models play an important role. They arrange the information of the environment into several layers with different levels of granularity, which permit solving the localization problem efficiently. Cebollada et al. [4] concentrate on this topic, and propose an algorithm to compress topological models in order to create the high-level layer of the map, by means of a clustering approach. They use omnidirectional images and global appearance descriptors. The experiments show the efficiency of the algorithm to create compact hierarchical models and to solve the problem of visual place recognition with a good balance between computation time and localization accuracy.

Continuing with the localization problem, another relevant issue in mobile robotics is the incremental estimation of egomotion from the exteroceptive sensors the robot is equipped with, typically visual sensors. This problem is commonly known as visual odometry (VO). In this regard, the work by Song et al. [5] presents a Visual-Inertial Odometry approach, based on the cubature information filter and H_∞ filter, namely, $MCH_\infty IF - VIO$, which uses a raw intensity-based measurement model for the camera. On the one hand, the measurements from the IMU (Inertial Measurement Unit), and the camera are fused by means of a hybrid information filter, which applies two cubature rules in the time-update and the measurement-update phases, to guarantee numerical stability. On the other hand, the H_∞ is used in the measurement-update phase to achieve robustness against non-Gaussian noises in the camera measurements. The authors validate their proposal experimentally, with a publicly available outdoors dataset, comparing its performance to other previous approaches.

Finally, disposing of versatile tools that enable researchers to quickly test and compare navigation algorithms in real operation conditions is key in autonomous mobile robotics. In this field, Muñoz-Bañon et al. [6] develop a framework for fast experimental testing of navigation algorithms in autonomous robotics, which is based on the Robot Operating System (ROS). They provide a basic structure arranged into a number of abstraction levels that allows researchers to implement and test their algorithms, focusing in any sub-problem of interest such as mapping or localization. The paper proves the validity of the framework by showing how to implement the localization module of a ground robot which uses global navigation satellite system positioning, and Monte Carlo localization with a Kalman filter, and is tested with large outdoor environments.

3. Path Planning and Motion Control

Once a local representation or a complete map of the environment is available, the robot can focus on the planning of optimal trajectories and the motion control to carry out a specific task or series of tasks, considering a set of constraints that will depend basically on the tasks, the architecture of the robot and the environment where the robot has to move.

Sun et al. [7] set their sights on the path planning problem, more concisely, on Artificial Potential Field (APF) approaches. They are an efficient alternative for motion planning in mobile robotics, but they are often limited by the presence of local minima in which the robot may get trapped. For this reason, they propose an improved version of this method (Dynamic APF (DAPF)), which uses a dynamic window approach to avoid local minima regions. Additionally, they address the problem of dynamic obstacles avoidance by means of a danger index which does not only consider the relative distance between robot and obstacle but also their relative velocity. The experimental section proves the ability of the algorithm to find optimal paths that avoid both local minima and moving obstacles.

Zeng et al. [8] present a two-level hierarchical framework for robot navigation in dynamic environments in a continuous way, named JPS-IA3C (Jump Point Search Improved Asynchronous Advantage Actor-Critic). On the one hand, the global planner JPS+ (P), which is a variant of JPS, efficiently computes a sequence of subgoals for the motion controller, which can eliminate first-move lag and avoid local minima. On the other hand, the low-level motion controller IA3C learns the control policies of the robots' local motion to satisfy the kinematic constraints and adapt to changing environments (moving obstacles). Additionally, IA3C builds a novel reward function framework, which avoids learning inefficiencies dues to sparse reward. The authors perform a set of simulation experiments that prove that this hierarchy is able to cope with incomplete and noisy information, and navigate robots in unseen and large environments with shorter path lengths and low execution time.

In some applications, the collaboration between the members of a team of robots can be of interest. Bae et al. [9] propose a multi-robot path planning algorithm that tries to overcome some of the shortcomings of conventional methods, such as the adaptation to complex and dynamic systems and environments. In multi-robot navigation, depending on the situation of the mission, each robot can be seen either as a moving obstacle which performs independent actions or as a cooperative robot that collaborates with other robots. To address these issues, the proposal of this paper consists in a framework based on the use of deep q learning combined with Convolutional Neural Networks, using visual information from the surrounding of the robots. The simulation results prove the flexible and efficient navigation provided by the method.

Liu et al. [10] present a method for path planning oriented to Unmanned Surface Vehicles (USV), which takes into account the risk of water depth. This is a crucial factor for the safe navigation in shallow waters. With this aim, the authors study the stability of USV's in a variety of situations and calculate the minimum safe water depth. To plan the path, a Water Depth Risk Level A* algorithm (WDRLA*) is proposed, and its performance is compared with the traditional A* shortest path and safest path. The authors use the depth point of the Electronic Navigation Chart (ENC) and a spline function interpolation algorithm to obtain a grid environment model considering water depth. The numerical simulations prove that the algorithm guarantees navigation safety in different conditions.

Zhao et al. [11] focus their work on space robotics, which are designed to work in outer space in a variety of tasks, such as assembly and maintenance of space stations. In this kind of robots, the Multitask-based Trajectory-Planning Problem (MTTP) is of utmost importance, as it would enable the robot to perform two or more tasks in each mission, what would suppose a save of energy.

The authors use piecewise continuous sine functions to create the trajectories along the waypoints and transform this problem into a parameter optimization, using an improved genetic algorithm to optimize the unknown parameters. Numerical simulations are carried out with a base spacecraft and a 7-degrees of freedom manipulator in two simulation cases, and they prove the efficiency of the approach.

Trajectory planning is also a relevant technology for autonomous Unmanned Aerial Vehicles (UAV). Majeed et al. [12] propose a flight path planning algorithm to find collision-free, minimum length and flyable paths for such vehicles in three-dimensional urban environments with fixed obstacles, for coverage missions. This problem consists in finding a low cost path that covers the free space of an area of interest with minimum overlapping. The authors address this problem based on a novel footprints' sweeps fitting method. They generate a sparse waypoint graph by connecting footprints' sweeps endpoints considering the obstacles, maneuverability constraints and footprints' sweeps visiting sequence. Simulation results prove the good performance of the algorithm in a variety of scenarios.

In the field of movement control of legged mobile robots, Yang et al. [13] aim attention at the problem of energy consumption, as it can be considered a performance index of quadruped robots. In their work, they model and analyze the energy consumption of the robot SCalf with a trot gait, and they study the effect of different gait parameters, such as step length, gait cycle, step height, and duty cycle. The experiments show which is the optimal choice of these relevant parameters, as far as energy consumption is concerned. To this purpose, the authors build the dynamics model of the robot, based on an analysis of the foot force distribution and derive a complete energy model which includes mechanical power and heat rate. Also, they use a foot trajectory based on cubic spline interpolation to describe the motions of the robot.

Finally, path planning is also a crucial problem in the field of industrial robotics. The work of Guo et al. [14] concentrates on the field of industrial manipulator robots, more concisely on machining and fabrication applications, among which deburring plays an important role. They develop a hybrid manipulator robot with five degrees of freedom. Also, they propose in the paper a deburring framework focusing on tool path planning (position and orientation) and robotic layered deburring planning; and a process parameter control based on fuzzy logic. A variety of experiments are performed to prove the dexterous manipulation and the orientation reachability of the manipulator and to verify the effectiveness of the two proposed methods in a deburring process.

Funding: This work was partially supported by the Spanish Government through the project DPI2016-78361-R (AEI/FEDER, UE) and the Generalitat Valenciana through the project AICO/2019/031.

Acknowledgments: This special issue would not have been possible without the valuable contributions of the authors, peer reviewers, and editorial team of Applied Sciences. We give our most sincere thanks to all the authors for their hard work, independently on the final decision about their submitted manuscripts. Also, all our gratitude to the peer reviewers for their help and fruitful feedback to authors. Finally, our warmest thanks to the editorial team for their untiring support and hard work during all the stages of development of this special issue and, in general, congratulations on the great success of the journal Applied Sciences.

Conflicts of Interest: The authors declare no conflicts of interest.

References

1. Qin, L.; Hu, Y.; Yin, Q.; Zeng, J. Speed Optimization for Incremental Updating of Grid-Based Distance Maps. *Appl. Sci.* **2019**, *9*, 2029. [CrossRef]
2. Zhu, Y.; Jia, C.; Ma, C.; Liu, Q. SURF-BRISK–Based Image Infilling Method for Terrain Classification of a Legged Robot. *Appl. Sci.* **2019**, *9*, 1779. [CrossRef]
3. Kamalova, A.; Navruzov, S.; Qian, D.; Lee, S. Multi-Robot Exploration Based on Multi-Objective Grey Wolf Optimizer. *Appl. Sci.* **2019**, *9*, 2931. [CrossRef]
4. Cebollada, S.; Payá, L.; Mayol, W.; Reinoso, O. Evaluation of Clustering Methods in Compression of Topological Models and Visual Place Recognition Using Global Appearance Descriptors. *Appl. Sci.* **2019**, *9*, 377. [CrossRef]

5. Song, C.; Wang, X.; Cui, N. Mixed-Degree Cubature H_∞ Information Filter-Based Visual-Inertial Odometry. *Appl. Sci.* **2019**, *9*, 56. [CrossRef]

6. Muñoz–Bañón, M.; del Pino, I.; Candelas, F.; Torres, F. Framework for Fast Experimental Testing of Autonomous Navigation Algorithms. *Appl. Sci.* **2019**, *9*, 1997. [CrossRef]

7. Sun, J.; Liu, G.; Tian, G.; Zhang, J. Smart Obstacle Avoidance Using a Danger Index for a Dynamic Environment. *Appl. Sci.* **2019**, *9*, 1589. [CrossRef]

8. Zeng, J.; Qin, L.; Hu, Y.; Yin, Q.; Hu, C. Integrating a Path Planner and an Adaptive Motion Controller for Navigation in Dynamic Environments. *Appl. Sci.* **2019**, *9*, 1384. [CrossRef]

9. Bae, H.; Kim, G.; Kim, J.; Qian, D.; Lee, S. Multi-Robot Path Planning Method Using Reinforcement Learning. *Appl. Sci.* **2019**, *9*, 3057. [CrossRef]

10. Liu, S.; Wang, C.; Zhang, A. A Method of Path Planning on Safe Depth for Unmanned Surface Vehicles Based on Hydrodynamic Analysis. *Appl. Sci.* **2019**, *9*, 3228. [CrossRef]

11. Zhao, S.; Zhu, Z.; Luo, J. Multitask-Based Trajectory Planning for Redundant Space Robotics Using Improved Genetic Algorithm. *Appl. Sci.* **2019**, *9*, 2226. [CrossRef]

12. Majeed, A.; Lee, S. A New Coverage Flight Path Planning Algorithm Based on Footprint Sweep Fitting for Unmanned Aerial Vehicle Navigation in Urban Environments. *Appl. Sci.* **2019**, *9*, 1470. [CrossRef]

13. Yang, K.; Rong, X.; Zhou, L.; Li, Y. Modeling and Analysis on Energy Consumption of Hydraulic Quadruped Robot for Optimal Trot Motion Control. *Appl. Sci.* **2019**, *9*, 1771. [CrossRef]

14. Guo, W.; Li, R.; Zhu, Y.; Yang, T.; Qin, R.; Hu, Z. A Robotic Deburring Methodology for Tool Path Planning and Process Parameter Control of a Five-Degree-of-Freedom Robot Manipulator. *Appl. Sci.* **2019**, *9*, 2033. [CrossRef]

Article

Speed Optimization for Incremental Updating of Grid-Based Distance Maps

Long Qin, Yue Hu, Quanjun Yin * and Junjie Zeng

College of Systems Engineering, National University of Defense Technology, Changsha 410073, China;
qldbx2007@sina.com (L.Q.); huyue.cse@gmail.com (Y.H.); zjjnudt@foxmail.com (J.Z.)
* Correspondence: yin_quanjun@163.com

Received: 2 April 2019; Accepted: 13 May 2019; Published: 16 May 2019

Abstract: In the context of robotics and game AI, grid-based Distance Maps (DMs) are often used to fulfill collision checks by providing each traversable cell maximal clearance to its closest obstacle. A key challenge for DMs' application is how to improve the efficiency of updating the distance values when cell states are changed (i.e., changes caused by newly inserted or removed obstacles). To this end, this paper presents a novel algorithm to speed up the construction of DMs on planar, eight-connected grids. The novelty of our algorithm, Canonical Ordering Dynamic Brushfire (CODB), lies in two aspects: firstly, it only updates those cells which are affected by the changes; secondly, it employs the strategy of Canonical Ordering from the fast path planning community to guide the direction of the update; therefore, the construction requires much fewer cell visits and less computation costs compared to previous algorithms. Furthermore, we propose algorithms to compute DM-based subgoal graphs. Such a spatial representation can be used to provide high-level, collision-free roadmaps for agents with certain safety radius to engage fast and rational path planning tasks. We present our algorithm both intuitively and through pseudocode, compare it to competing algorithms in simulated scenarios, and demonstrate its usefulness for real-time path planning tasks.

Keywords: distance map; incremental algorithms; canonical ordering; path planning; subgoal graph

1. Introduction

In the context of collision check and path planning in robotics and game AI, the Distance Map (DM) has been widely used as a consistent model to encode the search space [1–5]. In a grid-based environment with regions of blocked cells, a corresponding DM can be constructed to provide each cell a maximal clearance value, which registers the distance from itself to the nearest obstacle. Thus, a DM can help an agent (e.g., a Non-Player Character (NPC) in the video game or a robot in the real world) with a certain safety radius to efficiently search out collision-free paths and to avoid obstacles in motion. Figure 1 presents a DM constructed in an indoor environment.

In many practical applications, the underlying environments that an agent maneuvers in are often dynamic; therefore, it is necessary to reconstruct their corresponding DMs whenever changes of cell states are observed (e.g., an obstacle is inserted, removed, reshaped, or transferred). Since such changes usually occur within a relatively neighboring area around the agent, only portions of the previously constructed DM need repair. To make use of this localized feature, existing algorithms such as Dynamic Brushfire [6] and its subsequent variants [7,8] aim to speed up the reconstruction by launching a wavefront from the source of the state changes to incrementally repair the distance values, rather than reconstructing the whole DM from scratch. With such a localized mechanism, only those cells that are actually affected by the wavefront need to be handled; thus, in most cases, the computation costs can be efficiently reduced.

Figure 1. A distance map (DM) constructed from an indoor environment. The yellow regions consist of blocked cells (obstacles). Visually speaking, for each of these traversable cells in the unblocked areas, the father it is away from the nearest obstacle, the higher its brightness is.

However, for all of the previously proposed algorithms, the propagation of the wavefronts simply expands all the neighbors of a processed cell without preference, and then inserts the newly affected neighbors into a priority queue, so as to prepare for the next round propagation. Such indiscriminate expansion results in a much longer priority queue and thus becomes an efficiency bottleneck. In order to reduce the number of elements which need to be sorted by the priority queue, a searching strategy, Canonical Ordering, is introduced by us to systematically choose a single route from the equivalent propagation paths. Figure 2 shows how our algorithm propagates the wavefront form the source blocked cell (denoted as yellow tiles), visiting each affected cell only once (where the red arrows denotes the propagation directions of the wavefronts). We choose the successors of an expanded cell, s, by following two basic rules described in Nathan R. Sturtevant and Steve Rabin's paper [9]. That is, for a cell s, with the propagation direction c that was previously used to reach s, (1) If c to arrive at s is one of the four cardinal directions, the only legal direction at s for the next round propagation is c; (2) If the c to arrive at s is one of the four diagonal directions which can be decomposed into two perpendicular cardinal components c_1 and c_2, the legal directions at s for the next round propagation are c, c_1, and c_2. We name our algorithm Canonical Ordering Dynamic Brushfire (abbreviated to CODB in the rest of this paper).

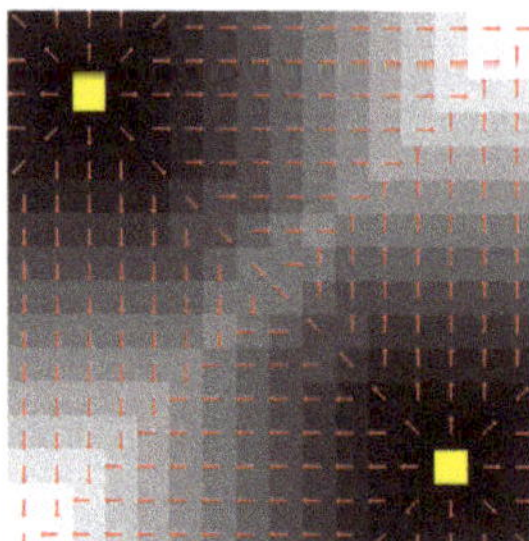

Figure 2. Two lower wavefronts started from the newly blocked cell denoted as yellow tiles.

Since a DM stores the maximal clearance for each traversable cell to its nearest obstacle, operations such as collision checks can be simplified to instant look-up queries (As shown in Equation (1)). For instance, if an agent with safety radius $R > 0$ is located on cell s, then the result of the collision check is determined by Equation (1) as below (where 1 denotes collision detected and 0 denotes collision-free):

$$C(R,s) = \begin{cases} 1 & if \ (R - DM(s)) \geq 0 \\ 0 & if \ (R - DM(s)) < 0 \end{cases} \tag{1}$$

To make better use of this feature, we furthermore propose algorithms to construct DM-based subgoal graphs. The resulting subgoal graphs are sparse but adequate high-level roadmaps to enable agents who possess safety radius to search out collision-free paths in real time. In order to reduce the possibility of replanning caused by dynamic terrain changes, we introduce Learning Real Time A* (LRTA*) [10], an algorithm for planning immediate moves at runtime, to drive the agents between subgoals connected by the high-level paths. Since each segment of the high-level paths is proved to be direct h-reachable, irrational behaviors such as trapping in local minima can be eliminated when LRTA* is iteratively applied between the direct-h-reachable subgoals (see the definition of direct h-reachable in Section 4.1.3).

There is another space representation; i.e., grid-based Voronoi diagrams can be used as a sparse model to help agents to maximize its distance to the obstacle cells. Actually, current algorithms for building grid-based Voronoi graphs can also obtain a corresponding distance map in which each cells keeps the distance to its nearest obstacle cell [11–13]. Although a Voronoi diagram can provide an agent with a sparser search space, its drawback is also prominent. For instance, in Multi-Agent Pathfinding Problems (MAPF) [14–19], a group of coordinated agents share the same Voronoi edges as their search space, thus a dense cluster of conflicts may occur, which needs to resolve during the planning process. Different from grid-based Voronoi Diagram, DM-based subgoal graphs don't conservatively compress the search space in unnecessary narrow channels. Furthermore, the search space between each pair of the direct-h-reachable subgoals commonly reserve more spaces than Voronoi edges for a group of coordinated agents to resolve conflicts.

We provide three main contributions in this paper. Firstly, we present an algorithm, Canonical Ordering Dynamic Brushfire (CODB), to speed up the incremental update of grid-based Distance Maps (DMs). Secondly, we propose algorithms to compute DM-based subgoal graphs which are used to provide high-level, collision-free roadmaps for agents with certain safety radius. Thirdly, we verify that under the guidance of the subgoal graphs, real-time search algorithms such as LRTA* can effectively avoid local minima; therefore, the resulting trajectories can successfully coincide with the optimal solutions searched by A*. We present our algorithms both intuitively and through pseudocode, compare them to current approaches on typical scenarios, and demonstrate their usefulness for fast path planning tasks.

The outline of this paper is as follow: Section 2 discusses related studies on DMs, Canonical Ordering, and subgoal graphs; Section 3 gives preliminaries and notations; Section 4 presents our algorithms both intuitively and through pseudocode; Section 5 compares CODB to other algorithms and tests the usefulness of DM-based subgoal graph for fast path planning tasks. This paper ends with conclusions in Section 6.

2. Related Work

2.1. Grid-Based Distance Maps

In the context of robotics and game AI, the grid-based DM is a popular spatial representation applied in navigation and motion planning tasks. The principal component of the recent approaches for constructing or reconstructing grid-based DMs is the well-known Brushfire algorithm [20]. Intuitively, Brushfire launches wavefronts to propagate changes of maximal clearance (i.e., changes caused by insertion or deletion of obstacle cells), updating distance values from the source of the change, and terminates when the change does not affect any more cells. Brushfire represents the OPEN list as a priority queue to incrementally record the affected cells and propagate the wavefronts. The priority of an element in the OPEN queue is determined by its newly updated distance and all these elements are popped up in increasing priorities. Sequentially, new cells which are adjacent to the popped one are tested, among which, newly updated cells will again be inserted into the OPEN list so that the propagation continues.

Kalra et al. [6], in their fundamental work, proposed a dynamic version of Brushfire algorithm, Dynamic Brushfire, to incrementally update grid-based DMs by propagating two kinds of wavefronts named "lower" and "raise" which start at newly blocked or freed cells, respectively; therefore, the update can be constrained within local areas. However, the wavefronts launched by Dynamic Brushfire roughly accumulate 8-connected grid steps to approximate maximal clearance, which overestimates the true Euclidean distances and would possibly lead to either a collision risk or overly conservative movements. To this end, Scherer et al. [21] proposed a method to propagate obstacle locations rather than counts of the grid steps, which reduces the absolute overestimation error below an upper bound of 0.09 pixel units. In the method proposed by Cuisenaire and Macq [22], the shortest distance at which this propagation error can occur is 13 pixels, which yields a maximum relative error of 0.69%. Regarding propagating obstacle references, Lau et al. proposed an approach to provide the location of the closest obstacle rather than just the distance to it, which can be appealing for collision check tasks [23]. Moreover, Lau et al. extended their method to 3D by adding the possibility to limit the propagated distances to maintain online feasibility in large open spaces and outdoors as proposed by Scherer et al.

Although these dynamic algorithms are fast and efficient for dealing with local changes, they just indiscriminately expand all the adjacent cells surrounding a currently processed cell, which results in a lot of redundant cell visits and scales up the size of the OPEN list and restricts the overall efficiency of the algorithm. We introduce the Canonical Ordering strategy in our work to prune the search space.

2.2. Canonical Ordering

The idea of applying Canonical Ordering as a speedup technique for real-time pathfinding systems that operate on regular grids was proposed by Daniel Harabor [24] and N Pochter [25]. As mentioned in the literature, searching in grids often becomes overwhelmed by a high degree of path symmetry, which accounts for a major part of the computational costs. Two paths are viewed as symmetric if (1) they have the same start and goal cells; (2) they are of the same length; and (3) their respective sequences of moves (i.e., cardinal or diagonal moves) can be reordered into the other. With symmetries in the grids, a search task will explore multiple cells for multiple times from those symmetric paths and this severely undermines the efficiency.

To break such symmetries, an online algorithm called Jump Point Search (JPS) [26] was presented by Daniel Harabor et al. to apply Canonical Ordering to recursively prune redundant successors and selectively expand only certain cells, called jump points. Canonical Ordering is essentially a special case of partial orderings among all the symmetric paths and prefers the diagonal-first ones to other alternatives. We say that a path has the diagonal-first property if there is no straight-diagonal turning point can be mutated into a diagonal-straight one of the same length constrained by the obstacles. By its virtue, JPS visits much fewer cells than traditional searching strategies; therefore, it answers a path query averagely faster than A* by an order of magnitude. After that, this algorithm's performance was further improved by a preprocessing based strategy and addition of Bounding Boxes, resulting in the algorithms JPS+ [26] and JPS+BB [27]. As an automatic move pruning technique for single-agent search [28], Canonical Ordering can not only be used in grids, but can also be built on general graphs and considerably reduce the number of cells generated by an A* search [29].

The outstanding performance of the Canonical Ordering strategy in compressing search space for real-time pathfinding algorithms provides us with a novel method to guide the direction of the wavefronts which propagate the distance changes, making it possible to speed up the construction of grid-based DMs.

2.3. Subgoal Graphs

A subgoal graph is a kind of sparse spatial representation which can be precomputed by abstracting the skeletons of the underlying grids into undirected graphs. The algorithm which accounts for computing subgoal graphs was proposed by Tansel Uras et al. [30]. It firstly introduces the basic version of the subgoal graph, called Simple Subgoal Graphs (SSGs). On grid maps, SSGs are constructed by placing subgoals at the convex corners of obstacles and connecting direct-h-reachable subgoals in the preprocessing stage. Two subgoals are mutually direct-h-reachable if all the optimal paths between them are valid and traverse no other subgoals.

SSGs reduce the search space and accelerate the process of finding shortest paths by abstracting the key points of the grid maps and reducing all symmetric paths between subgoals into only one edge. It can be proved that, for any given start and goal cells which are reachable, there is a shortest path that can be divided into segments between subgoals. Each segment connects direct-h-reachable subgoals. Therefore, when finding shortest paths on SSGs, one first connects the start and goal vertices to their direct-h-reachable subgoals in SSGs, and then searches the modified graphs via A* to get the shortest high-level path, which consists of a sequence of subgoals between the start and goal vertices. By refining the shortest high-level path, one can get the shortest path between start and goal on the grid map.

Moreover, a Two-level Subgoal Graph (TSG) is constructed from a SSG by partitioning the subgoals into global and local ones and only the global ones belong to the TSG. When removing the local subgoals, one has to add some extra edges so that the shortest paths between global subgoals remain the same. Finding the shortest path on a TSG is similar to finding a shortest path on a SSG. One first connects the start and goal vertices to their direct-h-reachable subgoals, among which the local ones should be temporarily connected to the TSG. Then, the search and refining processes are executed. It should be noted that the refining process of TSGs is slower than SSGs because TSGs connect h-reachable subgoals instead of direct-h-reachable subgoals. But, TSGs still find optimal paths faster due to the smaller search space. To reduce the search space even further, Tansel Uras et al. generalized the idea of partitioning [31,32], created a hierarchy among the vertices, and repeatedly divided the highest-level subgoals into global and local subgoals and increased the level of global subgoals by one. The resulting graphs are called N-Level Subgoal Graphs.

Subgoal Graphs can be used not only to find grid paths, but also to find any-angle paths which are shorter and more realistic [33–35]. Tansel Uras and Sven Koenig [36] exploited the similarities between Subgoal Graphs and visibility graphs and used Subgoal Graphs to quickly find any-angle paths with some small modifications. Their algorithm is up to two orders of magnitude faster than Theta*, a well-known any-angle path planning algorithm.

All the above algorithms only take regular grids as input and build subgoal graphs which do not take physical radius into considerations. Therefore, the paths planned from the resulting subgoal graphs would possibly be unable to meet the collision-free requirements in practical application. In this paper, we attempt to modify the existing algorithm so that it can make use of the maximal clearance values provided by the grid-based DMs and to build new type of subgoal graph that can efficiently search collision-free paths.

3. Preliminaries and Notation

All the algorithms studied in this paper work in planar, eight-connected grid maps with regions of obstacles consisting of blocked cells. A wavefront launched by a cell state change can propagate from one cell to its neighbor in any cardinal or diagonal directions, and the length of the cardinal and diagonal moves are 1 and $\sqrt{2}$, respectively. We follow the definition of octile distance to compute the heuristic distance between any two cells in a grid map, i.e., the distance between cells s and s' is computed by following Equation (2):

$$dist(s,s') = \sqrt{2} \times min(dx,dy) + |dx - dy| \tag{2}$$

In Equation (2), dx and dy denote the differences of the 2D coordinates of cell s and s'. For each cell s, $obst_s$ maintains coordinates of the obstacle cell s_o to which s is currently closest, and $dist_s$ maintains the distance between s and s_o. The notation dir_s maintains the direction along with which s would propagate the wavefront. The options of dir_s include four cardinal directions (denoted as left = 1, up = 3, right = 5, and down = 7), four diagonal directions (denoted as up-left = 2, up-right = 4, down-right = 6, and down-left = 8), and full directions (denoted as full-dir = 0, declaring that s is the source of the wavefront; thus, all the eight directions need to be considered in the next round propagation). The notation $raise_s$ shows if the wavefront on s is a raise wavefront or a lower one (the differences between raise and lower wavefronts are explained in Section 4.1.1). Given a cell s, a direction d, and an integer k, the notation $s' = s + kd$ denotes a cell s' that is reached from s by moving k steps along d. For two perpendicular directions c_1 and c_2, we have $d=c_1+c_2$ to denote that the sum of c_1 and c_2 results in their corresponding diagonal direction d.

4. The Methodology

4.1. Algorithm Intuition

Figure 3 shows the flowchart describing the main steps of CODB and its application in real-time pathfinding tasks.

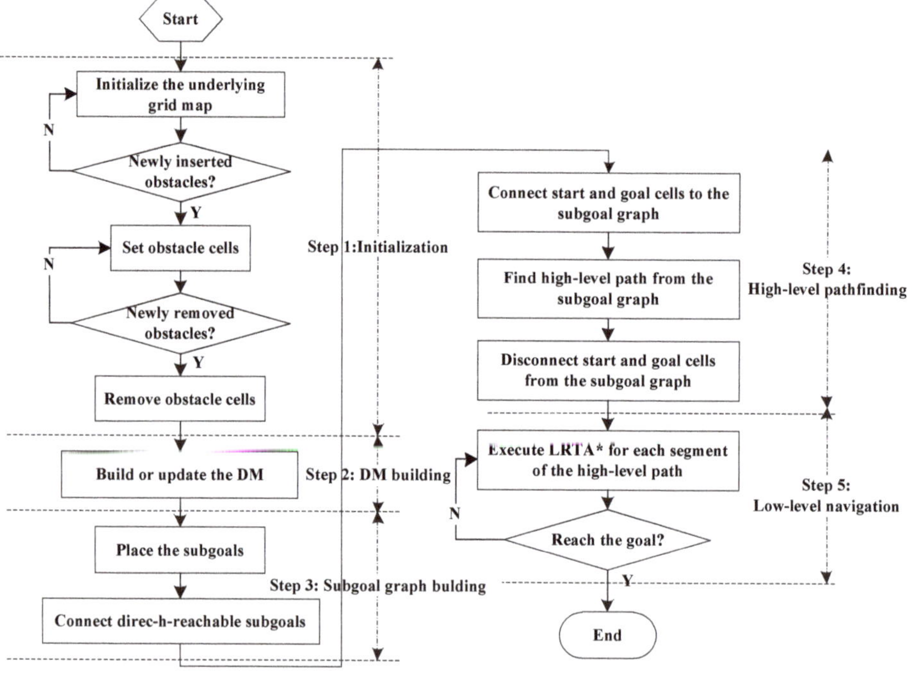

Figure 3. The flowchart describing the process of building a DM, building a DM-based subgoal graph, engaging a high-level path planning task, and executing Learning Real Time A* (LRTA*) to fulfill a low-level navigation task.

4.1.1. Lower and Raise Wavefronts

Following the basic ideas of Dynamic Brushfire algorithms, CODB employs two kinds of wavefront, lower and raise wavefronts, to incrementally update DMs. CODB keeps a priority queue (denoted as $OPEN$) to sort the cells to be explored by the wavefronts. A cell's priority is determined by its $dist_s$ value and is dequeued with increasing priority values. When a cell is popped from $OPEN$, the distance change on it will be propagated to its adjacent cells and any inconsistent cells (i.e., cells which are affected by the wavefront) are again put on $OPEN$, so as to prepare for the next round of propagation.

As shown in Figure 4a, when an obstacle cell (red dot denoted as s) is newly inserted into the center of the grids, it firstly sets $dist_s$ as 0 and $obst_s$ as its own coordinates; then, a so called "lower wavefront" is launched to propagate a distance reduction to its adjacent cells (as shown in Figure 4b). This propagation emanates from s, the source of the change, and terminates when the distance change no longer affects anymore cells (as shown in Figure 4c,d, the wavefront encounters cells which keeps equal clearance to other obstacle cells and thus failed to continue the propagation.). Finally, new boundaries of the updated DM are reached and the distance values of all affected cells are updated (As shown in Figure 4e).

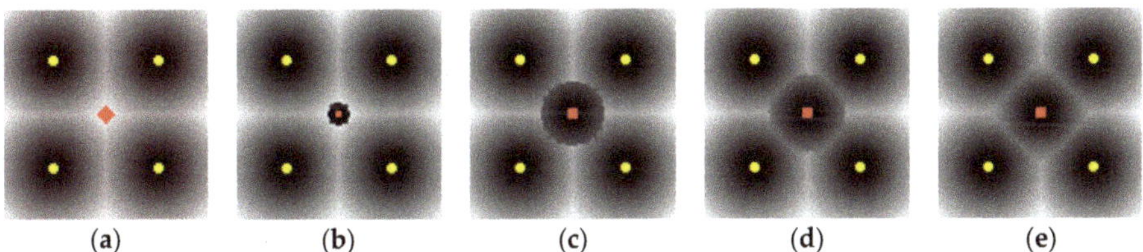

Figure 4. Lower wavefront propagation. (**a**) Insert a new obstacle cell; (**b**) Launch a lower wavefront; (**c**) Propagate the distance change; (**d**) Reach edges of affected area; (**e**) New DM is reconstructed.

On the other hand, when an obstacle cell is removed (e.g., as Figure 5a shows, we again remove the center obstacle cell s from the map), all the cells whose *obst* and *dist* value are computed based on s become invalid; therefore, a so called "raise wavefront" is launched to reset these invalid cells, declaring that they can be then updated by other lower wavefronts (as shown in Figure 5b). The "raise wavefront" terminates at the boundaries between s and other obstacle cells (as shown in Figure 5c), then lower wavefronts on the other sides of the boundaries are permitted to continue their propagation until the invalid region is again submerged (as shown in Figure 5d) and Figure 5e).

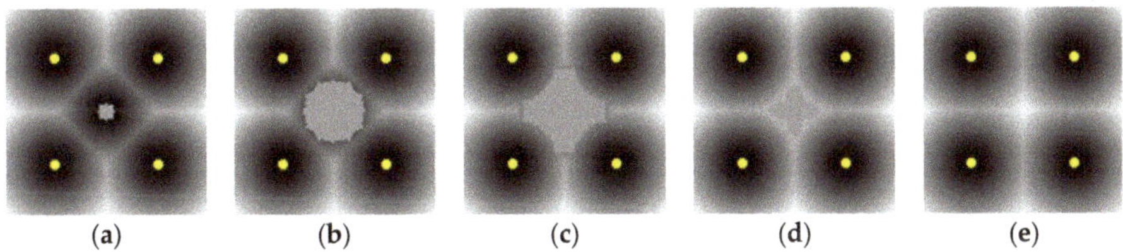

Figure 5. Raise wavefront propagation. (**a**) Remove the center obstacle; (**b**) Launch a raise wavefront; (**c**) Propagate the resetting to the invalid edges; (**d**) Continue the lower wavefront from other sides; (**e**)The invalid region is again submerged.

4.1.2. Propagation in Canonical Ordering

To speed up the construction efficiency, we introduce the concept of Canonical Ordering to guide the propagation of wavefronts. The application of Canonical Ordering has been recently discussed to speed up best-first search in grids-based pathfinding tasks in [24]. Different from the classical Dynamic Brushfire algorithm, which expands all the eight directions at every step, Canonical Ordering can eliminate redundant, symmetric paths from the source of state change, which accounts for the main part of the speed improvement.

As shown in Figure 6a, in an open grid map without obstacles, a wavefront launched by a cell (denoted as s) on the left top can follow three different optimal paths to propagate its distance change to the traversable cell (denoted as g) on the right bottom. The only difference between these paths relies on the permutation of actions of moving right one cell and moving down-right two cells. It is clear to see that the number of possible paths grows exponentially if we enlarge the grid map and keep g in the right bottom corner. Since the existing Dynamic Brushfire algorithm generates the same cell along different paths, such propagation would trigger large numbers of pushing and sort ascending operations in *OPEN* list, which can be computationally expensive.

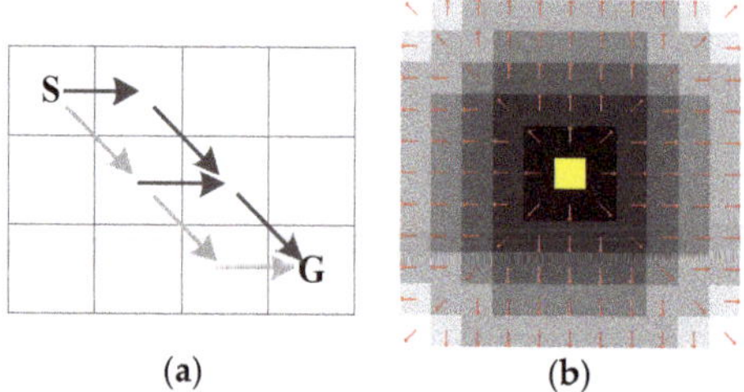

(a)   (b)

Figure 6. Canonical Ordering in wavefront propagation. (**a**) The possible paths to propagate a wavefront from cell S on left-top to cell G on bottom-right. Among these paths, only the gray one is tested when following the preference rule of Canonical Ordering; (**b**) By following the preference rule of Canonical Ordering, the expansion of a wavefront denoted as red arrows is launched from the center yellow cell and visits each affected cell only once.

To eliminate these redundancies, Canonical Ordering insists on a preference rule, i.e., distance changes first propagate diagonally before propagating cardinally whenever possible along a path (e.g., as shown in Figure 6a, by following this rule, only the gray path is tested.). With such a preference, a source cell in an open grid map returns a unique path from itself to each affected cell. As shown in Figure 6b a wavefront launched from the center cell (denoted as a yellow cell) draws lines diagonally and then branches vertically and horizontally, extending from the diagonals. Since every affected cell owns exactly one path from the source cell, the propagation is no longer a graph, but a tree.

4.1.3. DM-Based Subgoal Graph

In order to introduce the concept of maximal clearance provided by DMs into the precomputing algorithms of subgoal graphs, we extend the formal definitions of the traditional subgoal graph as follows:

Definition 1. *For two cells s and s', let dx and dy be the respective distances between s and s' along the x and y axes. The shortest trajectory between s and s' is a permutation of exactly min(dx,dy) diagonal and |dx-dx| cardinal moves, for a total of max(dx,dy) moves.*

Definition 2. *Given a safety radius R, a collision-free path between two cells s and s' is the shortest trajectory in which each cell ŝ is collision-free, i.e., C(R,ŝ)=0.*

Definition 3. *Given a safety radius R, an unblocked cell s is a collision-free subgoal if there are two perpendicular cardinal directions c_1 and c_2 such that C(R,s+c_1+c_2)=1, C(R,s+c_1)=0 and C(R,s+c_2)=0.*

Definition 4. *Given a safety radius R, two cells s and s' are h-reachable if there is a collision-free path of octile length h(s,s') between them. Two h-reachable cells are safe-h-reachable if all shortest trajectories between them are also paths. Two safe-h-reachable cell s and s' are direct-h-reachable if none of the shortest paths between them contains a subgoal s″ ∉ {s,s'}.*

Definition 5. *Given a safety radius R, an DM-based subgoal graph, Gs =(R,Vs,Es), on its corresponded grids, is a high-level, undirected graph, where Vs is the set of collision-free subgoals and Es is the set of edges connecting direct-h-reachable subgoals, and the length of the edges is the octile distances between the subgoals they connect.*

The process of constructing an DM-based subgoal graph can be intuitively divided into two steps: (1) placing collision-free subgoals at the corners of the expanded obstacle boundary to circumnavigate the collision regions (e.g., the orange circles shown in Figure 7a); (2) adding edges between those subgoals which are mutually direct-h-reachable (e.g., the green edges shown in Figure 7b). Given a start and a goal cell (e.g., the blue and red discs shown in Figure 7b), we first connect them to their respective direct-h-reachable subgoals and then plan a high-level shortest path by executing A* on the updated subgoal graph (e.g., the white trajectory shown in Figure 7b). We can refine each segment of the high-level path by arbitrarily selecting an h-reachable path between two connecting subgoals, and then orderly appending these refined paths together. Moreover, as shown in Figure 7c,d, for agents with different safety radius, their corresponding DM-based subgoal graph, according to Definition 5, can set aside adequate clearance; therefore, the paths found from the resulting graphs can be collision-free.

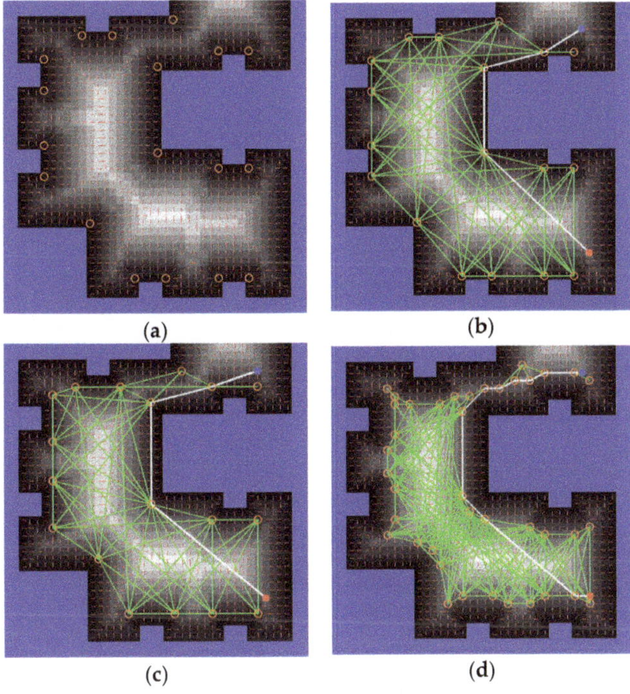

Figure 7. Steps of constructing DM-based subgoal graphs with different safety radius. (**a**) Step1: placing subgoals (R=1); (**b**) Step2: connecting direct-h-reachable subgoals (R=1); (**c**) a subgoal graph with R=1.5; (**d**) a subgoal graph with R=3.

4.2. Algorithm Pseudocode

4.2.1. Initiate the DM

Table 1 presents pseudocode for the initialization of a DM, including three functions, i.e., (1) **initialize()** to set aside certain space for required data structures and then initial their values; (2) **setObstacleCell(o)** to register a newly inserted obstacle cell; and (3) **removeObstacleCell(o)** to reset an removed obstacle cell. Upon initialization, all the cells are set to be traversable and undetermined, declaring that there is no obstacle cell, neither in the map, nor in finite distance (lines 1 to 7). When a cell o is marked as an obstacle cell by calling **setObstacleCell(o)**, it sets $dist_o$ as 0 and refers to itself as the closest obstacle cell, i.e., $obst_o = o$ (lines 8 and 9). Conversely, when o is freed by calling **removeObstacleCell(o)**, the function **resetCell(o)** resets it to the initial values, i.e., $dist_o = \infty$ and

$obst_o = \emptyset$ (line 11), and $raise_o$ is set as true (line 12). The function **insert** (*OPEN*, *c*, *d*) inserts a cell c into *OPEN* with a priority value d, or updates the priority if c is already in *OPEN*.

Table 1. The pseudocode for initialization.

initialize()		setObstacleCell (Cell *o*)		removeObstacleCell (Cell *o*)	
1.	*OPEN* ← ∅				
2.	**for each cell *s* in the grid map do**	8.	$dist_o \leftarrow 0$	11.	resetCell (*o*)
3.	$dist_s \leftarrow \infty$	9.	$obst_o \leftarrow o$	12.	$raise_o \leftarrow true$
4.	$obst_s \leftarrow \emptyset$	10.	**insert (*OPEN*,*o*,0)**	13.	**insert (*OPEN*, *o*, 0)**
5.	$voro_s \leftarrow false$				
6.	$raise_s \leftarrow false$				
7.	$dir_s \leftarrow 0$				

4.2.2. Update the DM

Table 2 presents pseudocode for updating the DM. The function **update()** orderly pops the next unprocessed cell s with the lowest $dist_s$ until the *OPEN* queue is empty (lines 14 and 15). If s is cleared and is not yet propagated by a raise wavefront, the function **raise()** is called (lines 16 and 17). However, if s has a valid closest obstacle cell, the function **lower()** is called (lines 18 to 20).

Table 2. The pseudocode for updating the Euclidean distance map.

update()		lower (cell s)		raise (cell o)	
14.	**while** *OPEN* ≠ ∅ **do**	21.	$N \leftarrow$ findSuccessors(s)	30.	$N \leftarrow$ findSuccessors(s)
15	$s \leftarrow$ pop(*OPEN*)	22.	**for each** $n \in N$ **do**	31.	**for each** $n \in N$ **do**
16.	**if** $raise_s$ **then**	23.	**if** $\neg raise_n$ **then**	32.	**if** $(obst_n \neq \emptyset \wedge \neg raise_n)$ **then**
17.	raise(s)	24.	$d \leftarrow$ distance $(obst_s, n)$	33.	**if** $\neg$ isOcc($obst_n$) **then**
18.	**else if** isOcc($obst_s$) **then**	25.	**if** $d < dist_n$; **then**	34.	clearCell(n)
19.	$voro_s \leftarrow false$	26.	$dist_n \leftarrow d$	35.	$raise_n \leftarrow true$
20.	lower(s)	27.	$obst_n \leftarrow obst_s$	36.	$dir_n \leftarrow$ direction(s,n)
		28.	$dir_n \leftarrow$ direction(s,n)	37.	**else** $dir_n \leftarrow full\text{-}dir$
		29.	insert(*OPEN*,n,d)	38.	insert (*OPEN*,n,$dist_n$)
				39.	$raise_s \leftarrow false$

findSuccessors(cell s)
40.   **if** dir_s is one of the four cardinal directions **then**
41.   **return** $N \leftarrow \{s + dir_s\}$
42.   **else**
43.   **return** $N \leftarrow \{s + dir_s, s + c_1, s+c_2 \mid$ where $dir_s = c_1+c_2\}$

Newly inserted obstacle cells call function **lower()** to launch a lower wavefront to propagate the reduction of *dist* and *obst* values from the currently popped cell s to its affected cells (lines 26 to 29). The lower wavefront continues when the distance value (denoted as d which is computed in line 24) between $obst_s$ and the newly expanded cell n holds the trend of distance reduction (being determined in line 25). Simultaneously, newly freed cells call function **raise()** to launch a raise wavefront, resetting the cells whose closest obstacle cell was the freed one (line 33 to 36). The raise wavefront terminates when it reaches those cells whose closest obstacle is valid; thus, a lower wavefront launched by n is generated (line 37). During the interwoven of these two wavefronts, inconsistent neighbors affected by the processed cell are again put on *OPEN* (line 29 and 38), thus the propagation continues.

Instead of propagating a wavefront along with full directions, we introduce a function **findSuccessors(s)** to filter cells by employing the rule of canonical ordering (line 21 and 30). As illustrated in Figure 6b, only the successors of a cell in the direction of the arrows are chosen as the candidates of the next round of propagation, while the others are ignored. For a cell n conducting a wavefront from one of its adjacent cell s, function **direction(s,n)** determines the direction from s to n and accordingly returns the value of dir_n (line 28 and 36).

4.2.3. Construct DM-Based Subgoal Graphs

By making use of the collision test equation shown in Equation (1), we modify the subgoal placement condition of the original construction algorithm; therefore. a collision-free subgoal graph for a certain safety radius R can be efficiently computed from the underlying DM.

Given an agent with safety radius $R \geq 0$, Table 3 shows how to construct a subgoal graph, $Gs := (Vs,Es)$, from the underlying DM. The entire construction process consists of two sequential phases, i.e., firstly placing collision-free subgoals on the corners of the expanded obstacle boundaries (line 45 to line 49), and secondly, adding edges to connect those subgoals which are mutually direct-h-reachable (line 50 to 53).

Table 3. The pseudocode for constructing DM-based subgoal graphs.

ConstrucDMBasedSubGraph(safetyRadius R)
44. $Vs \leftarrow \varnothing$, $Es \leftarrow \varnothing$
45.   **for each** *unblocked cells s* **do**
46.   **for each** pair of perpendicular cardinal directions c_1 and c_2 **do**
47.   **if** $C(R,s+c_1+c_2)=1$ **then**
48.   **if** $C(R,s+c_1)=0 \wedge C(R,s+c_2)=0$ **then**
49. $Vs \leftarrow Vs \cup \{s\}$
50.   **for each** $s \in Vs$ **do**
51. $S \leftarrow$ GetDirectHReachable(Vs, R,s)
52.   **for** *each $s' \in S$* **do**
53. $Es \leftarrow Es \cup \{(s,s')\}$
54. $Gs \leftarrow (Vs,Es)$

A time-consuming and important part in the algorithm shown in Table 3 is how to identify all direct-h-reachable subgoals from a given subgoal s (line 51), and this can be done by determining the direct-h-reachable area around s. As shown in Table 4 the algorithm proposed by Tansel Uras et al. [30] works in two steps. The first step (line 56 to 58) identifies the closest subgoals in each of the eight cardinal and diagonal directions, and the second step (line 59 to 70) incrementally finds out the other subgoals that can be reached via moves in two directions (i.e., either a diagonal direction or one of its two corresponded cardinal directions). In order to meet the strong requirements of Definition 5, we replace the Clearance(R, V, s, d) function in the original algorithm with SafeClearance(R, V, s, d), thus the resulting direct-h-reachable paths between two subgoals can be collision-free when taking safety radius R into consideration (e.g., the tests in line 73 ensure that every step of the incremental exploration is collision-free). For more detail about the function GetDirectHReachable, please refer to Tansel Uras' work [30].

Table 4. The pseudocode of determining the direct-h-reachable area of a given unblocked cell s.

GetDirectHReachable(subgoals *V*, safetyRadius *R*, cell *s*	SafeClearance(safetyRadius *R*, subgoals *V*, cell *s*, direction *d*)
55.　　$S \leftarrow \varnothing$	71.　　$i \leftarrow 0$
56.　　**for each** *directions d* **do**	72.　　**while** *true* **do**
57.　　　$s' \leftarrow (s + \text{SafeClearance}(R,V,s,d) \times d)$	73.　　　**if** $\neg(C(R,s+id) = 0 \wedge C(R,s+id+d)=0)$ **then**
58.　　　**if** $s' \in V$ **then** $S \leftarrow S \cup \{s'\}$	74.　　　　*return i*
59.　　**for each** *diagonal directions d* **do**	75.　　　$i \leftarrow i+1$
60.　　　**for each** *cardinal directions c associated with d* **do**	76.　　　**if** $(s+id) \in V$ **then** *return i*
61.　　　　$max \leftarrow \text{SafeClearance}(R,V,s,c)$	
62.　　　　$diag \leftarrow \text{SafeClearance}(R,V,s,d)$	
63.　　　　**if** $((s+max) \times c) \in V$ **then** $max \leftarrow max\text{-}1$	
64.　　　　**if** $((s+diag) \times d) \in V$ **then** $diag \leftarrow diag\text{ -}1$	
65.　　　　**for each** *i* from 1 *to diag* **do**	
66.　　　　　$j \leftarrow \text{SafeClearance}(R,V,s+id,c)$	
67.　　　　　**if** $j \leq max$ and $(s+id+c) \in V$ **then**	
68.　　　　　　$S \leftarrow S \cup \{s+id+c\}$	
69.　　　　　　$j \leftarrow j\text{-}1$	
70.　　　　　**if** $j < max$ **then** $max \leftarrow j$	

4.2.4. Find Paths in DM-Based Subgoal Graphs

The algorithms shown in Table 5 are proposed by Tansel Uras et al. [30], illustrating how to search for a high-level path between two cells in the subgoal graphs. The function **FindHighLevelPath** is called to connect *s* and *s'* to the underlying subgoal graph, engage an A* search to find a shortest collision-free path between *s* and *s'* from the updated subgoal graph (line 79), and to restore the original graph (line 80) before returning the resulting path (line 81). The function **ConnectToGraph** is called to identify all direct-h-reachable subgoals of *s* and *s'* (line 77 and line 78), and then add new edges between them into the graph (line 85)).

Table 5. The pseudocode for searching paths in DM-based subgoal graphs.

FindHighLevelPath(subgoals *V*, edges *E*, safetyRadius *R*, startCell *s*, *endCell s'*)
77.　　ConnectToGraph(*V,E,R,s*)
78.　　ConnectToGraph(*V,E,R,s'*)
79.　　$\Pi \leftarrow \text{A*}(V,E,s,s')$
80.　　*restore original graph*
81　　*return* Π

ConnectToGraph(subgoals *V*, edges *E*, safetyRadius *R*, cells *s*)
82.　　**if** $s \in V$ **then** *return*
83.　　　$Vs \leftarrow Vs \cup \{s\}$
84.　　　$S \leftarrow \text{GetDirectHReachable}(V,R,s)$
85.　　　**for each** $s' \in S$ **do** $Es \leftarrow Es \cup \{(s,s')\}$

According to Tansel Uras' work, the real-time requirement is easy to meet because each segment of the resulting high-level path can be quickly refined by arbitrarily choosing one of the symmetric paths between its two direct h-reachable subgoals. However, this global planning strategy often fails in some dynamic environments in which some other moving agents may block the preplanned routes. To solve this problem, we introduce Learning Real-time A* (LRTA*), a real-time heuristic search algorithm [10], to take local changes into consideration and repeatedly plan and execute actions within a constant time interval during the runtime.

We illustrate the key idea of LRTA* via pseudocode shown in Table 6. As long as the goal cell *s'* is not reached (line 87), the agent will follow the plan (line 88 and 89), learn (line 90), and execute the (91) cycle. The planning phase expands all the traversable, reachable cells within the range of a fixed look ahead range *R* and choose one cell *n* with the lowest $g(s,n)+h(n,s')$ as the immediate goal for the next move (line 89). During the learning part, the numeric value $h(s\hat{\ },s')$ is updated to approach the real total cost, denoted by $h^*(s\hat{},s')$ (line 90). Finally, the agent moves by changing its current position towards the most promising cell discovered in the planning phase (line 91).

Table 6. The pseudocode for learning real time A* search.

LRTA*(start s, end s', safety radius R)
86. $s\hat{} \leftarrow s$
87.   *while* $s\hat{}{\neq}s'$ *do*
88. $S \leftarrow frontier(s\hat{},R)$
89. $s'' \leftarrow argmin_{n \in S} \ (g(s,n)+h(n,s'))$
90. $h(s\hat{},s') \leftarrow g(s,s'')+h(s'',s')$
91. $s\hat{} \leftarrow s''$

A drawback that prevents the application of LRTA* is that agents who adopt this strategy could possibly be trapped in local minima [37]. As shown in Figure 8a, local minima often exist around certain terrain patterns such as concave regions, long distance barriers, and so on. Agents who use LRTA* to search paths in these patterns would meaninglessly move back and forth, visit the same cells several times to correct their heuristic values before escaping from these regions (As shown in Figure 8c). However, the high-level path planned from the DM-based subgoal graph can provide an agent with waypoints which circumnavigate the collision regions (As shown in Figure 8b). Therefore, it can efficiently evade local minima by iteratively popping subgoals from the high-level path as the next waypoint to head for (as shown in Figure 8d).

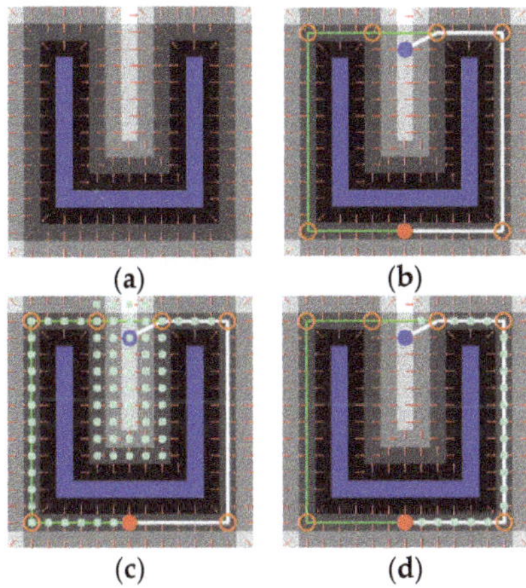

(a)   (b)

(c)   (d)

Figure 8. Examples of using LRTA* to search paths, with or without guidance of a high-level path provided by DM-based subgoal graph. (**a**) A DM with safety radius R=1.5 constructed on a concave blocked region; (**b**) The high-level path connecting the start and the goal cells can be searched from the DM-based subgoal graph; (**c**) The trajectory from the start cell to the goal cell searched by LRTA*, without guidance of the subgoal graph; (**d**) Making use of the subgoal graph, LRTA* visits much fewer cells when heading for the goal cell, and can efficiently deal with dynamical terrain changes which do not destroy connectivity.

5. Experiments and Results

In this section, we employ statistical methods to compare our algorithm with other competing methods on certain simulated scenarios. We also demonstrate the usefulness of the DM-based subgoal graphs to real-time path planning tasks. Our algorithms are implemented in C++, running on an Intel®Core(TM) i7-4790 CPU @ 3.60GHz.

5.1. Comparison to Other Algorithms

We compare CODB with other three competing algorithms (i.e., Brushfire, Dynamic Brushfire, and algorithms proposed by Boris Lau et al. [23] (we abbreviate it as BL below)) which are discussed in Section 2 in four typical scenarios as shown in Figure 9.

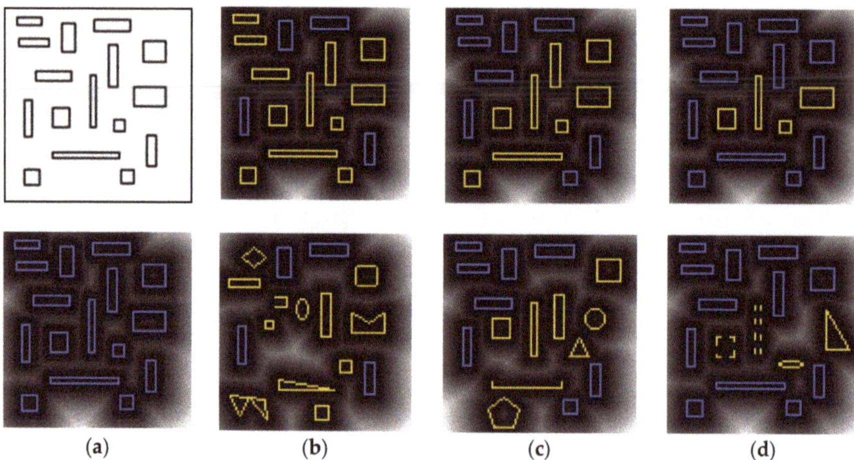

(a)   (b)   (c)   (d)

Figure 9. The grid maps used to compare performance of different algorithms. (**a**) Full reconstruction; (**b**) 75% changed; (**c**) 50% changed; (**d**) 25% changed. Within these figures, the obstacles marked in black and blue are static, while the obstacles marked in yellow randomly change their positions or shapes.

We set aside four grid maps (from Figure 9a to Figure 9d) which have the same size (100×100) while allowing different proportions of the obstacles (i.e., fully reconstruction, 75%, 50%, and 25% respectively) to randomly change their positions or shapes. After each set of changes, we use the above mentioned four algorithms to construct or reconstruct the DMs. We run each algorithm on each scenario 100 times, and the comparisons of the performances among these algorithms are shown in Table 7 and Figure 10 (i.e., the average computation time scaled in milliseconds and its variance) and Table 8 and Figure 11 (i.e., the maximal size of the priority queue $OPEN$ during construction and its variance).

Table 7. Comparison of the algorithms' efficiencies in the four scenarios (milliseconds).

Algorithm	Full Reconstruction		75% Dynamic		50% Dynamic		25% Dynamic	
	μ	σ	μ	σ	μ	σ	μ	σ
Brushfire	6.34	0.2533	6.53	0.1824	6.48	0.2316	6.32	0.2339
Dynamic Brushfire *	8.61	0.1753	6.19	0.2018	3.48	0.1899	1.39	0.2037
BL *	7.22	0.2451	5.24	0.2342	2.93	0.2171	1.12	0.1746
CODB *	4.26	0.2124	3.25	0.1937	1.67	0.2248	0.73	0.2102

* Dynamic algorithms that only repair the affected portions.

Table 8. Comparison of the maximal size of the priority queue $OPEN$ (Only dynamic algorithms are compared).

Algorithm	Full Reconstruction		75% Dynamic		50% Dynamic		25% Dynamic	
	μ	σ	μ	σ	μ	σ	μ	σ
Dynamic Brushfire	27,550	149	24,899	110	16,765	124	13,184	96
BL	12,523	78	10,268	106	8024	95	5327	77
CODB	9048	90	7611	121	6027	102	4116	88

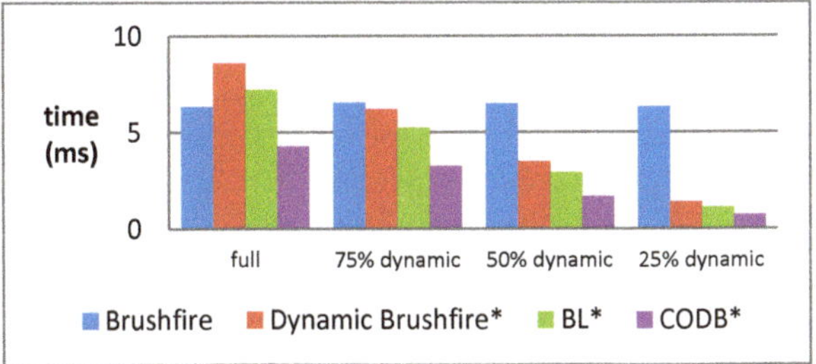

Figure 10. The average computation time provided by the four algorithms for updating DMs. The algorithms prefixed with "*" are dynamic algorithms and only update the affected areas.

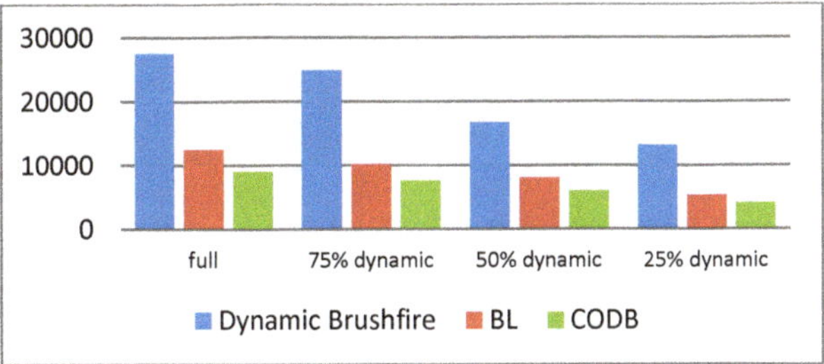

Figure 11. The average cell visits provided by the three dynamic algorithms.

For the first scenario in which full reconstructions are inevitable since all the obstacles are moved or reshaped, the extra operations that enable local repairs make Dynamic Brushfire (8.61 ms) and BL (7.22 ms) slower than their static counterpart, Brushfire (6.34 ms). However, due to the integration of Canonical Ordering strategy, CODB can substantially prune the search space; thus, it still maintains an obvious advantage over the other three algorithms in time efficiency (4.26 ms), even in the case of full reconstructions. As for the other three scenarios, as the proportion of the dynamic obstacles decreases, dynamic algorithms gradually manifest their superiority in speed. Among the three dynamic algorithms, CODB is also faster than the others due to fewer cell visits, reducing the computation time by at least 50%.

5.2. Application Tests on Real-Time Pathfinding

This work is motivated by the need to engage real-time and collision-free pathfinding tasks for agents with a certain safety radius. Such agents often maneuver in dynamic, crowded environments, and have to decide their next move in a limited amount of time. In order to demonstrate the usefulness of our algorithm in such scenarios, two agents (denoted as agent X and agent Y) working in a grid map of size 200×200 (as shown in Figure 12) are simulated. For each round of testing, both of the two agents are assigned the same start and goal cells located on the top right and left bottom of the grid map, respectively. The resulting paths of Agent X and Y are respectively shown in Figure 12a,b as light green trajectories; by contrast, the optimal path searched by the classical path planning algorithm, A*, is marked as red trajectories. Agent X simply adopts LRTA* to directly search and move in the underlying DM, while agent Y shown in Figure 12b would firstly search out a high-level path from the DM-based subgoal graph, and then engages LRTA* to search and move between the segments of the path. In order to gain the average performance, we run the test of each of the ten pairs of start and goal cells 100 times.

For agent X, applying LRTA* to engage a real-time search can help the agent successfully avoid collisions with obstacles, but it is very easy for the underlying DM to generate local minima around typical locations such as concaves, narrow channels, long distance barriers, and so on. To the cells within these locations, the occurrence of the local minima results in an increase in errors between the default heuristic values and the actual values. Therefore, LRTA* has to recheck these cells many times to incrementally correct their heuristic values (as shown in line 90, Table 6). Only in this way can the heuristic depressions be gradually filled up and finally drive the agent to escape from the local minima. Unfortunately, such runtime correction incurs meaningless movements (As the regions labeled as A, B, and C in Figure 12a) which are unacceptable to the requirements of engaging reasonable behavior.

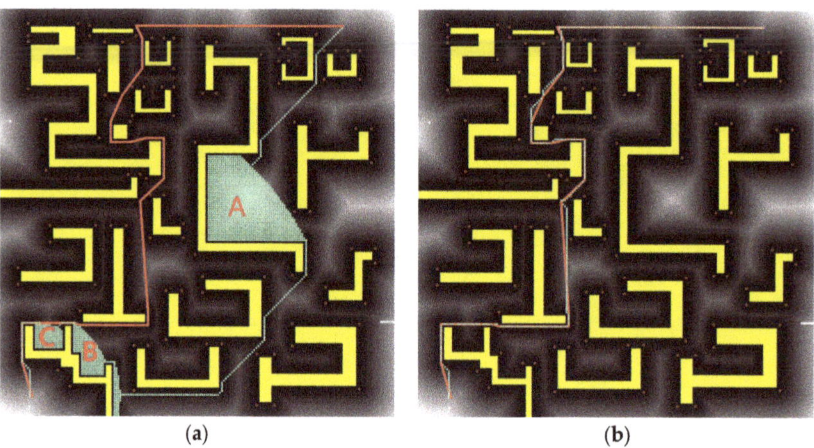

(a)   (b)

Figure 12. Performance comparison between agent X and Y. (**a**) The resulting path of agent X who simply uses LRTA* to engage a global search in a DM. As the light green trajectory shows, repeated heuristic corrections and meaningless movements caused by local minima (such as the regions denoted by identifiers A, B, and C) seriously reduce the rationality of the resulting trajectory when it is compared with the high level path searched by A* form the DM-based subgoal graph (denoted as the red trajectory); (**b**) The resulting path of agent Y who adopts LRTA* to refine the high-level path searched from the DM-based subgoal graph. Due to the sparse and h-direct reachable feature, Agent Y prefers to expand cells between consecutive subgoals and thus can efficiently avoid local minima. Therefore, the resulting path (shown as light green trajectory in (**b**)) is, to a large extent, close to the optimal path denoted as the red trajectory

As for agent Y, a high-level path (as the red path shown in both Figure 12a,b, is firstly searched out from the DM-based subgoal graph by executing the algorithms proposed in Table 5 (in order to make the figures clear, we only retain the subgoals and do not draw the direct-h-reachable edges). The resulting path provides a list of sparse and direct-h-reachable waypoints to circumnavigate the collision regions. Based on the path, agent B uses LRTA* at runtime to search and move along the path; therefore, it can efficiently avoid local minima and terminate meaningless movements.

6. Conclusions

In this paper, we present an algorithm, Canonical Ordering Dynamic Brushfire (CODB), to speed up the incremental updating of grid-based Distance Maps (DMs). CODB only updates those cells which are affected by local changes, and it employs the strategy of Canonical Ordering to guide the search direction; therefore, the algorithm requires much fewer cell visits and less computation costs compared to its competing approaches. Furthermore, we propose algorithms to compute DM-based subgoal graphs which are used to provide high-level, collision-free roadmaps for agents with certain safety radius to engage fast and rational path planning tasks. We present our algorithms both intuitively and through pseudocode, compare them to current approaches on typical scenarios, and demonstrate their usefulness for fast path planning tasks.

Appl. Sci. **2019**, *9*, 2029

Author Contributions: Conceptualization, L.Q., Y.H. and J.Z.; Data curation, L.Q.; Formal analysis, L.Q.; Funding acquisition, Q.Y.; Methodology, L.Q., Y.H. and J.Z.; Project administration, Q.Y.; Resources, Q.Y.; Software, L.Q.; Supervision, Q.Y.; Writing—original draft, L.Q.; Writing—review & editing, L.Q., Y.H. and J.Z.

Funding: This work described in this paper is sponsored by the National Natural Science Foundation of China under Grant No. 61473300.

Acknowledgments: This work described in this paper is sponsored by the National Natural Science Foundation of China under Grant No. 61473300. We appreciate the fruitful discussion with the Sim812 group: Qi Zhang, Kai Xu, Weilong Yang, and Cong Hu.

Conflicts of Interest: The authors declare no conflict of interest. The founding sponsors had no role in the design of the study; in the collection, analysis, or interpretation of data; in the writing of the manuscript, and in the decision to publish the results.

References

1. Chen, L.; Chuang, H.Y. A Fast Algorithm for Euclidean Distance Maps of a 2-D Binary Image. *Inf. Process. Lett.* **1994**, *51*, 25–29. [CrossRef]
2. Yazici, A.; Kirlik, G.; Parlaktuna, O.; Sipahioglu, A. A Dynamic Path Planning Approach for Multirobot Sensor-Based Coverage Considering Energy Constraints. *IEEE Trans. Cybern.* **2017**, *44*, 305–314. [CrossRef] [PubMed]
3. Borgefors, G. Distance Transformations in Digital Images. *Comput. Vis. Gr. Image Process.* **1988**, *34*, 344–371. [CrossRef]
4. Fabbri, R.; Costa, L.D.F.; Torelli, J.C.; Bruno, O.M. 2D Euclidean Distance Transform Algorithms: A Comparative Survey. *ACM Comput. Surv.* **2008**, *40*, 1–44. [CrossRef]
5. Charrow, B.; Kahn, G.; Patil, S.; Liu, S.; Goldberg, K.; Abbeel, P.; Michael, N.; Kumar, V. Information-Theoretic Planning with Trajectory Optimization for Dense 3D Mapping. *Robot. Sci. Syst.* **2015**, *11*.
6. Kalra, N.; Ferguson, D.; Stentz, A. Incremental Reconstruction of Generalized Voronoi Diagrams on Grids. *Robot. Auton. Syst.* **2009**, *57*, 123–128. [CrossRef]
7. Lau, B.; Sprunk, C.; Burgard, W. Efficient Grid-Based Spatial Representations for Robot Navigation in Dynamic Environments. *Robot. Auton. Syst.* **2013**, *61*, 1116–1130. [CrossRef]
8. Qin, L.; Yin, Q.; Zha, Y.; Peng, Y. Dynamic Detection of Topological Information from Grid-Based Generalized Voronoi Diagrams. *Math. Probl. Eng.* **2013**, *2013*, 1–11. [CrossRef]
9. Sturtevant, N.R.; Rabin, S. Canonical Orderings on Grids. In Proceedings of the Twenty-Fifth International Joint Conference on Artificial Intelligence, IJCAI 2016, New York, NY, USA, 9–13 July 2016; pp. 683–689.
10. Bulitko, V.; Lee, G. Learning in Real-Time Search: A Unifying Framework. *J. Artif. Intell. Res.* **2006**, *25*, 119–157. [CrossRef]
11. Lucet, Y. New Sequential Exact Euclidean Distance Transform Algorithms Based on Convex Analysis. *Image Vis. Comput.* **2009**, *27*, 37–44. [CrossRef]
12. Schouten, T.E.; Den Broek, E.L. Fast Exact Euclidean Distance (FEED): A New Class of Adaptable Distance Transforms. *IEEE Trans. Pattern Anal. Mach. Intell.* **2014**, *36*, 2159–2172. [CrossRef] [PubMed]
13. Tsardoulias, E.G.; Serafi, A.T.; Panourgia, M.N.; Papazoglou, A.; Petrou, L. Construction of Minimized Topological Graphs on Occupancy Grid Maps Based on GVD and Sensor Coverage Information. *J. Intell. Robot. Syst.* **2014**, *75*, 457–474. [CrossRef]
14. Wang, K.-H.C.; Botea, A. Scalable Multi-Agent Pathfinding on Grid Maps with Tractability and Completeness Guarantees. In Proceedings of the ECAI—European Conference on Artificial Intelligence DBLP, Lisbon, Portugal, 16–20 August 2017.
15. Sharon, G.; Stern, R.; Felner, A.; Sturtevant, N.R. Conflict-Based Search for Optimal Multi-Agent Pathfinding. *Artif. Intell.* **2015**, *219*, 40–66. [CrossRef]
16. Sigurdson, D.; Bulitko, V.; Yeoh, W.; Hernandez, C.; Koenig, S. Multi-Agent Pathfinding with Real-Time Heuristic Search. In Proceedings of the 2018 IEEE Conference on Computational Intelligence and Games (CIG), Maastricht, The Netherlands, 14–17 August 2018; pp. 1–8.
17. Li, J.; Harabor, D.; Stuckey, P.; Ma, H.; Koenig, S. Symmetry Breaking Constraints for Grid-based Multi-Agent Path Finding. In Proceedings of the National Conference on Artificial Intelligence, Honolulu, HI, USA, 27 January–1 February 2019.

18. Ma, H.; Wagner, G.; Felner, A.; Li, J.Y.; Kumar, T.K.S.; Koenig, S. Multi-Agent Path Finding with Deadlines. In Proceedings of the International Joint Conference on Artificial Intelligence, Wellington, New Zealand, 10–12 December 2018; pp. 417–423.
19. Boyarski, E.; Felner, A.; Stern, R.; Sharon, G.; Tolpin, D.; Betzalel, O.; Shimony, E. ICBS: Improved Conflict-Based Search Algorithm for Multi-Agent Pathfinding. In Proceedings of the International Conference on Artificial Intelligence, Buenos Aires, Argentina, 25–31 July 2015.
20. Rao, N.S.V.; Stoltzfus, N.; Iyengar, S.S. A "retraction" method for learned navigation in unknown terrains for a circular robot. *IEEE Trans. Robot. Autom.* **1991**, *7*, 699–707. [CrossRef]
21. Scherer, S.; Ferguson, D.; Singh, S. Efficient C-space and cost function updates in 3D for unmanned aerial vehicles. In Proceedings of the 2009 IEEE International Conference on Robotics and Automation, Kobe, Japan, 12–17 May 2009.
22. Cuisenaire, O.; Macq, B. Fast Euclidean Distance Transformation by Propagation Using Multiple Neighborhoods. *Comput. Vis. Image Underst.* **1999**, *76*, 163–172. [CrossRef]
23. Lau, B.; Sprunk, C.; Burgard, W. Improved Updating of Euclidean Distance Maps and Voronoi Diagrams. *Intell. Robot. Syst.* **2010**, 281–286.
24. Harabor, D.; Botea, A. Breaking Path Symmetries on 4-Connected Grid Maps. In Proceedings of the Sixth AAAI Conference on Artificial Intelligence and Interactive Digital Entertainment, AIIDE 2010, Stanford, California, CA, USA, 11–13 October 2010.
25. Pochter, N.; Zohar, A.; Rosenschein, J.S.; Felner, A. Search Space Reduction Using Swamp Hierarchies. In Proceedings of the Twenty-Fourth AAAI Conference on Artificial Intelligence, California, CA, USA, 11–15 July 2010.
26. Harabor, D.; Grastien, A. Improving Jump Point Search. In Proceedings of the Twenty-Fourth International Conference on Automated Planning and Scheduling, New Hampshire, NH, USA, 21–26 June 2014.
27. Rabin, S.; Sturtevant, N.R. Combining Bounding Boxes and Jps to Prune Grid Pathfinding. In Proceedings of the Thirtieth AAAI Conference on Artificial Intelligence, Phoenix, Arizona, 12–17 February 2016.
28. Harabor, D.; Grastien, A. Online Graph Pruning for Pathfinding on Grid Maps. In Proceedings of the Twenty-Fifth AAAI Conference on Artificial Intelligence, California, CA, USA, 7–11 August 2011.
29. Sturtevant, N.R. Generalizing JPS Symmetry Detection: Canonical Orderings on Graphs. In Proceedings of the Ninth Annual Symposium on Combinatorial Search, New York, NY, USA, 6–8 July 2016.
30. Uras, T.; Koenig, S.; Hernández, C. Subgoal Graphs for Optimal Pathfinding in Eight-Neighbor Grids. In Proceedings of the Twenty-Third International Conference on Automated Planning and Scheduling, Rome, Italy, 10–14 June 2013.
31. Uras, T.; Koenig, S. Identifying Hierarchies for Fast Optimal Search. In Proceedings of the Twenty-Eighth AAAI Conference on Artificial Intelligence, Quebec City, QC, Canada, 27–31 July 2014.
32. Uras, T.; Koenig, S. Understanding Subgoal Graphs by Augmenting Contraction Hierarchies. In Proceedings of the International Joint Conference on Artificial Intelligence, Macao, China, 28 September 2018; pp. 1506–1513.
33. Harabor, D.D.; Grastien, A.; Öz, D.; Aksakalli, V. Optimal Any-Angle Pathfinding in Practice. *J. Artif. Intell.* **2016**, *56*, 89–118. [CrossRef]
34. Hormazábal, N.; Díaz, A.; Hernández, C.; Baier, J.A. Fast and Almost Optimal Any-Angle Pathfinding Using the 2k Neighborhoods. In Proceedings of the Tenth Annual Symposium on Combinatorial Search, Pittsburgh, PA, USA, 16–17 June 2017.
35. Uras, T.; Koenig, S. An Empirical Comparison of Any-Angle Path-Planning Algorithms. In Proceedings of the Annual Symposium on Combinatorial Search, Ein Gedi, The Dead Sea, Israel, 11–13 June 2015.
36. Uras, T.; Koenig, S. Speeding-up any-angle path-planning on grids. In Proceedings of the Twenty-Fifth International Conference on Automated Planning and Scheduling, Jerusalem, Israel, 7–11 June 2015.
37. Hu, Y.; Zhang, Q.; Qin, L.; Yin, Q. Escaping Depressions in LRTS with Wall Following Method. In Proceedings of the 2017 9th International Conference on Intelligent Human-Machine Systems and Cybernetics (IHMSC), Hangzhou, China, 26–27 August 2017.

Article

SURF-BRISK–Based Image Infilling Method for Terrain Classification of a Legged Robot

Yaguang Zhu *, Chaoyu Jia, Chao Ma and Qiong Liu

Key Laboratory of Road Construction Technology and Equipment of MOE, Chang'an University, Xi'an 710064, China; jiachaoyu@chd.edu.cn (C.J.); machao@chd.edu.cn (C.M.); liuqiong@chd.edu.cn (Q.L.)
* Correspondence: zhuyaguang@chd.edu.cn; Tel.: +86-187-9285-2585

Received: 19 March 2019; Accepted: 25 April 2019; Published: 29 April 2019

Abstract: In this study, we propose adaptive locomotion for an autonomous multilegged walking robot, an image infilling method for terrain classification based on a combination of speeded up robust features, and binary robust invariant scalable keypoints (SURF-BRISK). The terrain classifier is based on the bag-of-words (BoW) model and SURF-BRISK, both of which are fast and accurate. The image infilling method is used for identifying terrain with obstacles and mixed terrain; their features are magnified to help with recognition of different complex terrains. Local image infilling is used to improve low accuracy caused by obstacles and super-pixel image infilling is employed for mixed terrain. A series of experiments including classification of terrain with obstacles and mixed terrain were conducted and the obtained results show that the proposed method can accurately identify all terrain types and achieve adaptive locomotion.

Keywords: terrain classification; image infilling method; multilegged robot

1. Introduction

Multilegged robot that origniated from reptile bionics has good walking stability and low energy consumption in its stationary state. It maintains good stability in complex environments owing to its redundant limb structure [1]. Compared with a wheeled robot, the multilegged robot can cross big obstacles and has many degrees of freedom. Its flexibility and adaptability on complex terrain allow the legged robot to have wide application. Researchers have designed different multilegged robots, such as mine-sweeping [2], volcano-detecting [3], underwater [4], strawberry-picking [5], and transfer robots, in addition to other prototypes. The autonomous mobile ability of multilegged robots is affected by how it perceives its surrounding environment. Multilegged robots mainly work in unstructured environments, so classifying various terrains, detecting obstacles, and localizing and recognizing complex terrain have become primary issues in the field.

For multilegged robots, environment perception is mainly related to accurate terrain identification and obstacle detection. The most normal way is to use image processing methods and classifiers. By extracting information from terrain images, such as spectra [6], color [7], texture [8,9], scale-invariant feature transform (SIFT) features [10], speeded up robust features (SURF) [11], and the DAISY descriptor [11], the terrain can be accurately identified. However, spectral-based methods concentrate on spatial frequencies of texture distributions and color-based methods have poor robustness and are easily affected by light and weather conditions. Among them, local features that are invariant in terms of scale, rotation, brightness, and contrast have been widely used in visual classification. Besides vision, legged robots are often equipped with other sensors, so information from multiple sensors for terrain recognition is also available. Kim [12] used the friction coefficient of different terrains to classify terrains using the Bayes classifier. Ojeda [13] proposed a terrain classification method based on an integration of information from multiple sensors (gyroscopes, accelerometers, encoders). Larson [14] proposed a model based on robot inclination angle obtained from an odometer. Hoepflinger [15] used current values of joint electricity and force sensor data to recognize terrain categories. Jitpakdee [16] proposed a neural network model for terrain classification based on robot body acceleration and angular velocity of the inertial measurement unit (IMU). These kinds of information are quite different from visual information and so special methods are needed.

Most of the existing methods have good classification accuracy on single-type terrain, but few are suitable for mixed terrain, which is common in the natural environment. To solve this problem, Filitchkin [17] used a sliding window technique for heterogeneous terrain images. Liang [18] compiled an algorithm for complex terrain classification based on multisensor fusion. Ugur [19] proposed a learning method to predict the environment by consecutive distance- and shape-related feature extraction. However, most of these methods have poor robustness because they require high-resolution images. Mixed terrain has different features than single-type terrain and sometimes the edges of the terrain cannot be recognized clearly, which makes identification of mixed terrain difficult. In order to enhance the recognition rate of detecting mixed terrain and terrain with obstacles, a systematic classification method for complex terrain is proposed in this paper. The following aspects were studied: Terrain information collected by a Kinect 3D vision sensor. Herein, we established a fast and effective terrain classifier based on speeded up robust features and binary robust invariant scalable key points (SURF-BRISK) features and support vector machine (SVM). A segmentation method for complex terrain images based on super-pixels is proposed, which can effectively segment complex terrain images into single terrain images. An image infilling method for terrain with obstacles and mixed terrain is also proposed. The local features are magnified to help the recognition of different complex terrains. Experiments on classifying terrain with obstacles and mixed terrain are conducted. The proposed system is validated by the multilegged robot.

This paper is organized as follows: In Section 2, the hexapod robot and SURF-BRISK–based image infilling method are introduced. In Section 3, the experimental results are presented and analyzed. Section 4 summarizes and concludes the paper.

2. Materials and Methods

2.1. Hexapod Walking Robot: SmartHex

In this paper, a six-legged robot with mammalian leg structure named SmartHex is used [20]. Each leg has three joints: base joint, hip joint, and knee joint. The robot is suitable for outdoor work due to its low energy consumption and large load characteristics [21]. The robot is controlled by a backward control based on a σ-Hopf oscillator with decoupled parameters for smooth locomotion [22]. The hardware structure of the robot is shown in Figure 1. BLi, HLi, and KLi (BRi, HRi, and KRi), respectively, indicate the base joint, hip joint, and knee joint of the left (right) leg. LF, LM, and LH (RF, RM, and RH), respectively, indicate the front leg, middle leg, and hind leg of the left (right) side. Each joint consists of a high-reduction-rate gear system and a DC servo motor with an integrated encoder, which is used to detect the position of the joint angle. In order to identify the environment around the robot with a high accuracy, Microsoft's Kinect 3D vision sensor is installed on the robot chassis to collect terrain information. The Kinect's red/green/blue (RGB) camera can collect color images with a resolution of 1920 × 1080 px. A complementary metal-oxide superconductor (CMOS) sensor is responsible for receiving and transmitting infrared signals. Meanwhile, current detection modules (CuSLi and CuSRi) are installed to record the energy consumption of each leg for different gaits. The arrows in Figure 1 indicate the directions of current flow. The robot's posture is monitored by the attitude sensor (AS). The data processed by the wireless module mounted on the control panel are transmitted to the host computer to create instructions [23].

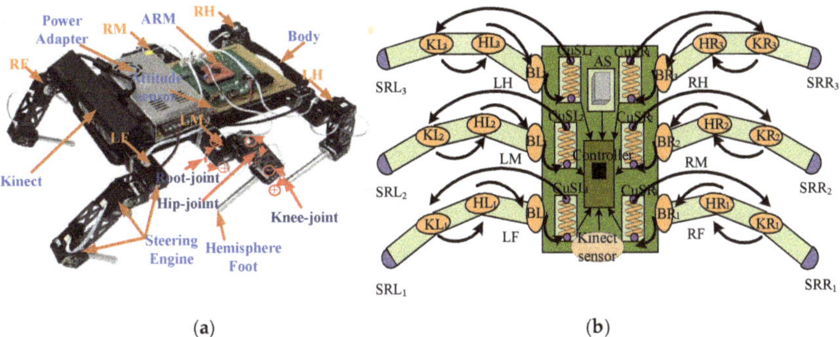

(a)   (b)

Figure 1. Architecture of robot: (**a**) hardware structure; (**b**) distribution of sensors. BLi, HLi, and KLi (BRi, HRi, and KRi), respectively, indicate base joint, hip joint, and knee joint of left (right) leg. LF, LM, and LH (RF, RM, and RH), respectively, indicate front leg, middle leg, and hind leg of left (right) side.

2.2. Terrain Classification Methodology

In robot navigation, terrain recognition can essentially be supposed as surface texture recognition. Terrain recognition based on local features is the most popular because of its robustness to illumination and weather and high recognition rate.

The terrain classification system proposed in this paper is depicted in Figure 2. The Kinect is installed on top of the robot to collect information on terrain (color, depth, and infrared images) and an obstacle detection module is established to detect obstacles in the front. If there are no obstacles, the information will be directly transmitted to the terrain classifier. Otherwise, the module will locate obstacles and identify their size. Meanwhile, a color image of the terrain is processed by the image infilling method to decrease the influence of obstacles on terrain identification. After classification, the confidence scores of each terrain are summarized into a pie chart. The analysis of the pie chart shows whether the terrain is mixed or not. If the terrain is mixed, the color image would be subjected to image segmentation and infilling. Then, the processed color image will be classified by the terrain classifier again, which provides an accurate identification of multiple areas of complex terrain. Finally, all terrain types can be predicted accurately, thus good performance by the robot is guaranteed. The function modules are described in detail in the following section.

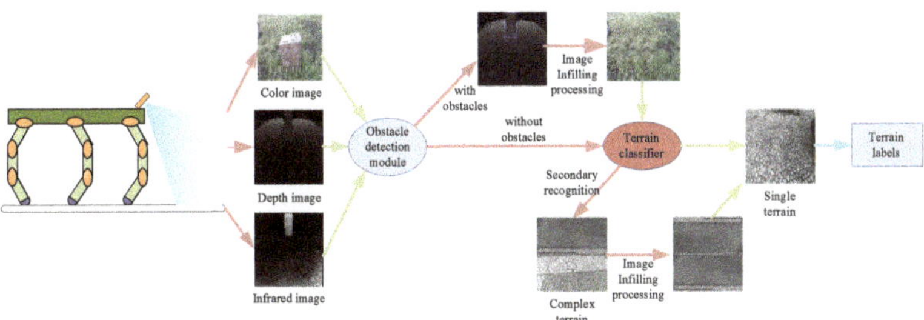

Figure 2. Terrain classification system.

2.2.1. Obstacle Detection Module

Detecting and localizing obstacles are important to realize autonomous motion and path planning. The sensors used for traditional obstacle detection mainly include laser radar sensors, ultrasonic sensors, infrared sensors, visual equipment, etc. [24]. In this paper, a fast and accurate detection method based on depth and infrared information is used [25]. The image is segmented by the mean-shift algorithm and the pixel gradient of the foreground is calculated. After pretreatment of edge detection and morphological operation, the depth and infrared information are fused. The characteristics of depth and infrared images are used for edge detection. Thus, the false rate of detection is reduced and detection precision is improved. Since depth images cannot be affected by natural sunlight, the influence of light intensity and shadow on obstacle recognition is effectively eliminated and the robustness of the algorithm is improved. This method can accurately identify the position and size of obstacles. In this paper, the results obtained by obstacle detection with this method are used as the input of the terrain image infilling method.

2.2.2. Terrain Classification Module

An online terrain classification system is needed to collect information on terrain through the Kinect sensor and then the key points are extracted from a color image of the terrain. Since the terrain classifier is based on the bag-of-words (BoW) module [26], all extracted features are processed by a clustering algorithm to ensure that the clusters have high similarity. These cluster centers are the visual vocabulary. Then, the terrain images are encoded to form the visual dictionary and a visual vocabulary frequency histogram corresponding to each terrain type. Finally, the information is used to train the support vector machine (SVM) [27] and an optimal hyperplane of each terrain type is divided to classify all terrain types. The algorithm can grasp the key samples and eliminate many redundant samples.

The main structure of the terrain classification system is demonstrated in Figure 3. The system is mainly divided into two steps: training and testing. In the first step, the information of all terrain types is collected and stored in memory and the data flow is presented as shown on the right in Figure 3. Then, local features of images in memory are extracted and extracted features are clustered by the k-means algorithm to generate a certain number of visual words [28]. Then, terrain images are encoded using the BoW module to form the visual dictionary and the visual vocabulary frequency histogram corresponding to each terrain type. Then, the information is used to train the SVM. With the aim of validating the terrain classification system established in the training part, the testing part is introduced, shown on the left in Figure 3. The local features from terrain images are extracted in the testing image set and the visual word dictionary is encoded. The images are converted to the frequency histograms that are input in the trained SVM to obtain the terrain label. This part is used by the hexapod robot for terrain recognition. The hexapod robot's gait transform algorithm is guided by terrain identification.

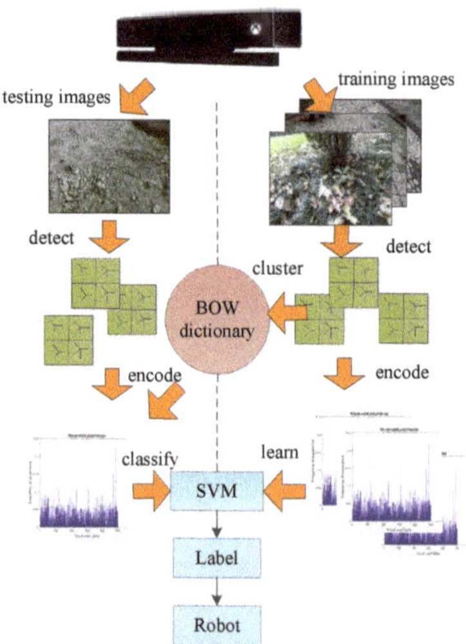

Figure 3. Diagram of terrain classification system. BOW: bag-of-words; SVM: support vector machine.

In this paper, a dataset was created using six common terrains: grass, asphalt, sand, gravel, tile, and soil. Each terrain image set contained 50 samples, which were acquired by a Kinect camera. A set of samples of terrain images with different illuminations and weather conditions is shown in Figure 4. The *K*-fold cross-validation was used and *K* = 5 [29]. All images were randomly partitioned into five equally sized groups. Each group was chosen as validation data for testing the classifier and other 4 groups for training set.

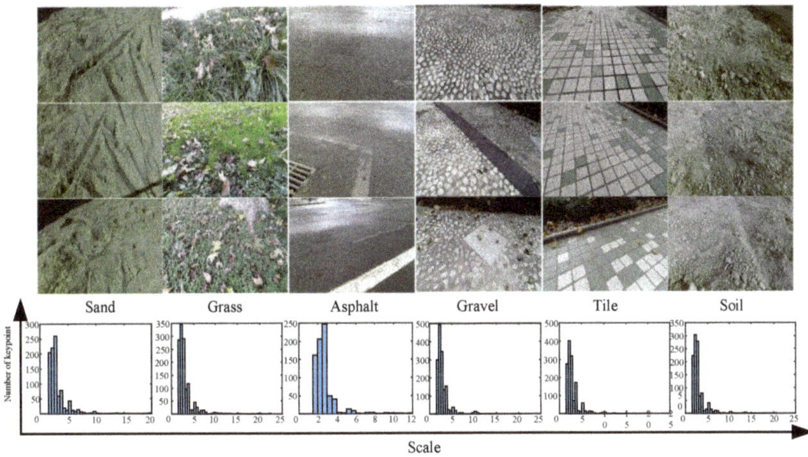

Figure 4. Different terrains and corresponding numbers of speeded up robust features (SURF) key points.

A. Point of Interesting Extracted by SURF

In the aspect of terrain image feature extraction, the SURF algorithm is a commonly used local feature extraction algorithm in image classification. The matching accuracy is high, but the real-time performance is generally poor. In recent years, many excellent algorithms have been proposed. BRISK [30], which combines detection of key points of features from accelerated segment test (FAST) and binary description can enhance the speed of the algorithm, but its classification performance is not ideal. Since the SURF algorithm with many feature points cannot satisfy real-time detection and the BRISK algorithm has fast computation speed but a low matching rate, a method for image matching based on the SURF-BRISK algorithm is proposed. The SURF-BRISK algorithm is established by combining the advantages of both algorithms. Points of interest are detected using the SURF algorithm, descriptors are calculated using the BRISK algorithm, and the Hamming distance is used [31] for similarity measurement, which enables not only high matching rates but also high calculation speed. The algorithm process is described below. In SURF, the criterion of feature points is the determinant of a Hessian matrix of pixel luminance. A pixel u(x, y) is given in image I. In this point, the scale σ of the matrix is defined by:

$$H(\mu, \sigma) = \begin{bmatrix} L_{xx}(u, \sigma) & L_{xy}(u, \sigma) \\ L_{xy}(u, \sigma) & L_{yy}(u, \sigma) \end{bmatrix}, \tag{1}$$

where $L_{xx}(u,\sigma)$ is the Gaussian second-order differential $\partial^2 g(\sigma)/\partial x^2$ convolution of image I at point u, and similarly for $L_{xy}(u,\sigma)$ and $L_{yy}(u,\sigma)$. In order to facilitate the calculation, the elements of the Hessian matrix are labeled as D_{xx}, D_{yy}, D_{xy}, and the weight of a square area is set to a fixed value. Hence, the approximate value of the Hessian matrix determinant H_{approx} is defined by:

$$\det(H_{approx}) = D_{xx}D_{yy} - (\omega D_{xy})^2, \tag{2}$$

where the correlation weight ω of the filter response is utilized to balance the expression of the Hessian determinant. In order to preserve the energy conservation of the Gauss kernel and approximate it, ω is usually set to 0.9. The Hessian matrix is used to calculate the partial derivative, which is usually obtained by a convolution of pixel light intensity and a certain direction of Gauss kernel partial derivative. In order to improve the speed of the SURF algorithm, the approximate box filter is used instead of the Gauss kernel with very little impact on precision. The convolution calculation can be used to optimize the integral image, which greatly improves the efficiency. It is necessary to use three filters to calculate Dxx, Dyy, and Dxy for each point. After filtering, a response graph of the image is obtained. The value of each pixel on the response graph is calculated by the determinant of the original pixel. The image is filtered with different scales and a series of responses of the same image at different scales is obtained. The detection method of feature points is if the value of det (Happrox) of a key point is greater than the value of 26 points in its neighborhood. The number of interest points sampled by SURF is shown in Figure 4.

B. Descriptors by BRISK

The BRISK descriptor adopts the neighborhood sampling model, which takes the feature points as the center of the circle. The points on the concentric circles of several radii are selected as the sampling points. In order to reduce the effect caused by sample image grayscale aliasing, the Gauss function can be used for filtering. The Gauss function of standard deviation sigma is proportional to the distance between the points on each concentric circle. Selecting a pair from the point pairs formed by all sampling points, denoted as (P_i, P_j), the gray values after treatment are $I(P_i, \sigma_i)$ and $I(P_j, \sigma_j)$, respectively. Hence, the gradient between two sampling points is

$$g(P_i, P_j) = (P_j - P_i) \cdot \frac{I(P_j, \sigma_j) - I(P_i, \sigma_i)}{\|P_j - P_i\|^2}. \tag{3}$$

Set A is a collection of all pairs of sampling points, S is a set containing all the short-range sampling pairs, and L is a set containing all the long-distance pairs of sampling points:

$$A = \left\{ (P_i, P_j) \in R^2 \times R^2 \middle| i < N \wedge j < i \wedge i, j \in N \right\}, \tag{4}$$

$$S = \left\{ (P_i, P_j) \in A \middle| \|P_j - P_i\| < \delta_{\max} \right\} \subseteq A, \tag{5}$$

$$L = \left\{ (P_i, P_j) \in A \middle| \|P_j - P_i\| < \delta_{\min} \right\} \subseteq A. \tag{6}$$

The general distance thresholds $\delta_{max} = 9.75\,t$, $\delta_{min} = 13.67\,t$, and t are characteristic point scales. The main direction for each feature point is specified by the gradient direction distribution characteristics of neighboring pixels of the feature point. In general, the BRISK algorithm can be used to solve for the direction g of the overall pattern according to the gradient between two sampling points:

$$g = \begin{pmatrix} g_x \\ g_y \end{pmatrix} = \frac{1}{L} \cdot \sum_{(P_i, P_j) \in L} g(P_i, P_j). \tag{7}$$

In order to achieve rotation and scale invariance, the sampling pattern is sampled again after the rotation angle $\theta = arctan2\,(g_y, g_x)$. The binary descriptor b can be constructed by performing Equation (8) on all pairs of points in set S by short-range sampling points.

$$b = \begin{cases} 1 & I(P_j^\theta, \sigma_j) > I(P_i^\theta, \sigma_i) \\ 0 & otherwise \end{cases} \qquad \forall (P_i^\theta, P_j^\theta) \in S \tag{8}$$

C. Local Feature Matching

After the feature descriptors extracted by SURF-BRISK are 512-bit binary bit strings consisting of 0 and 1, the Hamming distance is used to measure similarity. Assuming that there are two descriptors of S_1 and S_2, the Hamming distance is determined as

$$D_{kd}(S_1, S_2) = \sum_{i=1}^{512} (x_i \otimes y_i), \tag{9}$$

where $S_1 = x_1x_1 \dots x_{512}$, $S_2 = y_1y_2 \dots y_{512}$, x, y and the value of x and y is 0 or 1. The smaller the value of D_{kd}, the higher the matching rate, and vice versa. Therefore, the matching point pairs are obtained using the nearest-neighbor Hamming distance in the matching process.

Here, three descriptors, SURF, BRISK, and SURF-BRISK, are compared. Two tile images have been chosen for matching tests to compare the real-time and matching rate of these descriptors, as shown in Table 1 and Figure 5. Obviously, the SURF algorithm has the most matching points, the BRISK algorithm has the fastest matching, and the SURF-BRISK algorithm combines the advantages of both. The algorithm is faster than SURF and gets more matching points than BRISK.

Table 1. Detection times of different descriptors. Brisk: binary robust invariant scalable keypoints.

Descriptor Type	Detection Time (ms)	Matching Time (ms)
SURF	852	1251
BRISK	127	98
SURF-BRISK	765	142

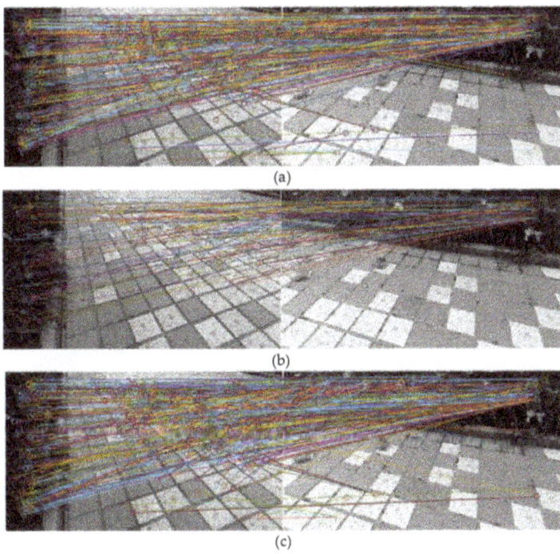

Figure 5. Matching results of different algorithms: (**a**) SURF; (**b**) BRISK; (**c**) SURF-BRISK.

D. BoW Model and SVM

Li et al. [32] first introduced the image method based on the BoW model. They believed that an image can be analogized to a document and the "words" of an image can be defined as feature vectors. The basic BoW model regards an image as a set of feature vectors and statistics of occurrence frequency of feature vectors, which are used for terrain classification. The BoW model can be set up by a clustering algorithm that is used to obtain the visual dictionary and the steps are as follows. Feature extraction: m images ($m \geq 50$) are collected for each terrain type and each image is extracted by SURF-BRISK to obtain $n(i)$ feature vectors. All terrain images form a total sum ($n(i)$) of feature vectors (words). Generation of dictionary/codebook: The feature vectors obtained from the previous step are clustered (here, the k-means clustering method is used [33]) to get k clustering centers in order to build the codebook. A histogram is generated according to the codebook. The nearest neighbor calculation of each word of the picture is used to find the corresponding words in the codebook in order to form the BoW model.

SVM is an excellent learning algorithm developed on the basis of statistics theory and is widely used in many fields, such as image classification, handwriting recognition, and bioinformatics. The input vector is mapped to a high-dimensional feature space by nonlinear mapping (a kernel function) and an optimal hyperplane is constructed in this space. Compared with the artificial neural network, which suffers from an overfitting problem, the support vector machine has better generalization ability for unknown samples [34]. SVM can be divided into three groups: linear separable, nonlinear separable, and kernel function mapping. Linear classifier performance is limited to linear problems, because in nonlinear problems constraints of excessive relaxation can lead to a large number of error samples. At this point, it can be transformed into a linear problem in a high-dimensional space using nonlinear transformation in order to obtain an optimal classification hyperplane.

2.3. Complex Terrain Recognition

In the field, terrain is usually complex. The accuracy of terrain with obstacles and mixed terrain, which is composed of two or more terrain types, is greatly reduced if traditional identification methods are used. This scenario affects the normal operation of the robot. Therefore, a systematic method based on image segmentation and infilling for recognition of terrain with obstacles and mixed terrain is introduced in this section.

2.3.1. Image Local Infilling for Terrain with Obstacles

Information on terrain with obstacles collected by Kinect is used as input data for the terrain classifier. It was found that large-volume obstacles cause low accuracy of final identification, because acquired features of terrain information are severely affected, since SURF-BRISK and SURF have the same points of interest. The distributions of points in different terrains with and without obstacles are shown in Figure 6. The obstacles greatly influence local feature extraction. In order to improve the accuracy of recognizing terrain with obstacles, a method of image local infilling (ILI) is presented. The errors for terrain with obstacles in the first round of recognition are shown in Table 2.

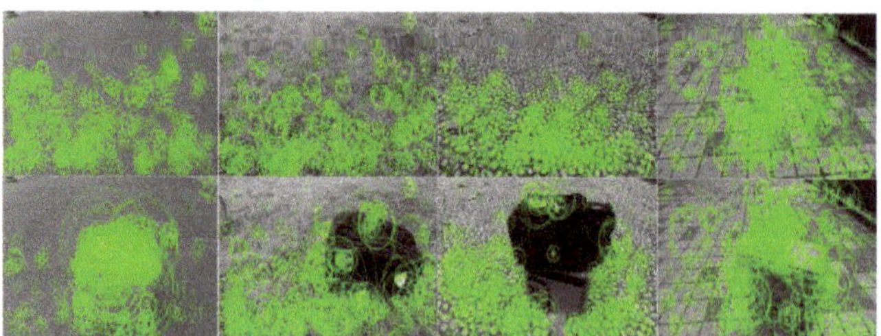

Figure 6. Distributions of feature points in terrain with and without obstacles.

The obstacle area of pixel matrix $I(m, n)$ is obtained using the obstacle detection method, as are the central pixel coordinates (u, v). Here, three infilling examples are illustrated for comparison. The obstacle area with a pixel value of $I = 255$ is presented as a white area in Figure 7b. The obstacle area with a pixel value of $I = 0$ is presented as a black area in Figure 7c. The obstacle area spliced by the no-obstacle sides of the background terrain image is presented in Figure 7d. Due to the use of both left and right sides of the terrain image for infilling, it only needs to compare the abscissa u of the obstacle area center and the abscissa u_c of the color image center. At the same time, according to the dimensions of $I(m, n)$, the size and orientation of obstacles are determined. If the width of the obstacle area, i.e., the number n of matrix $I(m, n)$, is too large, the image needs to be processed by multiple infilling. The classification and statistical results of the terrain classifier after ILI are also shown in Table 2. The first two methods do not improve the accuracy of image recognition, since white and black features do not contribute to the main feature points. However, the background terrain–based image infilling shows satisfactory results.

Table 2. Classification results of first round and after image local infilling (ILI).

Images with Obstacle					
First round Actual terrain	Asphalt Grass	Asphalt Grass	Asphalt Grass	Asphalt Grass	Asphalt Gravel
ILI White	Asphalt	Asphalt	Tile	Tile	Tile
Black	Asphalt	Asphalt	Asphalt	Tile	Tile
Terrain	Grass	Grass	Grass	Grass	Gravel

(a) (b) (c) (d)

Figure 7. ILI samples: (**a**) image sample; (**b**) local white; (**c**) local black; (**d**) local terrain.

2.3.2. Image Infilling for Mixed Terrain

After the first round of classification, both the terrain label and confidence score of the classified image are obtained. In SVM, the confidence score represents the geometric interval between the classified image and the hyperplane of each terrain type. Therefore, the confidence score needs to be normalized before conducting an analysis. The confidence score is adjusted to the interval [0, 1] to facilitate the comparison. Set S_d contains the confidence scores of all terrain types, and d_i is the confidence of the test image that corresponds to i terrain class before normalization. Set S_D contains confidence scores after normalization and D_i is normalized confidence. Therefore, after normalization we get

$$S_D = \{D_i | D_i = |d_i| / \sum_{i=1}^{6} |d_i|, i = 1, 2, \ldots, 6\}. \tag{10}$$

Moreover, a pie chart of confidence scores after normalization can clearly demonstrate the contribution of each terrain type. A pie chart of confidence scores after the first round of classification is shown in Figure 8. In the images of single terrain, the weight of single terrain is much higher than the weights of other terrains. A series of experiments demonstrated that if the highest terrain weight is larger than 30% and more than 10% higher than the second highest weight, the terrain can be considered as a single terrain. Otherwise, it is mixed terrain. For mixed terrain, it is difficult to identify the category from weights in the pie chart. In addition, it is important to note that mixed terrain usually appears at the intersection of different terrains. The traditional methods are not practical for images that contain two or more terrain types, because only one label will be notified. Obviously, some approaches can identify the boundaries of different terrains in an image and then make the decision. Actually, it is difficult to accurately determine terrain boundaries and the algorithm needs to do many computations, which causes poor real-time performance that affects the robot's outdoor walking. In the process of the robot moving in a forward direction, the terrain type is gradually changing. Different types of terrain appear in up and down form in the images. Taking this into consideration, a new method for identification of mixed terrain based on super-pixel image infilling (SPI) is presented.

Figure 8. Distribution of terrain types: (**a**) single terrain; (**b**) mixed terrain.

In the field of image segmentation, super-pixel has become a fast-developing image preprocessing technology. Ren et al. [35] first proposed the concept of super-pixels, which quickly divide images into a number of subareas that have image semantics. Compared with the traditional processing method, the extraction and expression of super-pixels are more conducive to collecting local characteristics of the image information. It can greatly reduce the calculation and subsequent processing complexity. Existing segmentation algorithms generally restrict the number of pixels, the compactness, the quality of segmentation, and the practicability of algorithms. Song et al. [36] evaluated the existing super-pixel segmentation algorithms. Their results indicate that the simple linear iterative cluster (SLIC) super-pixel segmentation algorithm has good performance in terms of the controllability of pixel numbers and the close degree of controllability. Aiming at segmentation, the SLIC algorithm is used for mixed terrain regions. The most super-pixels are selected as the target area in a multi-super-pixel area and the boundary pixels of the pixel coordinates of curve fitting are extracted as the terrain boundary segmentation of a complex terrain image. The procedure and results are shown in Figure 9.

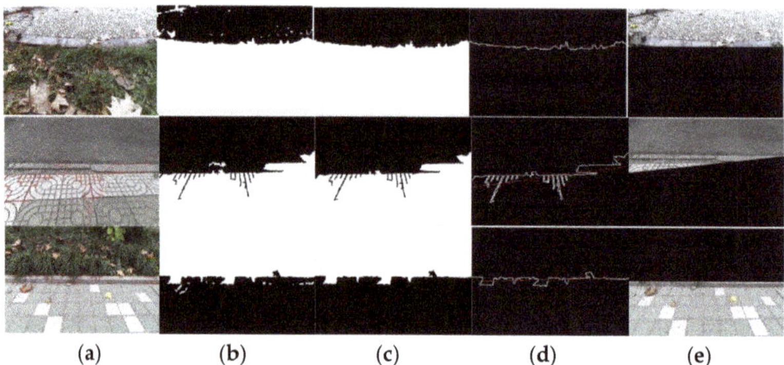

Figure 9. Segmentation result of mixed terrain images: (**a**) simple linear iterative cluster (SLIC) algorithm; (**b**) maximum super-pixel extraction; (**c**) filtering out smaller areas; (**d**) finding the boundary and fitting the line; (**e**) results.

Classification results after image segmentation are shown in Table 3. The output labels do not match actual terrain types. For this mismatch, the number of points of interest extracted from segmented images is shown in Figure 10. Compared with the original terrain image in Figure 4, the number of feature points of segmented images is still related to terrain type but much smaller. Obviously, it is impossible to realize an accurate prediction using the segmentation image, because the feature points are inadequate. Thus, the segmented images are spliced together to enhance the terrain features. Segmented color images would have only some of the pixels of the original color image collected by the Kinect camera and the blank pixels would be infilled by duplication of the segmented image. In the test, the rotation–inversion operation is used for image infilling. The results are shown in Figure 11.

Table 3. Image infilling and confidence scores.

Images										
Actual terrain	Tile		Grass		Tile		Grass		Grass	
Output label	Asphalt		Tile		Asphalt		Asphalt		Soil	
R-I	Tile		Grass		Tile		Grass		Grass	
Scores	Before	After	Before	After	Before	After	Before	After	Before	After
Sand	10.07	9.56	15.36	14.58	12.64	10.43	11.19	12.41	10.53	8.72
Grass	8.42	8.01	21.65	28.05	15.68	13.05	26.43	37.45	34.23	45.18
Asphalt	31.41	15.68	16.47	15.12	19.29	15.64	25.37	14.56	14.76	10.63
Gravel	14.55	16.56	15.83	16.45	15.67	13.54	14.36	13.22	11.7	8.35
Tile	22.01	36.05	17.26	14.67	22.36	31.73	10.27	8.68	18.29	15.64
Soil	13.54	14.14	13.43	11.13	14.37	15.62	12.38	13.69	10.5	11.48

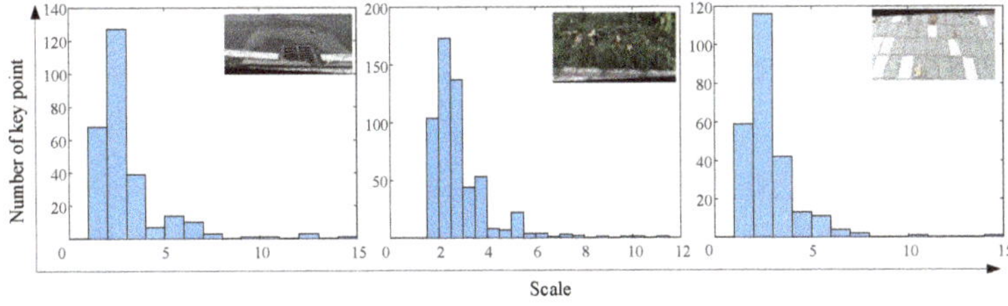

Figure 10. Number of feature points in segmented images.

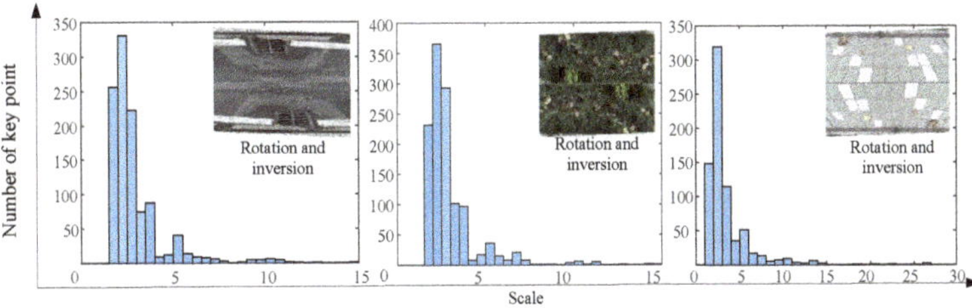

Figure 11. Number of feature points in spliced images.

The number of feature points in Figures 10 and 11 shows that the proposed method can enhance local features of segmented images. The classification results of a spliced image using this approach are shown in Table 3. Using the image infilling approach (rotation–inversion), the error results of the first-round classification can be corrected. It can be seen that confidence scores of the correct terrain type increased after image infilling. On the contrary, confidence scores of wrong terrain types decreased. That means the proposed method can effectively magnify image features for the classifier.

3. Results

3.1. Complex Terrain

In the experiments, the hexapod robot walked on six types of terrain without obstacles. Terrain images were collected by a Kinect camera installed on top of the robot. The inclination angle of the Kinect sensor is 40°. Images of terrain with obstacles were collected at different times and in different weather and light conditions. Obstacles mainly included cartons, trash, trees, and so on. There were 50 images collected for each terrain type. The collected images of terrain with obstacles were processed by the ILI method. Then, all images before and after ILI processing were classified by the presented terrain classifier. The recognition results for the two sets are shown in Figure 12a. The recognition rate of terrain with obstacles before ILI processing was relatively low. Since obstacles seriously affect local features of the terrain, error exists in most cases and average recognition accuracy is less than 75%. On the contrary, after the image infilling process, recognition accuracy was improved to above 85%.

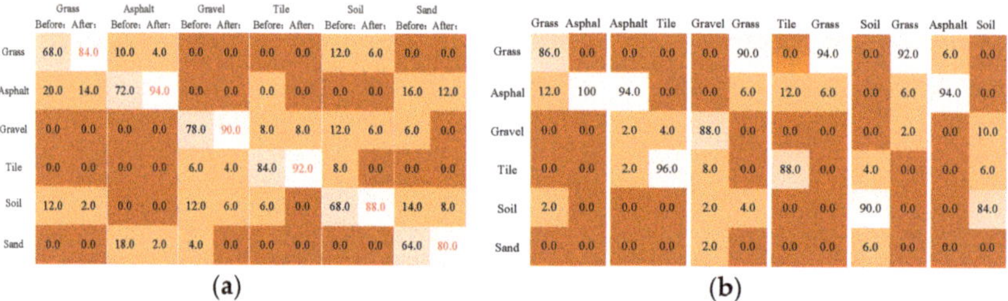

Figure 12. Recognition results: (**a**) terrain with obstacles; (**b**) mixed terrain.

Usually, mixed terrain appears at the intersection of different terrains. All mixed terrain images were collected at that moment. Specifically, 50 images were collected and processed by the SPI method for terrain recognition. The result is shown in Figure 12b. The classifier shows the labels of two terrain types for different subareas in the image. The average recognition accuracy reached 85% and the results show that the proposed method is effective in recognizing mixed terrain. At the same time, compared with a single label classifier, the SPI method has more practical significance for gait transition of the hexapod robot.

3.2. Robot Platform Application

When the robot walks on different terrains, different gaits have different effects on the robot's stability, performance, and energy consumption. The experiment showed that the gait can be changed based on the output of the terrain classifier. In the experiment, the hexapod robot walked for 30 s across three terrain types: asphalt, soil, and grass. The sampling period of the Kinect is 1 s. The gait pattern of the hexapod robot was set according to the output of the terrain classifier. The pseudocode of the gait transition algorithm is depicted in Table 4.

The value of G has a great influence on the smoothness and efficiency of motion on different terrains. The terrain classification results, including gait value, leg current from robot legs SRL1, and attitude angle, are shown in Figure 13. From 0–5 s, the terrain is supposed to be asphalt. Thus, the robot moves in tripod gait. From 5–6 s, the robot is in transition gait and ready to stride across mixed terrain consisting of asphalt and soil. The value of the terrain curve is nominal, showing that the terrain is complex, e.g., 1.3 means the terrain is changing from type 1 to type 3. From 6–21 s, the robot moves forward with its current gait. Then, from 22–23 s, it changes gait to get ready for another terrain. Finally, from 23–30 s, the terrain is grass and the robot continues to move in a wave gait.

The experimental results show that the robot can walk stably on a single-terrain type and transform its gait successfully according to different terrains. At the initial moment, captured images from the Kinect are classified by system. The confidence rating of the classified asphalt is greater than 30%, so the terrain image is judged by the system to be asphalt pavement in a single terrain. Similarly, at 14 s and 22 s, the system outputs a single terrain label for soil and grass. At 5 s and 22 s, the terrain image corresponds to the uncertain category and the highest confidence is less than 30%. Therefore, the terrain is supposed to be mixed, and image infilling method is used until the classification result meets recognition reliability requirements. The system outputs the labels of two terrains and causes the robot to make the corresponding gait transitions.

Table 4. Gait transition algorithm.

Initialize $G \in [0.5\ 1]$; $S_D \in \{D_{ij} = 0, j = 1, 2, \ldots, 6; i = 1, 2, \ldots, 2^n\}$; $B \in \{0, 1\}$; $n = 0$; $T_i \in \{1, 2, \ldots, 6\}$

  Repeat:

  (1) Collect terrain images: color, depth, and infrared;

  (2) Run the obstacle detection module and output B

       if $B = 1$ **then**

          Run image infilling processing I

          Jump to (2)

       else if $B = 0$ **then**

          continue

       end

  (3) Run the terrain classifier module and output S_D

       for $i = 1; i \leq 2^n; i{+}{+}$

          if max $\{D_{ij}, (j = 1, 2, \ldots, 6)\} < 0.3$ **then**

             n++

             Run image segmentation processing

             Run image infilling processing II

             Jump to repeat (3);

         Else if

             output the subscript j of max $\{D_{ij}, (j = 1, 2, \ldots, 6)\}$;

             $T_i = j$

         end

       end

  (4) Output classification results and gait G

       for $i = 1; i \leq 2n{-}1; i{+}{+}$

          if $T_i = 1$ or 2 **then** $G = 0.5$

             else if $T_i = 3$ or 4 **then** $G = 0.75$

             else if $T_i = 5$ or 6 **then** $G = 0.83$

         end

       end

       $T = T_1, T_2, T_3, \ldots, T_{2n-1}$

  (5) Run the robot

  Until: The robot is switched off.

  Note: G is the walking gait; typically, 0.5 for tripod gait, 0.75 for quadruped gait, and 0.83 for wave gait. S_D is the confidence score; j refers to terrain type: 1 for asphalt, 2 for tile, 3 for soil, 4 for gravel, 5 for sand, 6 for grass; i is the serial number of images; B refers to the result of obstacle detection: $B = 1$ means there is an obstacle, $B = 0$ means no obstacle. T_i is the output label of the terrain classifier. Image infilling processing I represents the ILI module, and image infilling processing II represents the SPI module.

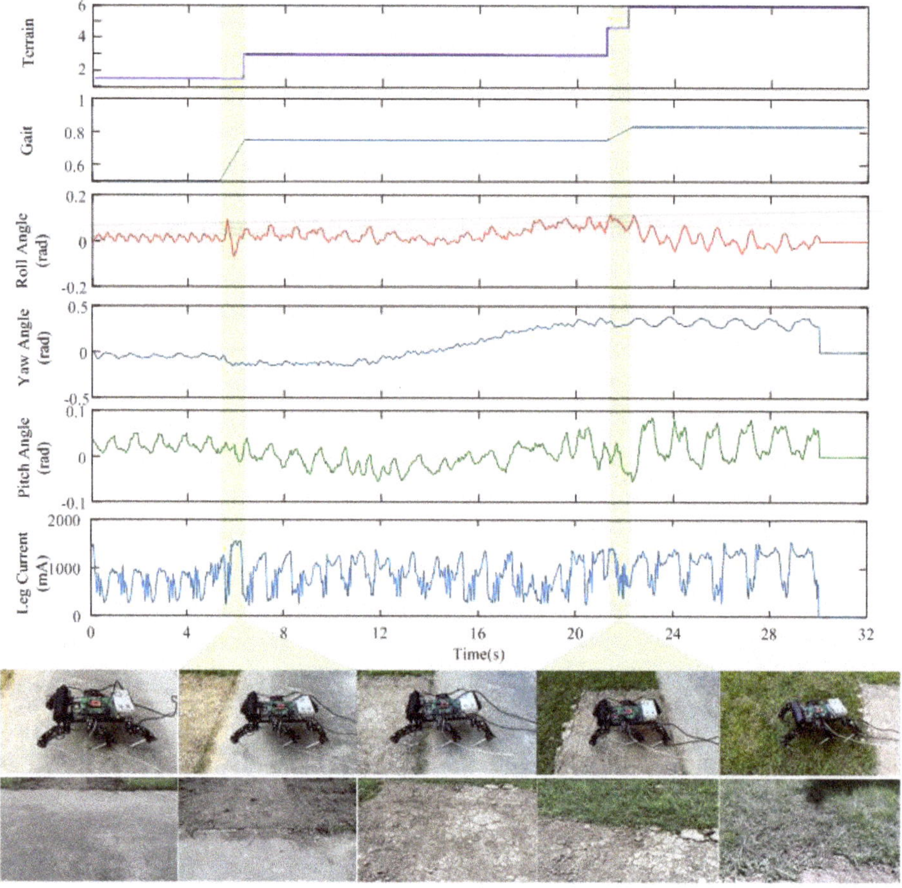

Figure 13. Gait transition.

4. Discussion

This paper describes a terrain classification system for a multilegged robot on complex terrain with obstacles. Several topographic classification methods are summarized in Table 5 [8,11,17,37–41]. With respect to single terrain recognition, several successful single terrain classification methods proved the effectiveness of this kind of methodology via local features, BoW model, and SVM. The common points (also advantages) of these works and our proposed algorithm include the following: Image features are extracted by selecting local features. Unlike color-based and spectra-based methods, local features are invariant to scale, rotation, brightness, and contrast and hence have become popular in image classification. In these methods, the SURF algorithm is used to extract local features of terrain images as input to the BoW model. The performance of SVM in classifying a small number of samples is also excellent. The characteristics of sensor-based information including frequency of leg current [38] and tactile data [40] are also used for terrain recognition. This kind of data is similar in the same terrain and has certain regularity in different terrains. The methods of building the classifier mainly focus on SVM [8,11,17], neural network [37–39], and mixtures of Gaussians [40,41]. Among them, SVM and neural network are the two main classification models. The application of terrain identification to legged robots is mainly concentrated on gait transition and path planning. For terrain classification with a multilegged robot, the precision requirement is low and a simple SVM is good enough for expected results. Our image infilling algorithm has the effect of magnifying local features of the image, which makes the classification more accurate. We also made an innovation in feature extraction: the SURF-BRISK algorithm is more suitable for real-time classification, its matching speed is much faster than the SURF algorithm alone, and its accuracy is also in line with the SURF algorithm.

Table 5. Comparison of recent terrain classification methods.

Author or Method (Year)	Feature	Classification Method	Number of Terrains	For Mixed Terrain?	Application
Khan (2011) [8]	SURF/DAISY	SVM	5	No	Visual terrain classification
Zenker (2013) [11]	SURF/SIFT	SVM	8	No	Energy-efficient gait
Filitchkin (2012) [17]	SURF	SVM	6	Yes	Selecting predetermined gaits
Lee (2011) [37]	SURF	ANN	5	No	Off-road terrain classification for UGV
Ordonez (2013) [38]	Characteristic frequency of leg current	PNN	4	No	Robot planning and motor control
Holder (2016) [39]	Convolutional encoder–decoder	CNN/SVM	12	Yes	Real-time road-scene understanding
Dallaire (2015) [40]	Tactile data	Mixture of Gaussians	12	No	Gait switching
Manduchi (2005) [41]	Color-based	Mixture of Gaussians	3	No	Recognizing different terrains and obstacles
This paper (2017)	SURF-BRISK	SVM	6	Yes	All kinds of outdoor robots

5. Conclusions

In this paper, a novel terrain classification system for accurate recognition of different terrains is proposed. By using Kinect, color, infrared, and depth images are acquired simultaneously. The infrared and depth information are fused and used for obstacle detection. The local feature extraction of terrain images is done by the SURF-BRISK algorithm. A terrain classifier based on the BoW model and SVM is employed. Using the proposed method, different terrains can be classified quickly and precisely. Complex terrain recognition is achieved by the local image infilling method for terrain with obstacles and mixed terrain. According to the experimental results, the proposed method greatly improves the accuracy of complex terrain recognition and plays an important role in locomotion guidance of multilegged robots.

The theoretical contributions and novelty of this paper can be summarized as follows:

(1) Images with obstacles are infilled by surrounding terrain parts in order to improve the classification accuracy. Thus, the local features of images are magnified and the method can achieve satisfactory results.

(2) A super-pixel image infilling method for mixed terrain classification is presented. The average classification accuracy of the proposed method for mixed terrain is over 80%. The proposed method can make acquired data more believable and reliable for locomotion planning and control of intelligent robots.

(3) Multiple terrain labels can be given instead of a single label, which indicates that the presented method is very practical for complex terrains.

This paper focuses on a complex terrain classification system and a combination of terrain classification and obstacle detection to complete the planning of a robot path. In the future, we will improve the rapid transformation of a robot's gait based on terrain information and make the robot more intelligent.

Author Contributions: Y.Z. designed the algorithm. Y.Z., C.M., C.J., and Q.L. designed and carried out the experiments. Y.Z. and C.J. analyzed the experimental data and wrote the paper. Q.L. gave many meaningful suggestions about the structure of the paper.

Funding: This research was funded by the National Natural Science Foundation of China (No. 51605039), the Thirteenth Five-Year Plan Equipment Pre-research Field Fund (No. 61403120407), the China Postdoctoral Science Foundation (No. 2018T111005 and 2016M592728), Fundamental Research Funds for the Central Universities, CHD (No. 300102259308, 300102258203 and 300102259401).

Conflicts of Interest: The authors declare no conflict of interest.

References

1. Yu, Z.; Chen, J.; Dai, Z. Study on Forces Simulation of Gecko Robot Moving on the Ceiling. *Adv. Intell. Soft Comput.* **2012**, *125*, 81–88. [CrossRef]
2. Abbaspour, R. Design and Implementation of Multi-Sensor Based Autonomous Minesweeping Robot. In Proceedings of the International Congress on Ultra Modern Telecommunications & Control Systems & Workshops, Moscow, Russia, 18–20 October 2010; IEEE: Piscataway, NJ, USA, 2010; pp. 443–449. [CrossRef]
3. Ayers, J. Localization and Self-Calibration of a Robot for Volcano Exploration. In Proceedings of the ICRA 04 2004 IEEE International Conference on Robotics and Automation, New Orleans, LA, USA, 26 April–1 May 2004; IEEE: Piscataway, NJ, USA, 2004; Volume 1, pp. 586–591. [CrossRef]
4. Zhao, S.D.; Yuh, J.K. Experimental Study on Advanced Underwater Robot Control. *IEEE Trans. Robot.* **2005**, *21*, 695–703. [CrossRef]
5. Cui, Y.; Gejima, Y.; Kobayashi, T. Study on Cartesian-Type Strawberry-Harvesting Robot. *Sens. Lett.* **2013**, *11*, 1223–1228. [CrossRef]
6. Semler, L.; Furst, J. Wavelet-Based Texture Classification of Tissues in Computed Tomography. In Proceedings of the IEEE International Conference on Image Processing, Atlanta, GA, USA, 8–11 October 2006; IEEE: Piscataway, NJ, USA, 2006; pp. 265–270.
7. Paschos, G. Perceptually uniform color spaces for color texture analysis: An empirical evaluation. *IEEE Trans. Image Proc.* **2001**, *10*, 932–937. [CrossRef]
8. Liu, X.; Wang, D. Texture classification using spectral histograms. *IEEE Trans. Image Proc.* **2003**, *6*, 661–670. [CrossRef]
9. Pietikäinen, M.; Mäenpää, T.; Viertola, J. *Color Texture Classification with Color Histograms and Local Binary Patterns*; IWTAS: New York, NY, USA, 2002; pp. 109–112.
10. Zenker, S.; Aksoy, E.E.; Goldschmidt, D. Visual Terrain Classification for Selecting Energy Efficient Gaits of a Hexapod Robot. In Proceedings of the IEEE/ASME International Conference, Wollongong, Australia, 9–12 July 2013; IEEE: Piscataway, NJ, USA; Advanced Intelligent Mechatronics (AIM): Marseille, France, 2013; pp. 577–584. [CrossRef]
11. Khan, Y.; Komma, P.; Bohlmann, K. Grid-Based Visual Terrain Classification for Outdoor Robots Using Local Features. In Proceedings of the IEEE Symposium on Computational Intelligence in Vehicles and Transportation Systems, Paris, France, 11–15 April 2011; IEEE: Piscataway, NJ, USA, 2011; pp. 16–22. [CrossRef]
12. Kim, J.; Kim, D.; Lee, D. Non-contact Terrain Classification for Autonomous Mobile Robot. In Proceedings of the IEEE International Conference on Robotics and Biomimetics, Guilin, China, 19–23 December 2009; IEEE: Piscataway, NJ, USA, 2009; pp. 824–829. [CrossRef]
13. Ojeda, L.; Borenstein, J.; Witus, G.; Karlsen, R. Terrain characterization and classification with a mobile robot. *J. Field Robot.* **2006**, *9*, 103–122. [CrossRef]
14. Larson, A.C.; Voyles, R.M.; Bae, J. Evolving Gaits for Increased Selectivity in Terrain Classification. In Proceedings of the IEEE/RSJ International Conference on Intelligent Robots and Systems, Sheraton Hotel and Marina, San Diego, CA, USA, 29 October–2 November 2007; IEEE: Piscataway, NJ, USA, 2007; pp. 3691–3696. [CrossRef]
15. Hoepflinger, M.A.; Remy, C.D.; Hutter, M. Terrain Classification for Legged Robots. In Proceedings of the IEEE International Conference on Robotics and Automation, ICRA 2010, Anchorage, AK, USA, 3–7 May 2010; IEEE: Piscataway, NJ, USA, 2010; pp. 2828–2833. [CrossRef]

16. Jitpakdee, R.; Maneewam, T. Neural Networks Terrain Classification Using Inertial Measurement Unit for an Autonomous Vehicle. In Proceedings of the SICE Annual Conference, Tokyo, Japan, 20–22 August 2008; IEEE: Piscataway, NJ, USA, 2008; pp. 554–558. [CrossRef]
17. Filitchkin, P.; Byl, K. Feature-Based Terrain Classification for LittleDog. In Proceedings of the 2012 IEEE/RSJ International Conference on Intelligent Robots and Systems (IROS), Vilamoura, Portugal, 7–12 October 2012; IEEE: Piscataway, NJ, USA, 2012; pp. 1387–1392. [CrossRef]
18. Zuo, L.; Wang, M.; Yang, Y. Complex Terrain Classification algorithm Based on Multi-Sensors Fusion. In Proceedings of the 32nd Chinese Control Conference (CCC), Xi'an, China, 26–28 July 2013; Inspec Accession Number: 13886652. pp. 5722–5727.
19. Ugur, E.; Dogar, M.R.; Cakmak, M.; Sahin, E. The learning and use of traversability affordance using range images on a mobile robot. In Proceedings of the 2007 IEEE International Conference on Robotics and Automation, Roma, Italy, 10–14 April 2007; IEEE: Piscataway, NJ, USA, 2007. [CrossRef]
20. Zhu, Y.G.; Jin, B. Trajectory Correction and Locomotion Analysis of a Hexapod Walking Robot with Semi-Round Rigid Feet. *Sensors* **2016**, *9*, 1392. [CrossRef] [PubMed]
21. Zhu, Y.G.; Jin, B. Compliance control of a legged robot based on improved adaptive control: Method and experiments. *Int. J. Robot. Autom.* **2016**, *5*, 366–373. [CrossRef]
22. Zhu, Y.G.; Wu, Y.S.; Liu, Q.; Guo, T.; Qin, R.; Hui, J.Z. A backward control based on σ -Hopf oscillator with decoupled parameters for smooth locomotion of bio-inspired legged robot. *Robot. Auton. Syst.* **2018**, *106*, 165–178. [CrossRef]
23. Zhu, Y.G.; Guo, T.; Liu, Q.; Zhu, Q.; Zhao, X.; Jin, B. Turning and Radius Deviation Correction for a Hexapod Walking Robot Based on an Ant-Inspired Sensory Strategy. *Sensors* **2017**, *17*, 2710. [CrossRef]
24. Discant, A.; Rogozan, A. Sensors for Obstacle Detection a Survey. In Proceedings of the 30th International Spring Seminar on the Electronics Technology, Cluj-Napoca, Romania, 9–13 May 2007; IEEE: Piscataway, NJ, USA, 2007; pp. 100–105. [CrossRef]
25. Zhu, Y.G.; Yi, B.M.; Guo, T. A Simple Outdoor Environment Obstacle Detection Method Based on Information Fusion of Depth and Infrared. *J. Robot.* **2016**, *9*, 1–10. [CrossRef]
26. Comaniciu, D.; Meer, P. Mean shift: A robust approach toward feature space analysis. *IEEE Trans. Pattern Anal. Mach. Intell.* **2002**, *5*, 603–619. [CrossRef]
27. Comaniciu, D. Rr Vision and Pattern Recognition. In Proceedings of the Conference on Computer Vision and Pattern Recognition, Hilton Head Island, SC, USA, 15 June 2000; IEEE: Piscataway, NJ, USA, 2000; pp. 142–149. [CrossRef]
28. Qin, L. Category Related BoW Model for Image Classification. *J. Inf. Comput. Sci.* **2015**, *9*, 3547–3554. [CrossRef]
29. Yadav, S.; Shukla, S. Analysis of k-Fold Cross-Validation over Hold-Out Validation on Colossal Datasets for Quality Classification. In Proceedings of the IEEE International Conference on Advanced Computing, Bhimavaram, India, 27–28 February 2016; IEEE: Piscataway, NJ, USA, 2016; pp. 71–83. [CrossRef]
30. Leutenegger, S.; Chli, M.; Siegwart, R. BRISK: Binary Robust Invariant Scalable Keypoints. In Proceedings of the 2011 IEEE International Conference on Computer Vision-ICCV, Barcelona, Spain, 6–13 November 2011; IEEE: Piscataway, NJ, USA, 2011; Volume 11, pp. 2548–2555. [CrossRef]
31. Cui, Z.; Li, Z. Two kinds of improved template matching recognition algorithm. *Comput. Eng. Des.* **2006**, *6*, 1083–1084.
32. Fei-Fei, L.; Perona, P. A Bayesian Hierarchical Model for Learning Natural Scene Categories. In Proceedings of the Conference on IEEE Computer Vision and Pattern Recognition, San Diego, CA, USA, 20–26 June 2005; IEEE Computer Society: Washington, DC, USA, 2005; pp. 524–531. [CrossRef]
33. Wang, Q.; Wang, C. Review of k-means algorithm for clustering. *Electr. Des. Eng.* **2014**, *6*, 479–484.
34. Chapelle, O. Training a Support Vector Machine in the Primal. *Neural Comput.* **2007**, *19*, 1155–1178. [CrossRef]
35. Ren, X.; Malik, J. Learning a Classification Model for Segmentation. In Proceedings of the IEEE International Conference on Computer Vision, Nice, France, 13–16 October 2003; IEEE: Piscataway, NJ, USA, 2003; pp. 11–17. [CrossRef]
36. Song, X.; Zhou, L.; Li, Z. Review on superpixel methods in image segmentation. *J. Image Gr.* **2015**, *20*, 599–608.

37. Lee, S.Y.; Kwak, D.M. A terrain Classification Method for UGV Autonomous Navigation Based on SURF. In Proceedings of the International Conference on Ubiquitous Robots & Ambient Intelligence, Incheon, Korea, 23–26 November 2011; IEEE: Piscataway, NJ, USA, 2011; pp. 303–306. [CrossRef]

38. Ordonez, C. Terrain identification for R Hex-type robots. *Unmanned Syst. Technol. XV* **2013**, *3*, 292–298. [CrossRef]

39. Holder, C.J.; Breckon, T.P. From On-Road to Off: Transfer Learning Within a Deep Convolutional Neural Network for Segmentation and Classification of Off-Road Scenes. *Springer Int. Publ.* **2016**, *9*, 149–162. [CrossRef]

40. Dallaire, P. Learning Terrain Types with the Pitman-Yor Process Mixtures of Gaussians for a Legged Robot. In Proceedings of the 2015 IEEE/RSJ International Conference on Intelligent Robots and Systems (IROS), Hamburg, Germany, 28 September–2 October 2015; IEEE: Piscataway, NJ, USA, 2015; pp. 3457–3463. [CrossRef]

41. Manduchi, R.; Castano, A.; Talukder, A. Obstacle Detection and Terrain Classification for Autonomous Off-Road Navigation. *Auton. Robots* **2005**, *1*, 81–102. [CrossRef]

Article

Multi-Robot Exploration Based on Multi-Objective Grey Wolf Optimizer

Albina Kamalova [1], Sergey Navruzov [1], Dianwei Qian [2] and Suk Gyu Lee [1,*]

[1]   Department of Electrical Engineering, Yeungnam University, Gyeongsan 38541, Korea
[2]   School of Control and Computer Engineering, North China Electric Power University, Beijing 102206, China
*   Correspondence: sglee@ynu.ac.kr; Tel.: +82-10-3060-2487

Received: 25 June 2019; Accepted: 17 July 2019; Published: 22 July 2019

Abstract: In this paper, we used multi-objective optimization in the exploration of unknown space. Exploration is the process of generating models of environments from sensor data. The goal of the exploration is to create a finite map of indoor space. It is common practice in mobile robotics to consider the exploration as a single-objective problem, which is to maximize a search of uncertainty. In this study, we proposed a new methodology of exploration with two conflicting objectives: to search for a new place and to enhance map accuracy. The proposed multiple-objective exploration uses the Multi-Objective Grey Wolf Optimizer algorithm. It begins with the initialization of the grey wolf population, which are waypoints in our multi-robot exploration. Once the waypoint positions are set in the beginning, they stay unchanged through all iterations. The role of updating the position belongs to the robots, which select the non-dominated waypoints among them. The waypoint selection results from two objective functions. The performance of the multi-objective exploration is presented. The trade-off among objective functions is unveiled by the Pareto-optimal solutions. A comparison with other algorithms is implemented in the end.

Keywords: multi-robot systems; multi-objective optimization; grey wolf optimizer; waypoints; exploration; uncertainties; unknown environment; mapping; grid map occupancy

---

1. Introduction

In robotics, exploration pertains to the process of scanning and mapping out an environment to produce a map, which can be used by a robot or group of robots for further work. Based on the type of environment, exploration can be one of the following: outdoor, indoor, or underwater, using a mobile-robot or multi-robot systems [1,2]. In this study, we focused on indoor exploration by robots, which are equipped with ranging sensors. It can be assumed that these robots with onboard sensors can scan an environment without any difficulties by walking randomly around. However, their motion is not efficient, which can result in an incomplete map coverage. As a solution to this issue, this paper proposes an algorithm that enhances the efficiency of the multi-robot exploration by using a multi-objective optimization strategy.

Naturally, all real-world optimization problems in engineering pursue multiple goals. They may differ from each other in various fields. However, it is common for all to optimize problems by maximization or minimization functions. In the past, multiple optimization problems were solved by one function because of the lack of suitable solution methodologies for solving a multi-objective optimization problem (MOOP) as a single-objective optimization problem. With the development of evolutionary algorithms, new techniques, which seek to optimize two or more conflicting objectives in one simulation run, have been applied to MOOPs. This new research area is named multi-objective optimization (MOO) [3].

In robotics, studies related to optimization have been gaining wide attention [4]. If we consider multi-robot systems [5], optimization is popularly applied in path planning [6], formation [7], exploration [8], and other fields where decision-making control needs to be optimized. Previous research conventionally found the optimal solutions as separated single-objective tasks: short path, obstacle-free motion, smoothest route, and constant search of uncertain terrain. The new impact of optimization in robotics is obtained due to the metaheuristics and its nature-inspired optimization techniques [9]. The nature-inspired algorithms are not only restricted to robotics but also have significant applications in different fields. Due to this, they have attracted the attention of scholars [10,11].

Metaheuristic algorithms are optimization approaches, which emulate the intelligence of various species of animals in nature. The number of agents classifies the metaheuristic algorithms into single-solution-based and population-based algorithms. In both classes, the solutions improve over the course of iterations with one single agent or an entire swarm of agents, respectively. The main advantage of population-based approaches is their ability to avoid stagnation in the local optima due to the number of agents. The swarm can explore search space more and faster than a single agent. Regardless of this, the benefit of one class over the other depends on its application in a certain problem. However, it is important to mention the No-Free-Lunch (NFL) theorem for optimization, which assures that there is no algorithm with universal optimal solution by all criteria and in all domains [12].

Despite the number of agents, the metaheuristic algorithms can be classified into single and multi-objective optimization techniques according to the number of objective functions. A multi-objective optimization is an extended approach to single optimization. It allows finding an optimal solution of two independent objectives simultaneously. In order to select just one best solution from the available ones, a trade-off should be considered among them. The Pareto-optimal front helps pick up one of the suitable solutions satisfying two objective functions.

Examples of the single-objective algorithms include the Particle Swarm Optimization (PSO) [13], Genetic Algorithm (GA) [14], Ant Colony Optimization (ACO) [15], Grey Wolf Optimizer algorithms (GWO) [16,17], which have the extended multi-objective optimization variations, namely MOPSO [18], MOGA [19], m-ACO [20], and MOGWO [21], respectively. In our previous study [8], we used the coordinated multi-robot exploration [22] and GWO algorithms together as a hybrid. In this study, our interest is to solve the multi-robot exploration problem as the MOOP using Multi-Objective Grey Wolf Optimizer (MOGWO).

Using the MOGWO exploration, we defined two objectives for optimization, namely the maximization of the search for new area and the minimization of the inaccuracy of the explored map. It can be said that the search process is divided into two stages (Figure 1). They switch during the simulation run depending on the value of the GWO parameter. If the value is greater than one, it searches occluded space. If the value is less than one, it increases the map accuracy by repeated visits in the explored space. It needs to be emphasized that the occupancy grid map with probabilistic values is used in this study [23].

The MOGWO exploration employs static waypoints in the simulation, which promotes the efficient exploration of an indoor simulated environment. It can be noted that the waypoints belong to the programmed logic of the algorithm and are not supposed to be used in the real environment [24]. The waypoints are grey wolf agents with some costs of probability values. In each iteration, the robots search alpha, beta, and gamma waypoints and save them in an archive wherein avoiding the selection the same non-dominated waypoints for several robots. After the selection, the robots can compute the next position, which is the closest to the average position of alpha, beta, and gamma waypoints, from among the frontier cells [25].

This paper is organized as follows: in Section 2, we briefly recall different algorithms of multi-robot exploration and evolutionary optimization techniques used in related works. In Section 3, the theory of GWO and MOGWO is presented. Sections 4 and 5 are dedicated to the proposed MOGWO exploration and its performance. Section 6 concludes the present study.

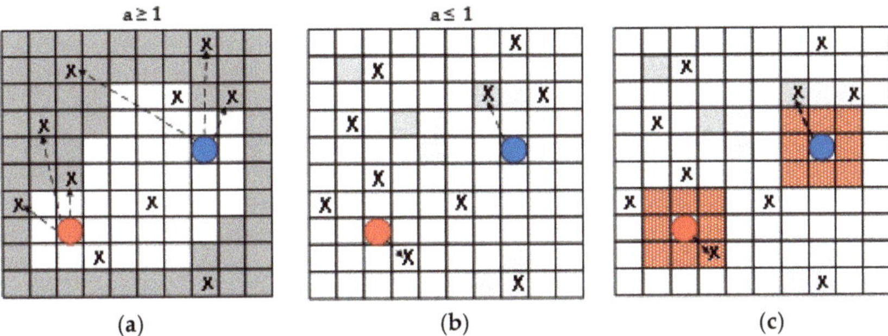

Figure 1. The proposed Multi-Objective Grey Wolf Optimizer (MOGWO) exploration in two stages:
(**a**) selecting the waypoints in unexplored space with high probability values of occupancy grid map;
(**b**) selecting the closest waypoints in explored space with high probability values proportional to the
distance; (**c**) the next position is one of the frontier points, which is the nearest to the selected waypoint.

2. Related Work

In the last two decades, many techniques have been proposed for robot exploration. Among
them, there are novel fundamental, hybridized, and modified methods. In this section, studies on
the different algorithms and the impact of the evolutionary optimization techniques in exploration
are discussed.

Considering exploration as one of the branches in robotics, Yamauchi et al.'s frontier-based method
is the pioneering work in this field [26]. From that time up to now, many frontier-based studies have
appeared, most of which were hybridized or modified with success.

The coordinated multi-robot exploration (CME) is frontier-based with the emphasis on the
cooperative work with a team of robots [22]. The robot's mission is to search for maximum utility
with the minimum cost that diverges the robots from each other keeping the direction to search
unexplored space. The alternative coordinated method is the randomized graph approach [27,28].
It builds a roadmap in an explored area that navigates robots to move through safe paths. Recently,
Alfredo et al. [29] introduced the efficient backtracking concept to the random exploration graph,
preventing the same robot in visiting the same place more than once. These above-mentioned methods
have the common idea of using frontier-based control.

Another approach in exploration that is completely different in theory and practice is artificial
intelligence (AI). Reinforcement Learning (RL) and convolutional neural network (CNN) are such
attempts, which have been proposed in previous studies [30–32]. The exploration-based approach
on neural networks differs considerably from the frontier-based approach in terms of environment
perception and control system. Visual sensors (cameras) scan a place for further computation using
image-processing algorithms [33]. The output of the calculation is the interaction of the robot with the
environment. Lei Tai et al. [34] conducted a survey of leading studies in mobile robotics using deep
learning from perception to control systems.

Recently, a novel branch of exploration that employs nature-inspired optimization techniques has
appeared. The approaches seek to enhance existing solutions to exploration. Sharma S. et al. [35] applied
clustering-based distribution and bio-inspired algorithms such as PSO, Bacteria Foraging Optimization,
and Bat algorithm. The clustering provides a direction of robot motion, while the nature-inspired
approaches involve exploring the unknown area. The study of [36] applied a combination of PSO,
fractional calculus, and fuzzy inference system. They compared their results with other six other
PSO variations that showed effective multi-robot exploration. A similar waypoint concept in our
study was performed in [37]. The artificial pheromone and fuzzy controllers help the multi-robot
systems to navigate efficiently by distributing the search between robots and avoiding repeated visits
in explored regions.

The study of [38] involved more than one optimization problem in the exploration, which is important to highlight here. The optimal solution seeks to minimize two objective functions: the variance of path lengths and the sum of the path lengths of all robots. Compared to our research, they applied the K-Means clustering algorithm instead of the bio-inspired technique used in our study.

The study of [39] presented the auto-adaptive multi-objective strategy for multi-robot exploration, where the multi-objective concept consisted of two missions: a search of uncertainties and stable communication. This work is closely related to the present work, but the focus of their research is an assessment of the communication conditions for providing efficient map coverage, which is a different perspective compared to our study.

In regard to multi-objective optimization in multi-robot systems, MOPSO [40] and multi-ACO [41] have already been applied to path planning problem. Broadly speaking, the metaheuristic algorithms are often applied in path planning problems compared to other issues, mainly, because optimization is the core study for finding a short and smooth path.

In general, MOGWO has never been applied before in mobile robotics studies.

3. Single and Multi-Objective Grey Wolf Optimizer

The section briefly describes the theories of GWO and MOGWO. The two techniques have the interconnection that one is inferred to another. Firstly, GWO will be presented, and then, the concept will be extended to the multi-objective optimization using MOGWO.

3.1. Grey Wolf Optimizer

GWO is a population-based metaheuristic algorithm, which mimics the wolf hunting process. Population- and single-based optimization algorithms differ from each other in the number of agents used to carry out a search of a global optimum. Each agent is a candidate for finding a global optimum. Figure 2 shows the GWO simulation. The search begins when all agents obtain random x, y values, where lower bound $\leq x, y \leq$ upper bound. Then, the cost function defines the best candidates α, β, γ among them in each iteration (Figure 2a) by equation:

$$f(x) = \sum_{i=1}^{n} x_i^2 \tag{1}$$

Every single agent needs to compute D_α, D_β, D_γ, and then, X_1, X_2, X_3 can be found. The random and adaptive vectors, $\vec{A}$ and $\vec{C}$ are upgraded in each iteration.

$$\vec{D}_\alpha = |\vec{C}_1 \cdot \vec{X}_\alpha - \vec{X}|, \vec{D}_\beta = |\vec{C}_2 \cdot \vec{X}_\beta - \vec{X}| \vec{D}_\gamma = |\vec{C}_3 \cdot \vec{X}_\gamma - \vec{X}| \tag{2}$$

$$\vec{X}_1 = \vec{X}_\alpha - \vec{A}_1 \cdot (\vec{D}_\alpha), \vec{X}_2 = \vec{X}_\beta - \vec{A}_2 \cdot (\vec{D}_\beta), \vec{X}_3 = \vec{X}_\gamma - \vec{A}_3 \cdot (\vec{D}_\gamma), \tag{3}$$

Finally, the next agent's position is mean value of $\vec{X}_1$, $\vec{X}_2$, $\vec{X}_3$, which Figure 2c is illustrated.

$$\vec{X}(t+1) = \frac{\vec{X}_1 + \vec{X}_2 + \vec{X}_3}{3} \tag{4}$$

The same calculation will be repeated in each run time for every search agent.

Depending on the values of the vectors, $\vec{A}$ and $\vec{C}$, also denoted as GWO parameters, the two phases make the transition between divergence (exploration) and convergence (exploitation) in the optimal solution search. The GWO parameters are calculated as follows:

$$\vec{A} = 2\vec{a} \cdot \vec{r}_1 - \vec{a},$$

$$\vec{C} = 2 \cdot \vec{r_2},$$

where the value of $\vec{a}$ decreases linearly from 2 to 0 using the update equation for iteration t:

$$\vec{a_t} = \vec{a_t} - \frac{a_0}{t}, \tag{5}$$

and $\vec{r}_1$ and $\vec{r}_2$ are random values ranging from 0 to 1.

In GWO, the parameter $\vec{A}$ determines the exploration and exploitation in searching behavior. Each agent of the population performs divergence when $\vec{A} > 1$ and executes convergence from α, β, γ agents when $\vec{A} < 1$. Figure 2 illustrates how the agent largely changes the next position at iterations t and $t+1$ by the divergence of the parameter $\vec{A}$, which is linearly decreasing.

The parameter $\vec{C}$ randomly determines the exploration or exploitation tendencies without dependency on the iterations. The stochastic mechanism in GWO allows the enhanced search of optimality by reaching different positions around the best solutions.

In recent years, the GWO algorithm has been widely modified in various studies. In study [42], the authors improved the convergence speed of GWO by guiding the population using the alpha solution. In another study [43], the new operator called reflecting learning was introduced in the algorithm. It mproves the search ability of GWO by the principle of light reflection in physics. The optimization was enhanced in studies [44–46] by random walk strategies and Levy flights distribution as well.

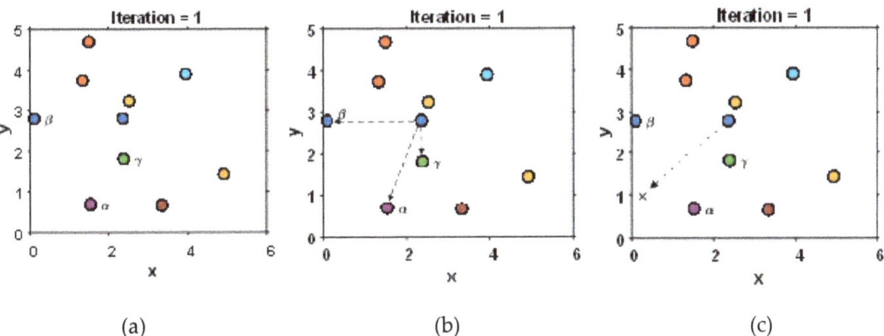

Figure 2. GWO simulation at first iteration: (a) initialization of grey wolf population and searching for the non-dominated agents by the cost sphere function of Equation (1); (b) calculating the divergence of each agent positions to α, β, γ agents in Equation (2); and (c) calculating the next agent position by Equations (3) and (4). The position of agent $x(t) = 2.3531$, $y(t) = 2.8036$ will change the position in the next iteration to $x(t+1) = 0.2914$, $y(t) = 0.9727$, where $t = 1$.

3.2. Multi-Objective Grey Wolf Optimizer

Two new components were integrated into MOGWO for performing multi-objective optimization: an archive of the best non-dominated solutions and a leader selection strategy of alpha, beta, and gamma solutions. The archive is needed for storing non-dominated Pareto solutions through the course of iterations. The archive controller has dominant sorting rules for entering new solutions and for archive states. The size of the archive is closely related to the number of objective functions, which is named segments or hypercubes. Figure 3a shows the archive of three hypercubes with the non-dominated solutions for the t-iteration.

The second component is a leader selection, which chooses the least crowded hypercube of the search space and offers available non-dominated solutions from the archive (Figure 3b). In case there are only two solutions, the third one can be taken from the second least crowded hypercube.

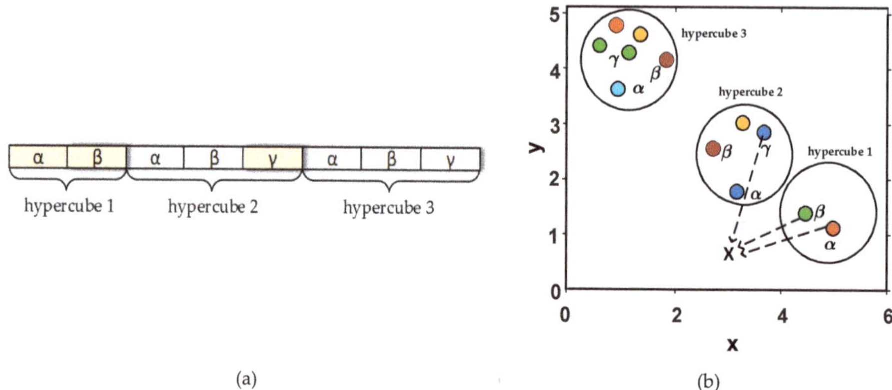

(a)   (b)

Figure 3. Two modules of MOGWO: (**a**) the archive, which consists of three hypercubes, and (**b**) the leader selection mechanism.

Generally, it can be said that the archive stores the best solutions for each objective function. It saves them not only as alpha, beta, and gamma agents, but also with the segment priorities, which are defined by the number of total solutions. Thus, the global best solution can be chosen among the local ones in the archive. This selection mechanism in MOGWO prevents the picking of the same leaders. In other words, it avoids stagnation in local optimal points. Figure 4 shows the full algorithm of MOGWO.

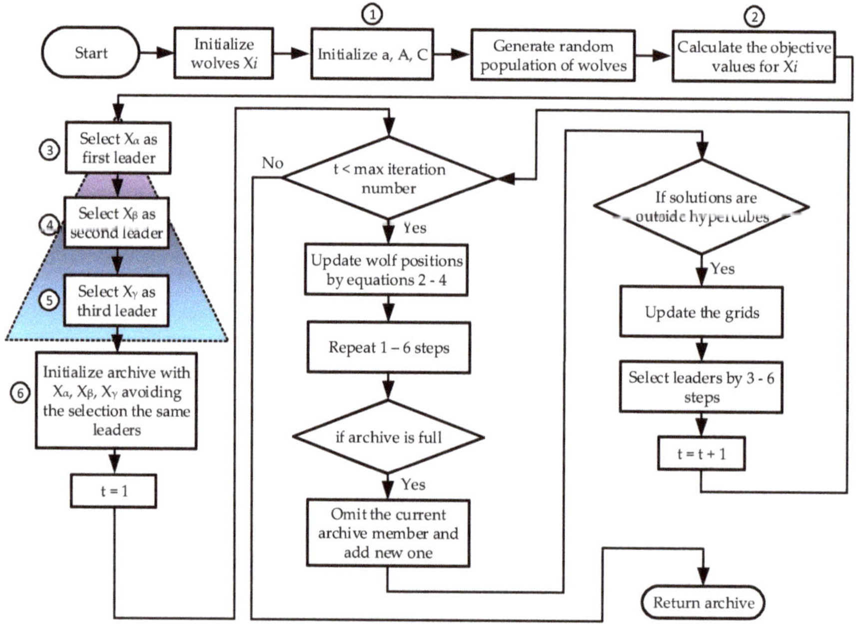

Figure 4. MOGWO algorithm.

MOGWO finds application in cloud computing for virtual machine placement [47], medicine for preventing cervical cancer by scanning images [48], wind power for speed forecasting [49], and energy-efficient scheduling [50]. However, it has not been used in the robotic field up to this time.

4. MOGWO Exploration for Multi-Robot System

In this section, we describe the proposed multi-robot exploration based on MOGWO optimization. First, we define the optimization problems in the exploration. As mentioned above, there are two objective functions for which the study tries to find an optimal solution. Then, the second subsection presents the approach for solving the problems using the MOGWO exploration algorithm.

4.1. Mathematical Formulation of MOOPs in the Multi-Robot Exploration

The process of searching uncertainties by a team of robots can be considered a multi-tasking system. Each robot receives the sensor reading data and upgrades the probabilities of the grid occupancy in the map. One robot should have the same task as another robot in the multi-robot system, wherein the task can be any of the following: scanning the environment using sensors, avoiding obstacles and collisions with other robots, seeking to explore new terrain, and increasing the accuracy of the map. It means that together, each single robot should provide good implementation as a multi-robot system satisfying the multi-objective functions for obtaining the best solutions.

In this paper, we formulated the objective functions of the exploration as follows:

$$\text{Maximize}: \ f_1 \ \to \ number\ of\ explored\ cells, \tag{6}$$

$$\text{Minimize}: \ f_2 \ \to \ probability\ values\ P\big(occ_{x,y}\big). \tag{7}$$

Subject to:

$$i \geq (total\ number\ of\ cells - total\ number\ of\ obstacle\ cells)\,/\,number\ of\ robots, \tag{8}$$

$$wp_w \ \geq \ number\ of\ robots^3, \tag{9}$$

$$wp_w \ \leq \ total\ number\ of\ cells - total\ number\ of\ obstacle\ cells, \tag{10}$$

where,

$$i - number\ of\ iterations,$$

$$wp_w - number\ of\ waypoints.$$

The first objective function in Equation (6) tries to maximize the search space by visiting various numbers of cells in the map. In a good scenario, robots should avoid explored cells. The waypoints in MOGWO exploration allow saving the direction to the unexplored part of the map. However, there are some constraints for a successful search. For example, the number of waypoints should not be too small and big. In the case when it is small, robots will stay in one point, because they do not have the tasks to drive next waypoints. If it is bigger than the total number of cells in a map, then a robot will drive around one place longer than it is needed.

After the map is explored, the second objective function tries to improve the map accuracy by reducing the probability values of the grid cells. It means once a sensor beam touches a grid cell, the cell is marked as explored. However, the signal strength projected on the cell is not identical. In the robot position, the probability value has the lowest value. In frontier cells, the values are higher according to the power of a signal.

In the subsection below, the MOGWO exploration algorithm is described extensively.

4.2. The Proposed MOGWO Exploration Algorithm

Equations (6) and (7) define the objectives of the exploration in this study. For such problems, there is no single solution that satisfies all objectives simultaneously at one time. It is not possible to explore new cells (f_1) and to revisit explored cells (f_2) at the same time. Based on the GWO parameter a in Equation (5), the search process is divided into two parts. When $a > 1$, it searches new waypoints.

When $a < 1$, the process switches to revisiting already explored areas to improve the map accuracy. Thus, the approach serves two MOOPs in a single run-time.

Algorithm 1 demonstrates the MOGWO exploration for the multi-robot system. The process begins with the random initialization of waypoints in the search space. Their positions are set only once in the first iteration and will not be upgraded throughout the whole exploration. In line 1, it was noted that the number of waypoints should be higher than $nRbt^3$ because each robot needs at least three of the best solutions α, β, γ for the search.

Algorithm 1. The pseudocode of the proposed MOGWO exploration
1: Set waypoints wp_w randomly in unknown space ($wp > nRbt^3$)
2: Set the archive is empty
3: Set initial robot position r_j ($j = nRbt$)
4: Initialize a, A, C
5: **while** t is not over
6:     Update A, C
7:     Set the archive is empty
8:     **for** j = 1: nRbt
9:         Find current position x, y of r_1, r_2, r_3
10:         Find frontier point $Vn_{x,y}$ (n = 1, … ,8) of r_j
11:         Insert rays to the map from x, y position
12:         Calculate the distances by the objective function (f_1) wp_w
13:         Calculate the probability values by objective function (f_2) for wp_w
14:         if a $\geq$ 1
15:             **if** wp_w is explored
16:                 Then, to increase wp_w cost
17:             **if else** wp_w is unexplored
18:                 Then, wp_w cost
19:             **end if**
20:             Find minimum wp_α, wp_β, wp_δ costs and save in archive
21:             Find D_α, D_β, D_δ of r_j by Equation (2)
22:             Find X_1, X_2, X_3 and $X(i+1)$ by Equations (3) and (4)
23:             Find $r_j(t+1) = min\left(eucl\_dist\left[Vn_{x,y}, X(t+1)\right]\right)$
24:         **end if**
25:         if a $\leq$ 1
26:             Divide probability cost by distance cost
27:             Find maximum wp_α, wp_β, wp_δ costs and save in archive
28:             Find $r_j(t+1) = min\left(eucl\_dist\left[Vn_{x,y}, wp_\alpha\right]\right)$
29:         **end if**
30:     **end for**
31: Reduce a
32: **show** map
33: **end while**

The proposed algorithm uses the archive for the same purposes as MOGWO does. It allows the storage of the non-dominated solution, which prevents the repeated selection of the same waypoint by the robots. For each iteration in the loop, the robots upgrade the positions, the GWO parameters, and the positions and probability costs of the frontier cells (lines 9, 10).

Two objective functions are used for all stages (lines 12, 13). The first one calculates the distances between waypoints and robots. The second objective function computes the probability values in the waypoint positions. Thus, the waypoints have distance costs of f_1 and probability costs f_2.

Lines 14–24 show the exploration stage for $a \geq 1$. At first, the algorithm needs to divide the waypoints into explored and unexplored ones due to the probability costs in lines 15–19. Then, it can select the unexplored wp_α, wp_β, and wp_γ according to the distance costs (Figure 5). In lines 21 and 22, it computes the position $X(t+1)$. However, the robots cannot jump physically to the position, thus, the frontier cell, which is the closest to $X(t+1)$, is selected for the next robot position.

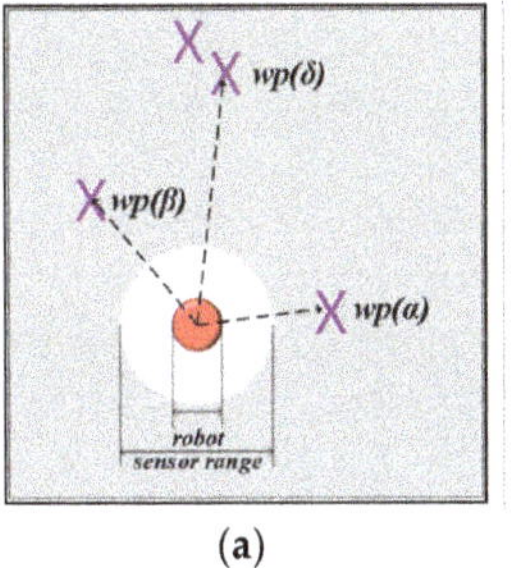
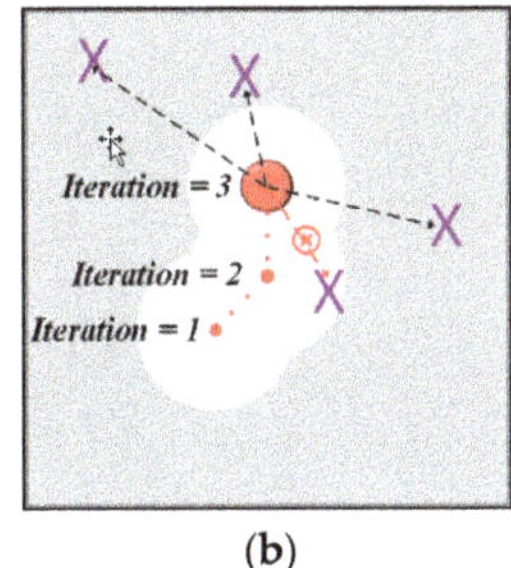
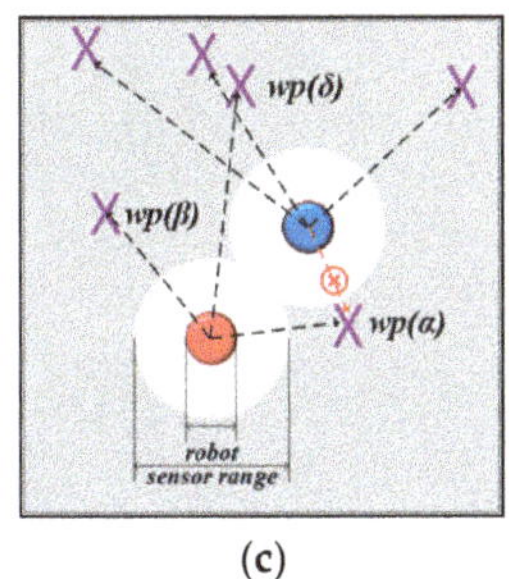

(a)   (b)   (c)

Figure 5. MOGWO exploration: (**a**) selection of the closest waypoints $wp_\alpha, wp_\beta, wp_\gamma$ among wp_w; (**b**) the waypoints that are located in the explored space have the lowest probability to be selected by the robot than waypoints in unknown space; (**c**) the best waypoints for one robots should not be duplicated for another robot.

Lines 25–28 illustrate the exploitation stage for $a \leq 1$. It divides the probability cost f_2 by the distance cost f_1. Afterward, it finds the maximum value of the result in line 26 and saves it in the archive. The next robot position is one of the frontier cells, which is closest to the alpha waypoint.

In line 30, the a parameter is reduced from 2 to 0 iteratively. Lastly, the map is upgraded by all robots in the end of iteration.

The implementation results are performed in the next section. The multi-robot exploration by MOGWO algorithm obtained notable results, which can be observed in various conditions by adjusting the numbers of waypoints and iterations.

5. Simulation Results and Analysis

In this section, we implemented the proposed MOGWO exploration and analyzed the obtained results. As alternatives, there are varying parameters for testing the simulation performance such as the number of waypoints and the number of iterations. The parameters are used to test the algorithm performance in several ways. However, some parameters were maintained as constant throughout all the simulation runs (see Table 1).

Table 1. Experimental parameters.

Parameters	Value
Initial poses	$r_1 = (5,5)$, $r_2 = (7,9)$, $r_3 = (4,9)$
Map size	15×15
Obstacle Width	0.5
Ray Length	1.5
Probabilities of occupancy cells	$P(robotR_{x,y}R) = 0.0010$ $P(obstacleR_{x,y}R) = 0.9990$ $P(unexploredR_{x,y}R) = 0.5000$ $P(unexploredR_{x,y}R) > P(exploredR_{x,y}R) \geq P(robotR_{x,y}R)$

The major goal, which we seek to attain, is to know how many iterations and how many waypoints are needed for efficient exploration. If the number of iterations is too low, the robots do not have time to explore the entire map physically considering that the size of the step in each iteration is unchanged. The same is true for the number of waypoints. They should be enough for free robot driving in the environment. In the next subsection, the experiments with certain constraints are presented, and the Pareto optimal set is proposed for the selection of the optimal solution of the environment.

5.1. Simulation Results

The analysis of the MOGWO exploration algorithm takes into consideration two aspects of the objective function: how it explores and how it improves the accuracy of the map. The experiment constraints influence the performance of the algorithm. Due to the GWO stochastic parameters, the decision-making process can be different in each simulation run. It leads us to test the algorithm performance several times with the same constraints. Based on the experiment parameters in Table 1, it can be calculated using Equation (8) that the iteration number should not be less than 60 and more than 120. In addition, for the waypoints, the range should be from 60 to 150 for three robots in a certain map size (Equations (9) and (10)). In this study, we selected the parameters as 60, 80, 100, and 120 iterations and 60, 80, 100, and 150 waypoints. Table 2 shows the results of map coverage in percentage, which is computed using the following equation:

$$Map\ coverage\ (\%) = \left(100 - \frac{sum\ of\ cell\ probability\ values\ after\ run\ time}{sum\ of\ cell\ probability\ values\ before\ run\ time}\right) \times 100. \qquad (11)$$

Table 2. Simulation results of the MOGWO exploration in several constraints.

Number of Waypoints	Number of Iterations			
	60	80	100	120
60	90.17 **92.36** 82.08	92.42 91.16 88.27	91.51 85.93 93.54	**98.21** 95.81 95.38
	85.55 89.10 85.36	90.20 87.91 87.90	95.92 96.07 93.47	89.61 89.62 89.03
	86.58 85.52 90.62	90.72 91.70 94.03	92.07 92.30 190.33	94.26 98.02 96.59
	86.79 90.01 82.14	91.11 89.21 88.30	87.92 95.22 91.70	92.89 95.22 93.38
	87.78 88.84 88.00	93.94 88.85 93.18	95.30 88.25 95.76	96.33 91.68 94.10
	86.03 87.62 89.68	90.65 85.15 92.61	88.89 93.60 93.46	95.99 91.63 93.65
	81.42 88.60 92.15	94.39 89.23 93.05	92.21 90.34 92.77	91.60 91.22 97.20
	88.48 85.58 84.89	90.43 91.99 89.39	92.81 91.31 94.86	94.28 91.69 95.57
	84.16 86.74 86.61	93.40 94.95 **95.50**	93.50 88.99 91.83	95.84 92.64 95.48
	83.59 91.63 84.70	93.14 92.40 95.23	**96.57** 94.61 92.46	94.63 97.47 97.33
80	90.53 89.00 89.24	96.28 93.19 94.32	90.16 95.48 94.30	96.68 94.28 86.57
	90.01 90.24 89.59	93.73 94.72 92.70	90.94 96.16 86.94	97.28 95.38 94.13
	88.92 88.62 86.90	95.38 96.69 91.98	96.42 96.67 94.17	**98.40** 94.95 97.86
	87.66 84.95 91.15	89.37 90.90 92.56	94.85 93.84 95.92	94.07 95.16 96.54
	89.38 87.89 88.41	92.05 93.81 92.11	96.61 94.65 94.66	96.14 95.16 97.52
	88.39 90.47 87.61	90.56 **97.01** 94.86	96.51 95.20 92.80	94.90 94.76 97.84
	82.78 87.21 82.00	95.15 94.39 90.49	94.89 93.81 96.76	96.92 95.63 94.72
	82.86 **91.21** 84.37	93.43 91.49 95.56	93.81 **96.72** 95.43	92.89 95.91 96.74
	89.07 89.44 88.21	93.47 90.93 91.19	92.19 95.63 91.21	94.38 94.67 98.01
	89.54 87.64 89.23	94.18 92.93 88.49	94.38 93.71 92.50	97.78 98.45 92.14
100	90.12 87.49 90.16	95.34 89.59 93.45	95.11 94.45 96.69	97.78 97.05 95.14
	88.19 87.59 90.10	92.88 95.47 95.37	97.29 95.81 96.48	97.48 95.13 97.11
	89.32 92.26 88.19	**97.58** 96.16 91.33	93.90 96.97 96.57	96.86 98.48 96.17
	88.81 87.18 89.89	93.93 93.17 95.23	96.03 94.61 97.52	97.07 96.51 98.17
	80.68 87.74 89.14	93.77 93.65 95.73	92.49 97.32 95.99	93.09 96.31 97.09
	89.01 89.23 **91.39**	95.55 93.25 94.18	96.97 97.49 97.72	96.16 98.40 96.43
	84.98 88.36 90.95	94.71 95.08 92.78	97.27 95.42 95.84	97.24 96.49 96.67
	90.45 87.77 89.86	92.91 91.44 86.62	94.40 97.11 92.47	96.23 95.15 97.56
	88.52 86.66 88.18	92.31 93.42 92.71	94.00 93.77 95.49	95.75 95.89 96.38
	91.14 86.81 89.48	92.62 94.45 94.70	**97.95** 95.40 97.09	**99.03** 94.30 98.03
150	91.15 92.52 92.83	96.26 90.48 94.41	98.48 98.06 98.41	99.06 98.94 98.96
	91.41 91.97 86.52	95.21 96.27 95.72	96.60 **98.84** 97.24	95.48 98.97 97.40
	92.14 85.72 92.02	95.38 93.91 96.80	95.91 95.21 96.83	98.21 98.15 97.13
	87.60 89.30 91.56	95.70 95.75 94.94	97.53 97.71 97.34	98.33 98.11 97.54
	89.73 91.20 84.98	95.82 93.62 96.76	98.52 95.53 98.54	98.31 97.61 96.17
	88.67 90.40 88.26	91.84 95.58 95.02	96.67 97.64 97.26	98.19 98.69 98.11
	90.52 80.15 88.25	96.50 95.98 93.46	93.97 96.51 95.28	97.26 98.63 94.85
	88.63 92.41 **93.22**	96.55 95.52 95.62	97.56 96.43 96.52	98.13 98.26 98.92
	93.22 92.24 88.31	95.38 **97.78** 94.24	96.97 95.96 98.27	98.49 98.52 97.71
	90.85 91.21 86.41	93.77 95.94 91.67	98.78 96.67 97.27	**99.47** 98.94 98.04

In Table 2, it was emphasized that, for example, the maximum map coverage with a certain sequence of decisions and the constraints, 60 iterations by 60 waypoints, is 92.36%. Furthermore, the highest result among all the set of constraints used is 99.47% at the maximum allowable set of constraints.

Figure 6 shows one of the simulation-runs with the constraints: 120 iterations and 150 waypoints. In the $a \geq 1$ stage, 87.57% of the environment was explored in half of the total number of iterations (61). For the map in Figure 6b, it can be concluded that the robots touched all the waypoints with the sensor rays. This means that the exploration ability of the algorithm is satisfied in this stage. Figure 6c demonstrates the completed result for the $a \leq 1$ stage with a total of 99.06% map coverage. The trajectories of the robots can be observed through the blue, red, and green colored lines in Figure 6d.

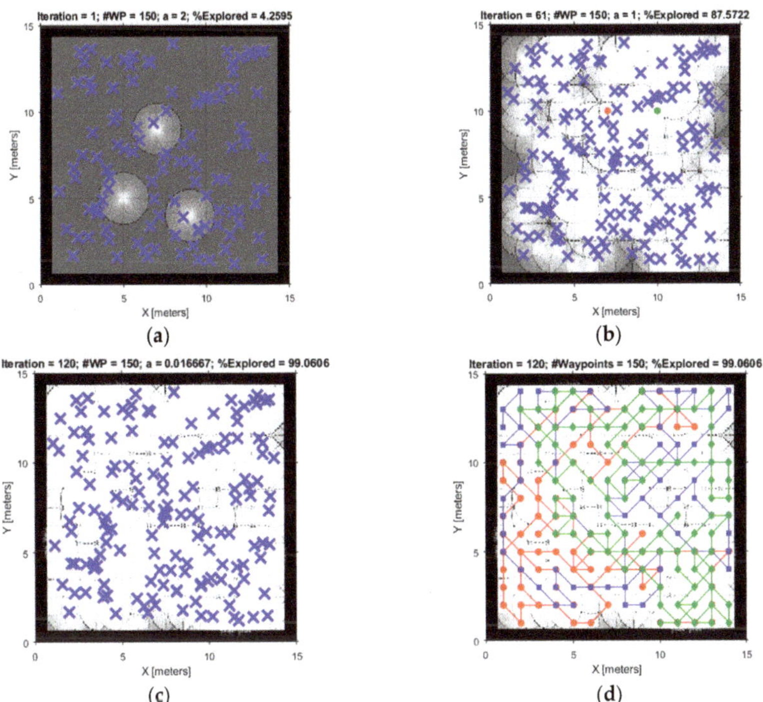

Figure 6. The simulation of MOGWO exploration algorithm in (**a**) iteration 1 with a = 2; (**b**) iteration 61 with a = 1; (**c**) the last iteration, 120, with a = 0.016; and (**d**) the trajectories of three robots.

The decision-making process of each robot is presented in Figure 7. The values of alpha solutions in the simulation above (Figure 6) vary for the two stages: exploration and exploitation. It can be seen that when $a \geq 1$, the trend goes up to maximum values, and when $a \leq 1$, the simulation tries to achieve the minimum values.

The simulation of MOGWO exploration algorithm was implemented in MATLAB using the OccupancyGrid class of the Robotic System Toolbox [51,52]. The video of the simulation can be seen here [53].

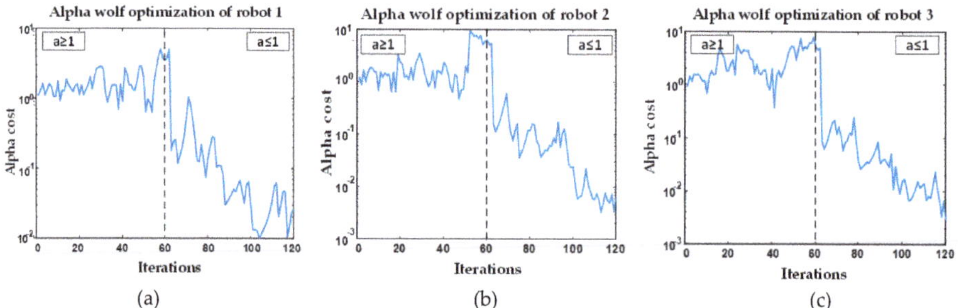

Figure 7. Decision-making process of the alpha solutions: (**a**) robot 1, (**b**) robot 2, and (**c**) robot 3.

5.2. The Pareto Optimality Analysis for MOGWO Exploration Algorithm

Table 2 shows numerous exploration results, which are categorized by constraints. Considering only the percentage of map coverage, it is obvious that 99.47% is the best performance. However, the exploration with 120 iterations and 150 waypoints as constraints takes the longest time, which means it is not the optimal solution.

In this study, we take two factors that are important for the exploration: map coverage and time. In Figure 8, we searched for the trade-offs between the minimum number of iterations and the maximum number of map coverage. The plot was made based on the data in Table 2. The results for 120 and 60 iterations, which can be considered too long and too short for exploration, respectively, are extreme solutions. Thus, the solutions belong to the Pareto optimal front, which lies between two lines.

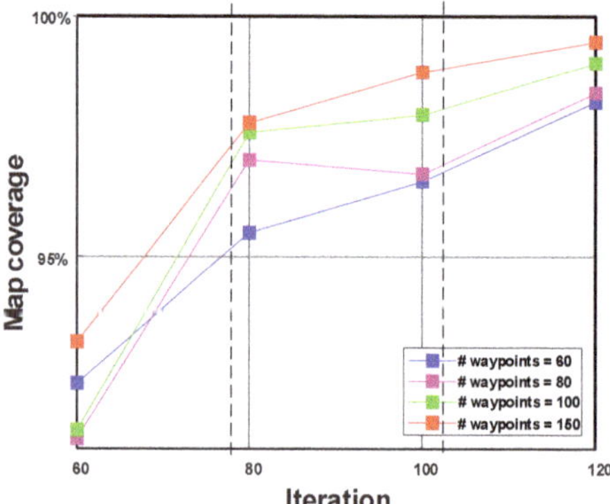

Figure 8. Pareto optimal set of the results of the MOGWO exploration algorithm.

It can be concluded here that the optimal set lies between 80 and 100 iterations with 150 waypoints. In the next subsection, the MOGWO exploration algorithm using 150 waypoints will be compared to two other algorithms using the same environment and map coverage computation using Equation (11).

5.3. Comparison

In this subsection, the proposed MOGWO exploration was compared with the original deterministic CME algorithm [22] and the hybrid stochastic exploration algorithm based on the GWO and CME [8]. The same map and experiment parameters (Table 1) were selected for the two algorithms with 60, 80, 100, and 120 iterations as it was implemented in the MOGWO exploration algorithm in Section 5.1. It should be noted that waypoints were not applied for the other two algorithms used in the comparison.

In the experiment, the CME algorithm was run only once for each iteration class (60, 80, 100, and 120) because it does not generate any random values even when tested multiple times. By its deterministic nature, CME implements differently for every modification in the environment. For instance, the simulation runs were aborted after the 98th iteration during our experiment when one of the robots got stuck next to the wall obstacle. For the purpose of completing the exploration, the initial position of robot 2 or r_2 (from Table 1) was changed from (7,9) to (6,5). From these results, we can conclude that the exploration by the deterministic CME algorithm requires fine-tuning the map parameters for successful map coverage.

The hybrid stochastic algorithm is a stochastic approach, which uses the single-based GWO algorithm. During our experiments, the simulation-runs were aborted several times due to the robot's selection of inappropriate positions among the frontier cells. This situation occurs when the GWO parameters, A and C, oblige a robot to move into the wrong places, such as obstacles or another robot position. Fortunately, in this algorithm, the A and C parameters vary in each simulation-run, which allows us to obtain successful results.

Figure 9 shows the comparison of the results obtained using the original CME, the hybrid stochastic exploration, and the MOGWO exploration with 150 waypoints. It can be seen that the deterministic CME approach has the lowest values of map coverage among all the algorithms. The proposed MOGWO algorithm does not outperform the hybrid stochastic exploration algorithm in iteration classes, 60, 80, and 100. However, it surpasses the original CME in all iteration categories and the hybrid stochastic exploration algorithm in the category of 120 iterations. Additionally, aborted simulation-runs, which are drawbacks of the CME and hybrid stochastic exploration, did not occur in the MOGWO exploration. Thus, the MOGWO exploration proved more efficient and stable compared to the other algorithms studied in this subsection.

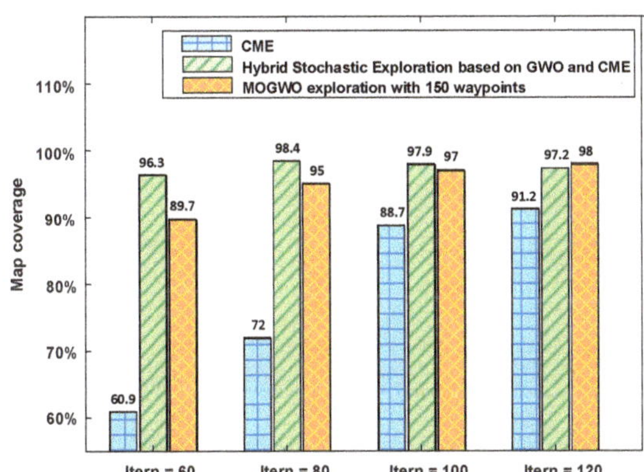

Figure 9. Comparison of three algorithms for the multi-robot exploration. The bars of the hybrid stochastic exploration algorithm based on GWO and coordinated multi-robot exploration (CME) are average values of 30 simulation-runs. The bars of the MOGWO exploration algorithm are average values taken from the row corresponding to 150 waypoints in Table 2.

6. Conclusions

This paper proposed a new method of solving the multi-robot exploration problem as a multi-objective problem. Two objective functions were formed: to search new terrain and to enhance the map accuracy. The use of the MOGWO algorithm enabled us to obtain high percent values of the map coverage without any aborted simulation-runs. The simulation results successfully demonstrated the capability of the MOGWO algorithm to build complete maps, which were completed within certain constraints: the number of waypoints and the number of iterations. Based on the results, the optimal solution was defined by the Pareto optimal set. Furthermore, the proposed MOGWO exploration algorithm was compared with the deterministic exploration and the hybrid stochastic exploration algorithms. The comparison showed that the proposed MOGWO exploration technique outperforms the deterministic exploration in all set of constraints and the hybrid stochastic exploration algorithm at 120 iterations and 150 waypoints.

Author Contributions: A.K. conceived and designed the algorithm. A.K., S.N. and D.Q. designed and performed the experiments. A.K., S.N. and D.Q. wrote the paper. A.K., D.Q. and S.G.L. formulated the mathematical model. S.G.L. supervised and finalized the manuscript for submission.

Funding: This work was supported by the Basic Science Research Program, through the National Research Foundation of Korea, Ministry of Science, under Grant 2017R1D1A3B04031864.

Conflicts of Interest: The authors declare no conflict of interest.

References

1. Limosani, R.; Esposito, R.; Manzi, A.; Teti, G.; Cavallo, F.; Dario, P. Robotic delivery service in combined outdoor–indoor environments: Technical analysis and user evaluation. *Robot. Auton. Syst.* **2018**, *103*, 56–67. [CrossRef]
2. Vidal, E.; Hernández, J.D.; Palomeras, N.; Carreras, M. Online Robotic Exploration for Autonomous Underwater Vehicles in Unstructured Environments. In Proceedings of the IEEE 2018 OCEANS-MTS/IEEE Kobe Techno-Oceans (OTO), Kobe, Japan, 28–31 May 2018; pp. 1–4.
3. Deb, K. Multi-objective optimization. In *Search Methodologies*; Springer: Boston, MA, USA, 2014; pp. 403–449.
4. Amorós, F.; Payá, L.; Marín, J.M.; Reinoso, O. Trajectory estimation and optimization through loop closure detection, using omnidirectional imaging and global-appearance descriptors. *Expert Syst. Appl.* **2018**, *102*, 273–290. [CrossRef]
5. Rizk, Y.; Mariette, A.; Edward, W.T. Cooperative Heterogeneous Multi-Robot Systems: A Survey. *ACM Comput. Surv. (CSUR)* **2019**, *52*.
6. Zhang, Y.; Gong, D.W.; Zhang, J.H. Robot path planning in uncertain environment using multi-objective particle swarm optimization. *Neurocomputing* **2013**, *103*, 172–185. [CrossRef]
7. Pang, B.; Song, Y.; Zhang, C.; Wang, H.; Yang, R. A Swarm Robotic Exploration Strategy Based on an Improved Random Walk Method. *J. Robot.* **2019**, *2019*. [CrossRef]
8. Kamalova, A.; Lee, S.G. Hybrid Stochastic Exploration Using Grey Wolf Optimizer and Coordinated Multi-Robot Exploration Algorithms. *IEEE Access* **2019**, *7*, 14246–14255.
9. Fong, S.; Suash, D.; Ankit, C. A review of metaheuristics in robotics. *Comput. Electr. Eng.* **2015**, *43*, 278–291. [CrossRef]
10. Wadood, A.; Khurshaid, T.; Farkoush, S.G.; Yu, J.; Kim, C.-H.; Rhee, S.-B. Nature-Inspired Whale Optimization Algorithm for Optimal Coordination of Directional Overcurrent Relays in Power Systems. *Energies* **2019**, *12*, 2297. [CrossRef]
11. Kim, C.H.; Khurshaid, T.; Wadood, A.; Farkoush, S.G.; Rhee, S.B. Gray wolf optimizer for the optimal coordination of directional overcurrent relay. *J. Electr. Eng. Technol.* **2018**, *13*, 1043–1051.
12. Wolpert, D.H.; Macready, W.G. No free lunch theorems for optimization. *IEEE Trans. Evolut. Comput.* **1997**, *1*, 67–82. [CrossRef]
13. Kennedy, J. Particle swarm optimization. *Encycl. Mach. Learn.* **2010**, 760–766.
14. Gen, M.; Lin, L. Genetic Algorithms. *Wiley Encyc. Comput. Sci. Eng.* **2007**, 1–15.
15. Dorigo, M.; Di Caro, G. Ant colony optimization: A new meta-heuristic. In Proceedings of the 1999 Congress on Evolutionary Computation-CEC99, Washington, DC, USA, 6–9 July 1999.
16. Seyedali, M.; Lewis, A. Grey wolf optimizer. *Adv. Eng. Softw.* **2014**, *69*, 46–61.
17. Mirjalili, S.; Aljarah, I.; Mafarja, M.; Heidari, A.A.; Faris, H. Grey Wolf Optimizer: Theory, Literature Review, and Application in Computational Fluid Dynamics Problems. In *Nature-Inspired Optimizers*; Springer: Cham, Switzerland, 2020; pp. 87–105.
18. Coello, C.A.; Lechuga, M. MOPSO: A proposal for multiple objective particle swarm optimization. In Proceedings of the 2002 Congress on Evolutionary Computation. CEC'02 (Cat. No. 02TH8600), Washington, DC, USA, 12–17 May 2002.

19. Konak, A.; Coit, D.W.; Smith, A. Multi-objective optimization using genetic algorithms: A tutorial. *Reliab. Eng. Syst. Saf.* **2006**, *91*, 992–1007. [CrossRef]

20. Alaya, I.; Solnon, C.; Khaled, G. Ant colony optimization for multi-objective optimization problems. In Proceedings of the 19th IEEE International Conference on Tools with Artificial Intelligence (ICTAI 2007), Washington, DC, USA, 29–31 October 2007.

21. Mirjalili, S.; Saremi, S.; Mirjalili, S.M.; Coelho, L.D.S. Multi-objective grey wolf optimizer: A novel algorithm for multi-criterion optimization. *Expert Syst. Appl.* **2016**, *47*, 106–119. [CrossRef]

22. Burgard, W.; Moors, M.; Stachniss, C.; Schneider, F.E. Coordinated multi-robot exploration. *IEEE Trans. Robot.* **2005**, *21*, 376–386. [CrossRef]

23. Thrun, S. A probabilistic on-line mapping algorithm for teams of mobile robots. *Int. J. Robot. Res.* **2001**, *20*, 335–363. [CrossRef]

24. Mirjalili, S.; Dong, J.S.; Lewis, A. Ant Colony Optimizer: Theory, Literature Review, and Application in AUV Path Planning. In *Nature-Inspired Optimizers*; Springer: Cham, Switzerland, 2020; pp. 7–21.

25. Kulich, M.; Kubalík, J.; Přeučil, L. An Integrated Approach to Goal Selection in Mobile Robot Exploration. *Sensors* **2019**, *19*, 1400. [CrossRef]

26. Yamauchi, B. A frontier-based approach for autonomous exploration. *Cira* **1997**, *97*.

27. Franchi, A.; Freda, L.; Oriolo, G.; Vendittelli, M. A randomized strategy for cooperative robot exploration. In Proceedings of the 2007 IEEE International Conference on Robotics and Automation, Roma, Italy, 10–14 April 2007.

28. Franchi, A.; Freda, L.; Oriolo, G.; Vendittelli, M. The sensor-based random graph method for cooperative robot exploration. *IEEE/ASME Trans. Mechatron.* **2009**, *14*, 163–175. [CrossRef]

29. Palacios, A.T.; Sánchez, L.A.; Bedolla Cordero, J.M.E. The random exploration graph for optimal exploration of unknown environments. *Int. J. Adv. Robot. Syst.* **2017**, *14*, 1729881416687110. [CrossRef]

30. Tai, L.; Liu, M. Mobile robots exploration through cnn-based reinforcement learning. *Robot. Biomim.* **2016**, *3*, 24. [CrossRef] [PubMed]

31. Tai, L.; Li, S.; Liu, M. Autonomous exploration of mobile robots through deep neural networks. *Int. J. Adv. Robot. Syst.* **2017**, *14*. [CrossRef]

32. Caley, J.A.; Lawrance, N.R.; Hollinger, G.A. Deep learning of structured environments for robot search. In Proceedings of the 2016 IEEE/RSJ International Conference on Intelligent Robots and Systems (IROS), Daejeon, South Korea, 9–14 October 2016.

33. Papoutsidakis, M.; Kalovrektis, K.; Drosos, C.; Stamoulis, G. Design of an Autonomous Robotic Vehicle for Area Mapping and Remote Monitoring. *Int. J. Comput. Appl.* **2017**, *167*, 36–41. [CrossRef]

34. Tai, L.; Liu, M. Deep-learning in mobile robotics-from perception to control systems: A survey on why and why not. *arXiv* **2016**, arXiv:1612.07139.

35. Sharma, S.; Shukla, A.; Tiwari, R. Multi robot area exploration using nature inspired algorithm. *Biol. Inspir. Cognit. Archit.* **2016**, *18*, 80–94. [CrossRef]

36. Wang, D.; Wang, H.; Liu, L. Unknown environment exploration of multi-robot system with the FORDPSO. *Swarm Evolut. Comput.* **2016**, *26*, 157–174. [CrossRef]

37. De Almeida, J.P.L.S.; Nakashima, R.T.; Neves-Jr, F.; de Arruda, L.V.R. Bio-inspired on-line path planner for cooperative exploration of unknown environment by a Multi-Robot System. *Robot. Auton. Syst.* **2019**, *112*, 32–48. [CrossRef]

38. Puig, D.; García, M.A.; Wu, L. A new global optimization strategy for coordinated multi-robot exploration: Development and comparative evaluation. *Robot. Auton. Syst.* **2011**, *59*, 635–653. [CrossRef]

39. Benavides, F.; Ponzoni Carvalho Chanel, C.; Monzón, P.; Grampín, E. An Auto-Adaptive Multi-Objective Strategy for Multi-Robot Exploration of Constrained-Communication Environments. *Appl. Sci.* **2019**, *9*, 573. [CrossRef]

40. Thabit, S.; Mohades, A. Multi-Robot Path Planning Based on Multi-Objective Particle Swarm Optimization. *IEEE Access* **2019**, *7*, 2138–2147. [CrossRef]

41. Chen, X.; Zhang, P.; Du, G.; Li, F. Ant colony optimization based memetic algorithm to solve bi-objective multiple traveling salesmen problem for multi-robot systems. *IEEE Access* **2018**, *6*, 21745–21757. [CrossRef]

42. Hu, P.; Chen, S.; Huang, H.; Zhang, G.; Liu, L. Improved alpha-guided Grey wolf optimizer. *IEEE Access* **2018**, *7*, 5421–5437. [CrossRef]

43. Long, W.; Wu, T.; Cai, S.; Liang, X.; Jiao, J.; Xu, M. A Novel Grey Wolf Optimizer Algorithm with Refraction Learning. *IEEE Access* **2019**, *7*, 57805–57819. [CrossRef]

44. Han, T.; Wang, X.; Liang, Y.; Wei, Z.; Cai, Y. A Novel Grey Wolf Optimizer with Random Walk Strategies for Constrained Engineering Design. In Proceedings of the International Conference on Information Technology and Electrical Engineering, Bandung, Padang, Indonesia, 22–25 October 2018.

45. Heidari, A.A.; Pahlavani, P. An efficient modified grey wolf optimizer with Lévy flight for optimization tasks. *Appl. Soft Comput.* **2017**, *60*, 115–134. [CrossRef]

46. Gupta, S.; Deep, K. A novel random walk grey wolf optimizer. *Swarm Evolut. Comput.* **2019**, *44*, 101–112. [CrossRef]

47. Fatima, A.; Javaid, N.; Anjum Butt, A.; Sultana, T.; Hussain, W.; Bilal, M.; Ilahi, M. An Enhanced Multi-Objective Gray Wolf Optimization for Virtual Machine Placement in Cloud Data Centers. *Electronics* **2019**, *8*, 218. [CrossRef]

48. Sahoo, A.; Chandra, S. Multi-objective grey wolf optimizer for improved cervix lesion classification. *Appl. Soft Comput.* **2017**, *52*, 64–80. [CrossRef]

49. Wu, C.; Wang, J.; Chen, X.; Du, P.; Yang, W. A novel hybrid system based on multi-objective optimization for wind speed forecasting. *Renew. Energy* **2020**, *146*, 149–165. [CrossRef]

50. Qin, H.; Fan, P.; Tang, H.; Huang, P.; Fang, B.; Pan, S. An effective hybrid discrete grey wolf optimizer for the casting production scheduling problem with multi-objective and multi-constraint. *Comput. Ind. Eng.* **2019**, *128*, 458–476. [CrossRef]

51. Available online: https://www.mathworks.com/help/robotics/ref/robotics.occupancygrid-class.html1U30T (accessed on 20 July 2019).

52. Kumar, N.; Vámossy, Z.; Szabó-Resch, Z.M. Robot path pursuit using probabilistic roadmap. In Proceedings of the 2016 IEEE 17th International Symposium on Computational Intelligence and Informatics (CINTI), Budapest, Hungary, 17–19 November 2016; pp. 139–144.

53. Available online: https://youtu.be/b_IiUjwM-bQ (accessed on 15 July 2019).

Article

Evaluation of Clustering Methods in Compression of Topological Models and Visual Place Recognition Using Global Appearance Descriptors

Sergio Cebollada [1,*,†], Luis Payá [1,†], Walterio Mayol [2,†] and Oscar Reinoso [1,†]

[1]   Department of Systems Engineering and Automation, Miguel Hernández University, 03202 Elche, Spain; lpaya@umh.es (L.P.); o.reinoso@umh.es (O.R.)
[2]   Department of Computer Science, University of Bristol, Bristol BS81TH, UK; wmayol@cs.bris.ac.uk
*   Correspondence: sergio.cebollada@umh.es
†   These authors contributed equally to this work.

Received: 11 December 2018; Accepted: 17 January 2019; Published: 22 January 2019

Abstract: This paper presents an extended study about the compression of topological models of indoor environments. The performance of two clustering methods is tested in order to know their utility both to build a model of the environment and to solve the localization task. Omnidirectional images are used to create the compact model, as well as to estimate the robot position within the environment. These images are characterized through global appearance descriptors, since they constitute a straightforward mechanism to build a compact model and estimate the robot position. To evaluate the goodness of the proposed clustering algorithms, several datasets are considered. They are composed of either panoramic or omnidirectional images captured in several environments, under real operating conditions. The results confirm that compression of visual information contributes to a more efficient localization process through saving computation time and keeping a relatively good accuracy.

Keywords: mapping; localization; clustering; omnidirectional images; global appearance descriptors

---

1. Introduction

The presence of mobile robots in many kinds of environments has increased substantially during the past few years. Robots need a high degree of autonomy to develop their tasks. In the case of autonomous mobile robots, this means that they must be able to localize themselves and to navigate through environments that are a priori unknown. Hence, the robot will have to carry out the mapping task, which consists of obtaining information from the environment and creating a model. Once this task is done, the robot will be able to address the localization task, i.e., estimating its position within the environment with respect to a specific reference system.

Vision sensors have been widely used for mapping, navigation, and localization purposes. According to the number of cameras and the field of view, different configurations have been proposed. Some authors (such as Okuyama et al. [1]) have used monocular configurations. Others proposed stereo cameras by using binocular (such as Yong-Guo et al. [2] or Gwinner et al. [3]) or even trinocular systems (such as Jia et al. [4]).

Despite stereo cameras permitting measuring depth from the images, these systems present a limitation related to their field of view. In order to obtain complete information from the environment, several images must be captured. In this respect, omnidirectional cameras constitute a good alternative. They can provide a big amount of information with a field of view of 360 deg. around them, and their cost is relatively low in comparison with other kinds of sensors. Furthermore, omnidirectional vision systems present further advantages. For instance, the features in the images are more stable (because they stay longer as the robot moves), and they permit estimating both the position and the orientation of the robot. Omnidirectional cameras have been successfully used by different authors for mapping and localization [5–9]. A wide study was carried out by Payá et al. [10], who introduced a state-of-the-art of the most relevant mapping and localization algorithms developed with omnidirectional visual information. An example of a mobile robot that has an omnidirectional camera mounted on it is shown in Figure 1a, and an example of an omnidirectional image is shown in Figure 1b.

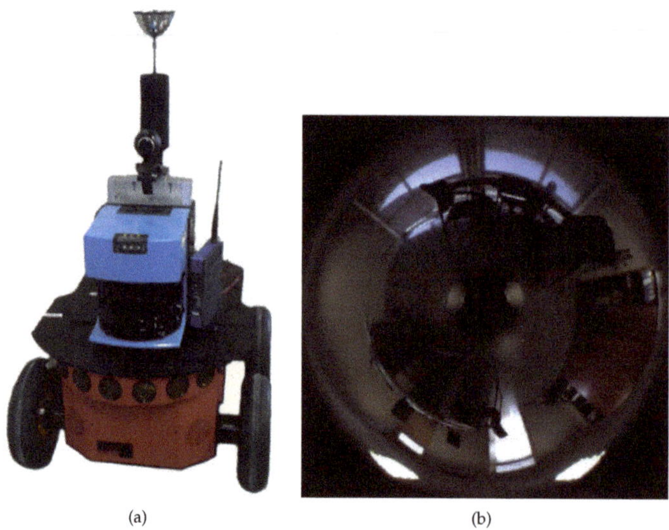

(a)   (b)

Figure 1. (**a**) Example of a robot Pioneer P3-AT® equipped with an omnidirectional vision system and a laser range finder. In this work, only the omnidirectional camera is used. (**b**) Example of an omnidirectional image captured from one office.

In the related literature, two main frameworks have been proposed in order to carry out the mapping task: the metric maps, which represent the environment with geometric accuracy; and the topological maps, which describe the environment as a graph containing a set of locations with the related links among them. Regarding the second option, some authors have proposed to arrange the information in the map hierarchically, into a set of layers. The way a robot solves the localization task efficiently in hierarchical maps is as follows: first, a rough, but fast localization is carried out using the high-level layers; second, a fine localization is tackled in a local area using the low-level layers. Therefore, in order to address the mapping and localization issue, hierarchical maps constitute an efficient alternative (like the works [11–13] show).

Visual mapping and localization have been solved mainly by using two main approaches to extract the most relevant information from scenes; either by detection, description, and tracking of some relevant landmarks or working with global appearance algorithms, i.e., building a unique descriptor per image. On the one hand, the methods based on local features consist of extracting some outstanding points from each scene and creating a descriptor for each point, using the information around it (Figure 2a). The most popular description methods used for this purpose are SIFT (Scale-Invariant Feature Transform) [14] and SURF (Speeded-Up Robust Features) [15]. More recently, descriptors such as BRIEF (Binary Robust Independent Elementary Features) [16] or ORB (Oriented FAST and Rotated BRIEF) [17] have been proposed, trying to overcome some drawbacks such as the computational time and invariance against rotation. These descriptors have become very popular in visual mapping and localization, and many authors have proposed methods that use them, such as Angeli et al., who employed SIFT [18], or Murillo et al., who used SURF [8]. Nonetheless, these methods present some disadvantages. For instance, to obtain reliable landmarks, the environments must be rich in details. Furthermore, keypoints' detection is not always robust against changes in the environments (e.g., changes of lighting conditions), and sometimes, the description is not totally invariant to changes in the robot position. Moreover, these approaches might be computationally complex; hence, in those cases, it would not be possible to build models in real time. On the other hand, the methods based on the global appearance of scenes consist of treating each image as a whole. Each image is represented by a unique descriptor, which contains information about its global appearance (Figure 2b). These methods lead to simpler mapping and localization algorithms, due to the fact that each scene is described by only one descriptor. Hence, mapping and localization can be carried out by just storing and comparing the descriptors pairwise. Besides, they could be more robust in dynamic and unstructured environments. However, as drawbacks, these methods present a lack of metric information (they are commonly employed to build topological maps). Visual aliasing also might have a negative impact on the mapping and localization tasks, due to the fact that indoor environments are prone to present repetitive visual structures. Additionally, modelling large environments would require a big amount of images, and this can introduce serious issues when these techniques have to be used in real-time applications. Therefore, global appearance is an intuitive alternative to solve the mapping and localization problem, but its robustness against these issues must be tested. Many authors have addressed mapping and localization using global appearance descriptors (Figure 2b). For instance, Menegatti et al. [19] used the Fourier signature in order to build a visual memory of a relatively small environment from a set of panoramic images. Liu et al. [20] proposed a descriptor based on colour features and geometric information. Through this descriptor, a topological map can be built. Payá et al. [21] proposed a mapping method from global appearance and solved the localization in a probabilistic fashion, using a Monte Carlo approach. Furthermore, they developed a comparative analysis of some description methods. Rituerto et al. [22] proposed the use of the descriptor *gist* [23,24] to create topological maps from omnidirectional images. More recently, Berenguer et al. [6] proposed the Radon transform [25] as the global appearance descriptor of omnidirectional images and a hierarchical localization method. Through this method, first, a rough localization is obtained; after that, a local topological map of a region is created and used to refine the localization of the robot.

In light of the previous information, in the present paper, the use of hierarchical models is proposed to solve the localization task efficiently. In this sense, compression methods are used as a solution to generate the high-level layers of the hierarchical model. Some authors have used clustering algorithms to carry out the compression task. For instance, Zivkovic et al. [26] used spectral clustering to obtain higher level models, which improved the efficiency of the path-planning. Grudic and Mulligan [27] built topological maps through the use of an unsupervised learning algorithm, which worked with spectral clustering. Valgren et al. [28] tackled an on-line topological mapping through the use of incremental spectral clustering. Štimec et al. [29] used an unsupervised clustering based on the multiple eigenspaces algorithm to carry out topological mapping hierarchically using omnidirectional images. More recently, Shi et al. [30] proposed the use of a differential clustering method to improve the compression of telemetry data.

We propose a method to build hierarchical maps through a combination of clustering methods and global appearance descriptors. We compare the performance of spectral and self-organizing maps' clustering. In addition, an exhaustive experimental evaluation is carried out to assess the performance of the method in mapping and localization tasks, and we evaluate the influence of the most relevant parameters in the results. This is an interesting problem in the field of mobile robotics because, as pointed out before, global appearance descriptors are a straightforward way of describing visual information, but they contain no metric information, comparing to local-features' descriptors. Additionally, no deep study to assess the performance of global-appearance descriptors in hierarchical mapping can be found in the literature. The experiments show that the proposal that we present is a feasible alternative to build robust compact maps, despite the phenomenon of visual aliasing, which is present in the sets of images that we have used in the experiments.

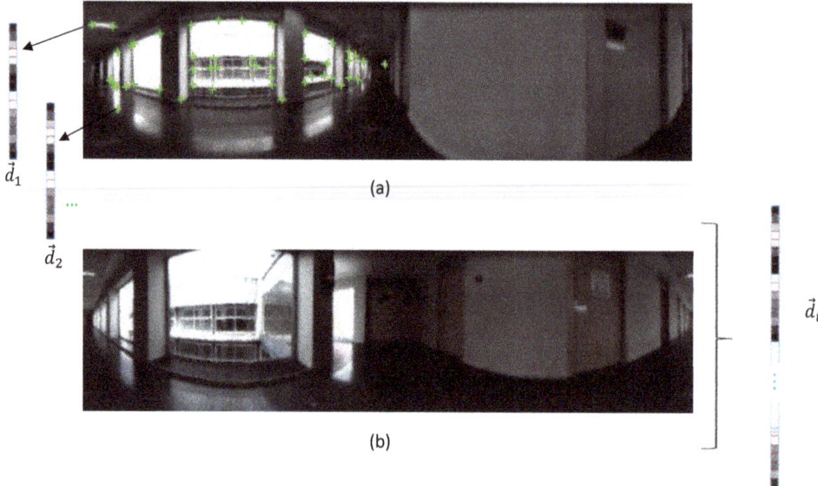

Figure 2. Two main methods to extract the most relevant information from the images for mapping and localization purposes. (**a**) Detection, description, and tracking of some relevant landmarks along a set of scenes. (**b**) Building a unique descriptor per image that contains information on its global appearance.

The present paper continues and extends the study presented in [31], which is a comparative evaluation in which the performance of some descriptors was assessed to create compact models and estimate the position of the robot. The contributions of the present paper are the following: (a) a new method to compact the visual model is proposed; (b) the trade-off compactness-accuracy-computational cost is addressed, and the performance of the compact models is compared to raw models (with no compaction); (c) a comparison between compression through direct methods and compression through clustering methods to solve the localization task is evaluated; and (d) new indoor environments with different topologies are included in the experimental section.

The remainder of the paper is structured as follows: Section 2 outlines the global appearance descriptors that will be tested throughout the paper. After that, Section 3 shows the clustering approaches used to compress the models. Next, Section 4 presents the method to obtain the localization within the compact models. Section 5 presents the experimental results of clustering and localization and also the discussions about the results. Finally, Section 6 outlines the conclusions and future research lines.

2. Global Appearance Descriptors

As mentioned in the previous section, global appearance descriptors constitute an interesting alternative for mapping and localization. In this work, the robot moves along the floor plane, and it captures images using a hyperbolic mirror, which is mounted over a camera along the vertical axis. This section details three methods to describe the global appearance of a set of panoramic scenes $IM = \{im_1, im_2, ..., im_N\}$ where $im_j \in \mathbb{R}^{M_x \times M_y}$. After using each description method, a descriptor for each image is calculated; thus, the database is composed of a set of descriptors, $D = \{\vec{d_1}, \vec{d_2}, ..., \vec{d_N}\}$ where each descriptor is $\vec{d_j} \in \mathbb{C}^{l \times 1}$ and corresponds to the image im_j.

The remainder of the section presents the global appearance techniques used throughout the paper and the homomorphic filter, which is used as a pre-treatment for the images.

2.1. Fourier Signature Descriptor

The Fourier signature descriptor was firstly used by Menegatti et al. [19] to create an image-based memory for robot navigation. Payá et al. [21] studied the computational cost and the error in localization by using Fourier Signature (FS) and proposed a Monte Carlo approach to solve the localization problem in indoor environments.

This description method is based on the use of the Discrete Fourier Transform (DFT). After calculating the FS of a panoramic image, a new complex matrix is obtained $IM(u,v)$. It can be decomposed into two real matrices, one containing the magnitudes and the other the arguments. The steps to obtain a global appearance descriptor from a panoramic image through the Fourier Signature (FS) are: First, departing from the intensity matrix of the original image, the DFT of each row is calculated. The result is a complex matrix with the same size as the original image ($IM(u,v) \in \mathbb{C}^{N_x \times N_y}$). Second, only the k_1 first columns of this matrix are retained since the main information is in the low frequency components. Third, the resultant matrix ($IM(u,v) \in \mathbb{C}^{N_x \times k_1}$) is decomposed into the magnitudes and arguments matrices. The matrix of magnitudes ($A(u,y) \in \mathbb{R}^{N_x \times k_1}$) is invariant against changes of the robot orientation in the movement plane if the image is panoramic. In the last step, the global appearance descriptor is obtained by arranging the k_1 columns of the magnitudes matrix in one single column ($\vec{d} \in \mathbb{R}^{N_x \cdot k_1 \times 1}$).

2.2. Histogram of Oriented Gradients Descriptor

The Histogram of Oriented Gradients (HOG) is a description method used in computer vision to detect objects. This descriptor is remarkable due to the fact that it is easy to build, leads to successful results in detection tasks, and also requires a low computational cost. It is built from the orientation of the gradient in localized parts of the panoramic image. The development consists of dividing the image into small regions (k_2 horizontal cells in this work) and compiling a histogram with b bins for the pixels, which are included inside each cell using their gradient orientation. The combination of this information provides the desired descriptor ($\vec{d} \in \mathbb{R}^{b \cdot k_2 \times 1}$). This method has been used by some authors such as Mekonnen et al. [32] to develop a person detection tool, or Dong et al. [33], who proposed an HOG-based multi-stage approach for object detection and pose recognition in the field of service robots. This method was firstly used in mobile robotics by Dalal and Triggs [34] to solve people detection task. Zhu et al. [35] presented an improved version with respect to computational time and efficiency to detect people.

The HOG version proposed in this work is described in detail in [36].

2.3. Gist Descriptor

The *gist* description was introduced by Oliva et al. [37], and it has been commonly used to recognize scenes. Since then, several versions can be found, which work with different features from the images, such as colour, texture, orientation, etc. [38]. Some researchers have used *gist* in mobile robotics. For instance, Chang et al. [39] used this global appearance descriptor for localization and navigation. Murillo et al. [40] also used the *gist* descriptor to solve the localization problem, but in this case, the *gist* descriptor was a reduced version obtained with Principal Components Analysis (PCA).

The version we use throughout this paper is described in [36] and works with the orientation information obtained through a set of Gabor filters. From the panoramic image, m different resolution levels are obtained. Then, n_{masks} orientation filters are applied over each level. Finally, the pixels of every image are grouped into k_3 horizontal blocks, and the information is arranged in a vector ($\vec{d} \in \mathbb{R}^{n_{masks} \cdot m \cdot k_3 \times 1}$).

2.4. Homomorphic Filter

In order to solve the localization task, typical situations may happen such as lighting variations and changes in the position of some objects (chairs, tables, open doors, etc.). Hence, descriptors must be robust against these circumstances.

Fernandez et al. [41] showed that some pre-treatments could improve the localization accuracy in indoor environments with different lighting levels. Among the studied techniques, the use of the homomorphic filter [42] can be highlighted. The homomorphic filter permits filtering the luminance and reflectance components from an image separately.

The use of this filter has proven to provide especially good results when it is used in combination with the HOG descriptor [31], whereas in the FS and *gist* cases, the results were similar to or worse than without this pre-treatment filter. Hence, in the present paper, the following configurations will be used throughout the experiments: FS without filter, HOG with filter, and *gist* without filter.

3. Clustering Methods to Compact the Visual Information

In this section, the creation of topological models and how to compact them will be addressed. Subsequently, these models will be utilized to solve the localization problem. Only visual information and global appearance descriptors will be used in both tasks. This way, the problem will be addressed through the next two steps.

1. Learning: creating a map of the environment and compacting it. A set of omnidirectional images is captured from different positions, and a global appearance descriptor for each image is calculated. After that, a clustering method is used to determine the structure and compact the model.
2. Validation: Once the map is built, the robot obtains a new image from an unknown position, calculates the descriptor, and compares it with the set of descriptors obtained in the learning step. Through this comparison, the robot must be able to estimate its position.

Focusing on the learning step, the robot moves around the environment and captures some images from different positions to cover the whole environment. This way, a set of omnidirectional images is collected $I = \{im_1, im_2, ..., im_N\}$ where $im_j \in \mathbb{R}^{N_x \times N_y}$. After that, a global appearance descriptor is calculated for each image; hence, a set of descriptors is obtained $D = \{\vec{d_1}, \vec{d_2}, ..., \vec{d_N}\}$ where $\vec{d_j} \in \mathbb{C}^{l \times 1}$.

This set of descriptors can be considered as a straightforward model of the environments [43,44], as some previous works do. However, in this mapping strategy, important problems appear when the environment has considerable dimensions. The larger the environment is, the more images have to be captured to model it completely. This leads to the requirement of more computational time and also more memory space in order to process and collect the information related to each captured image and to solve the subsequent localization problem. This way, the model should be compacted in such a way that it retains most of the visual information and permits solving the localization problem efficiently.

In this work, we propose a clustering approach to compact the model, with the objective of creating a two-layer hierarchical structure. The low-level layer is composed of a set of descriptors and, to obtain the high-level layer, this set will be compacted via clustering. Each cluster is characterized by the common attributes of the instances that form that group. This way, the dataset $D = \{\vec{d_1}, \vec{d_2}, ..., \vec{d_N}\}$ is divided into n_c clusters $C = \{C_1, C_2, ..., C_{n_c}\}$ under the conditions:

$$C_i \neq \emptyset, i = 1, ..., n_c$$

$$\bigcup_{i=0}^{m} C_i = D \tag{1}$$

$$C_i \cap C_j = \emptyset, i \neq j, i, j = 1, ..., n_c.$$

After this, each cluster is reduced to a unique representative descriptor, which is obtained in this work as the average of all the descriptors that compose that cluster. A set of representatives is obtained $R = \{\vec{r_1}, \vec{r_2}, ..., \vec{r_{n_c}}\}$, and therefore, the model is compacted. This set of representatives composes the high-level layer.

Figure 3 shows how a sample map is compacted. Figure 3a shows the positions where panoramic images were captured to cover the whole environment. The result of the clustering process is presented in Figure 3b, and then, one representative per cluster is obtained (Figure 3c). The representative descriptor is obtained as the average descriptor among those grouped in the same cluster. Additionally, the position of this representative descriptor is calculated as the average position of the capture points of the images included in the same cluster. These positions are calculated just as a ground truth to test the performance of the compact map in a localization process, but they are not used either to build the map, nor to localize the robot. Only visual information is used with these aims. Different clustering methods will be analysed. These methods will only use visual information, and ideally, the objective is to group images captured from near positions despite visual aliasing. To evaluate the correctness of the approach, the geometrical compactness of the clusters and their utility to solve the localization task will be tested in Section 5.

Regarding the clustering process to compact the visual models, two methods are studied: spectral clustering and self-organizing maps.

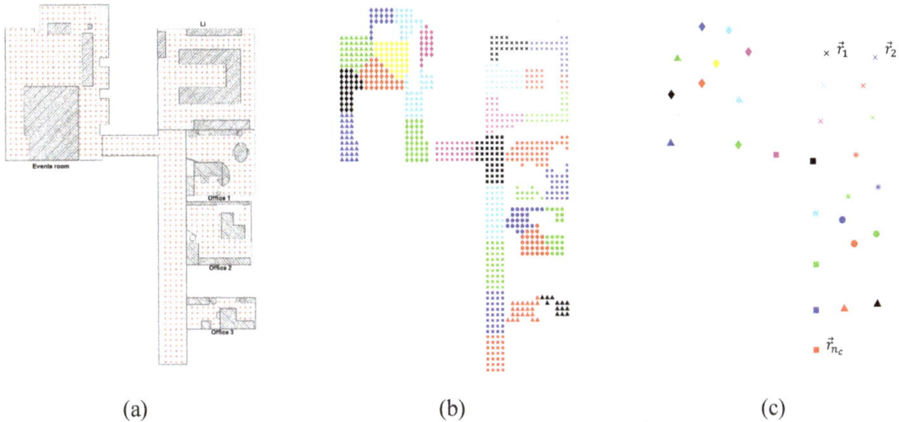

(a)	(b)	(c)

Figure 3. Example of an indoor map and a compression of the information. (**a**) Positions where the images were captured. (**b**) Result of the clustering process. (**c**) Each cluster is reduced to one representative.

3.1. Spectral Clustering Algorithm

Spectral clustering algorithms [45] have proven to be suitable to process highly-dimensional data. In this work, a spectral normalized clustering algorithm is used as was introduced by Ng et al. [46]. This algorithm has been already used for mapping along with local features extracted from the scenes [29,47].

In our work, the algorithm departs from the set of global appearance descriptors $D = \{\vec{d_1}, \vec{d_2}, ...\vec{d_N}\}$ obtained from the images collected in the environment, and the parameter n_c is the desired number of clusters. Initially, the similitude between descriptors is calculated. This parameter is calculated for each pair of descriptors; hence, a matrix of similitudes S is obtained as $S_{ij} = e^{-\frac{|\vec{d_i}-\vec{d_j}|^2}{2\sigma^2}}$ where σ is a parameter that controls the rapidity of the reduction of the similitude when the distance between $\vec{d_i}$ and $\vec{d_j}$ increases. The steps to carry out the clustering are the following:

1. Calculation of the normalized Laplacian matrix:

$$L = I - D^{-1/2}SD^{1/2} \tag{2}$$

where D is a diagonal matrix $D_i = \sum_{j=1}^{N} S_{ij}$.

2. Calculation of the n_c main eigenvectors of L, $\{\vec{u_1}, \vec{u_2}, ..., \vec{u_{n_c}}\}$. Arranging these vectors by columns, the matrix $U \in \mathbb{R}^{N \times n_c}$ is obtained.

3. Normalization of the matrix U to obtain the matrix $T \in \mathbb{R}^{N \times n_c}$.

4. Extraction of vector $\vec{y_i} \in \mathbb{R}^{n_c}$ from the ith row of the matrix T. $i = 1, ..., N$.

5. Clustering of the $\vec{y_i}$ vectors by using a simple clustering algorithm (such as k-means or hierarchical clustering). Through this, the clusters $A_1, A_2, ..., A_{n_c}$ are obtained.

6. Obtaining the clusters with the original data as $C_1, C_2, ..., C_{n_c}$ where $C_i = \vec{d_j} \mid \vec{y_j} \in A_i$.

If the number of instances N or the dimension l is high, the computation of the n_c eigenvectors (third step) will be computationally expensive. To solve this issue, Luxburg [45] proposed cancelling some components of the similitude matrix. This way, in the matrix S, only the components S_{ij} so that j is among the t nearest neighbours of i are retained. After this, the n_c first eigenvectors of the Laplacian matrix L are calculated by using the Lanczos/Arnoldi factorization [48].

Finally, for each cluster, a representative is obtained as the average visual descriptor of the set of descriptors that compose that cluster.

Spectral clustering may result in being more efficient than traditional methods such as k-means or hierarchical clustering in large environments due to the fact that spectral clustering considers the mutual similitude between the instances.

3.2. Cluster with a Self-Organizing Map Neural Network

As a second alternative, Self-Organizing Maps (SOM) have been chosen to carry out the clustering evaluation in this work. This algorithm was introduced by Kohonen [49], and it is an effective option to carry out a mapping distribution when the data present a high dimensionality [50]. This algorithm has been commonly used for clustering or reducing the dimensionality of data. Therefore, in this work, the input data are the set of global appearance descriptors calculated with one of the methods described in Section 2. The size of the neural network map ($W_{SOM} \times H_{SOM} = n_c$) is chosen. After the training step, the data will be grouped into n_c different clusters.

Self-organizing maps automatically learn to classify input vectors according to their similarity and topology in the input space. They differ from competitive layers in that neighbouring neurons in the SOM learn to recognize neighbouring sections of the input space. Thus, self-organizing maps learn both the distribution (as the competitive layers do) and topology of the input vectors with which they are trained. The neurons can be arranged in a grid, hexagonal, or random topology. The self-organizing map network identifies a wining neuron i^* using the same procedure as employed by the competitive layer, but instead of updating only the winning neuron, all neurons within a certain neighbourhood $N_{i*}(d)$ of the winning neuron are updated.

4. Using the Compact Topological Maps to Localize the Robot

At this point, the robot is provided with a model of the environment, which, in this case, is a hierarchical map. From it, the robot firstly uses the high-level layer to carry out a rough localization, and secondly, a fine localization is tackled through the use of the low-level layer. The visual localization problem has been solved by many authors through local features by using probabilistic approaches such as particle filters or Monte Carlo localization [51,52]. Nevertheless, the works developed with global appearance descriptors are scarce. Hence, this paper presents a comparison of this kind of descriptor to estimate hierarchically the position of the robot within a hierarchical map in a specific time instant. In order to test the accuracy of the localization method proposed in this work, the coordinates where the images were captured within the environment are known (ground truth). Nevertheless, they are not used to estimate the position of the robot since, as mentioned before, the presented method only considers visual information. This decision permits studying the feasibility of visual sensors as the only source of information to create a compact topological map and, more concisely, of global appearance descriptors. Therefore, not using the position information in the mapping and localization algorithms permits isolating the effect of the main parameters of these descriptors and knowing the performance of this kind of information. The remainder of this section is structured as follows: Section 4.1 outlines the types of distances that have been used to calculate how different the global appearance descriptors are. Section 4.2 explains the localization step within maps that have not been compacted previously, i.e., no clustering has been carried out (the full information about the environment is provided). Finally, Section 4.3 explains the localization task within hierarchical topological maps.

4.1. Distance Measures between Descriptors

In order to know how similar two panoramic images are through their global appearance descriptors, some distance measurements have been used. This way, a comparison can be carried out by calculating the distance between the descriptors of two images captured from different positions of the environment. The lower the distance between those images is, the more similar they are. This kind of distance is used in the localization step. We consider two descriptors $\vec{a} \in \mathbb{R}^{l \times 1}$ and $\vec{b} \in \mathbb{R}^{l \times 1}$, where a_i and b_i are the ith components of $\vec{a}$ and $\vec{b}$ with $i = 1, ..., l$. The distances used in this work are:

- Euclidean distance: This a particular case of the the weighted metric distance and is defined as:

$$dist_{euclidean}(\vec{a}, \vec{b}) = \sqrt{\sum_{i=1}^{l}(a_i - b_i)^2} \tag{3}$$

- Cosine distance: Departing from a similitude metric, which is defined as the scalar product between two vectors, the distance is defined as:

$$dist_{cosine}(\vec{a}, \vec{b}) = 1 - sim_{cosine}(\vec{a}, \vec{b})$$
$$sim_{cosine}(\vec{a}, \vec{b}) = \frac{\vec{a}^T \cdot \vec{b}}{|\vec{a}||\vec{b}|} \tag{4}$$

- Correlation distance: Again, departing from a similitude metric, which is defined as a normalized version of the scalar product between two vectors, the distance is defined as:

$$dist_{correlation}(\vec{a}, \vec{b}) = 1 - sim_{correlation}(\vec{a}, \vec{b})$$
$$sim_{correlation}(\vec{a}, \vec{b}) = \frac{(\vec{a} - \bar{a})^T(\vec{b} - \bar{b})}{\sqrt{(\vec{a} - \bar{a})^T(\vec{a} - \bar{a})}\sqrt{(\vec{b} - \bar{b})^T(\vec{b} - \bar{b})}} \tag{5}$$

where:

$$\bar{a} = \frac{1}{l}\sum_{i=1}^{l} a_i; \qquad \bar{b} = \frac{1}{l}\sum_{i=1}^{l} b_i \tag{6}$$

Previous research works [21,36] have evaluated the relation between the distance between the global appearance descriptors and the geometrical distance between capture points. These works show that even if the robot moves a short distance, the descriptor changes. Therefore, global appearance descriptors can be used to detect even small movements.

4.2. Resolution of the Localization Problem in a Model That Has Not Been Compacted

In this case, the map is composed of a straightforward set of descriptors (i.e., this set has not been treated to create a hierarchical map through any clustering process). Once this straightforward map is available, the localization process starts:

1. The robot captures a new image at time instant t from an unknown position (im_t).
2. It calculates the global appearance descriptor of the captured image $\vec{d_t}$.
3. The distances between this new descriptor and the set of descriptors in the map are obtained. The comparison between descriptors is carried out through one of the distance metrics presented in Section 4.1.
4. A distance vector $l_t = \{l_{t1}, ..., l_{tN}\}$ is obtained where $l_{tj} = dist\{\vec{d_t}, \vec{d_j}\}$ according to any distance measure.
5. Considering the position of the robot as the position of the closest neighbour within the map (the problem known as image retrieval [53]), the corresponding position of the robot is the position in the map that minimizes the distance $arg\min_j l_{tj}$. This way, the position (x, y) of the robot in the instant t is estimated.

4.3. Resolution of the Localization Problem in a Compact Model

Image retrieval is an inefficient process due to the fact that the maps are usually composed by a huge number of images and the descriptors have a high dimensionality. Therefore, the computational cost could be a problem. In this case, clustering is used to compact the map. Additionally, indoor environments may present visual aliasing. As explained in Section 3, after clustering, the map $\mathfrak{M}$ will be formed only by a set of clusters $C = \{C_1, ..., C_{n_c}\}$, where n_c is the number of clusters. For each cluster, a representative descriptor was calculated as the average of the descriptors in it and the coordinates of those representatives as the average coordinates of the descriptors that compose that cluster. Thus, a set of cluster representatives $\{\vec{r_1}, ..., \vec{r_{n_c}}\}$ and the coordinates of each representative $\{(x, y)_{r_1}, ..., (x, y)_{r_{n_c}}\}$ are known (ground truth).

The localization in this hierarchical map is carried out as follows. (1) The robot captures a new image im_t from an unknown position (x_t, y_t), which must be estimated, and (2) the descriptor corresponding to the new captured image is obtained ($\vec{d_t}$) by using any of the description algorithms explained in Section 2 (FS, HOG, or *gist*). (3) The distance vector is obtained $\vec{l_t} = \{l_{t1}, , l_{tn_c}\}$ where $l_{tj} = dist\{\vec{d_t}, \vec{r_j}\}$ is the distance (one of the three types explained in Section 4.1) between the descriptor $\vec{d_t}$ and each representative $\vec{r_j}$. Finally, (4) the estimated position of the robot (x_e, y_e) is the position associated with the nearest neighbour $d_t^{nn}|t = arg\min_j l_{tj}$.

The coordinates of the representatives are not used in the localization step. However, to measure the goodness of the estimation, the geometric distance between (x_t, y_t) and the centre of the corresponding cluster (obtained as the average position among the positions of the images that belong to that cluster) is calculated: $error = \sqrt{(x_e - x_t)^2 + (y_e - y_t)^2}$. Furthermore, the required computational cost to estimate the localization is calculated.

5. Experiments

5.1. Datasets

Two different types of datasets were used to develop the experiments; QuorumV, which contains grid-distributed visual data, and the COsy Localization Database (COLD), which contains visual information along a trajectory. On the one hand, Quorum V is a publicly-available dataset [54], which consists of a set of omnidirectional images that have been captured in an indoor environment at Miguel Hernandez University (Spain). The database includes 3 offices, a library, a meeting room, and a corridor. It is composed by two datasets; the first one is a training dataset, and it is composed of 872 images, which were captured on a dense 40 × 40 cm grid of points. As for the second dataset, the test dataset, it is composed of 77 images, which were captured in different parts of the environment, in half-way positions among the points of the training dataset, and including changes in the environment (e.g., people walking, position of furniture, etc.). Figure 4 shows the bird's eye view of theQuorum V database and the grid points captured by the robot for the training dataset.

On the other hand, COLD (COsy Localization Database) [55] (also publicly available) contains several sets of images captured in three different indoor environments, which are located in three different cities: Ljubliana (Slovenia), Saarbrücken, and Freiburg (Germany). This database contains omnidirectional images captured while the robot traversed several paths within the environments under real operating conditions (with people that appear and disappear from scenes, changes in the furniture, etc.). In the present work, we use the two longest paths: Saarbrücken and Freiburg. Both datasets include several rooms such as corridors, personal offices, printer areas, kitchens, bathrooms, etc. In order to represent the same distance between images as the distance presented in the Quorum V database, a downsampling is carried out to obtain an acquisition distance between images of 40 cm approximately. Therefore, two training datasets are generated: $Freiburg_{training}$ and $Saarbrucken_{training}$, with 519 and 566 images, respectively. Moreover, from the remaining images, test datasets were created. Figure 5 shows the bird's eye view of the environments and the path that the robot traversed to obtain the images. To summarize, Table 1 shows the datasets used for this work and the number of images that each of them contains.

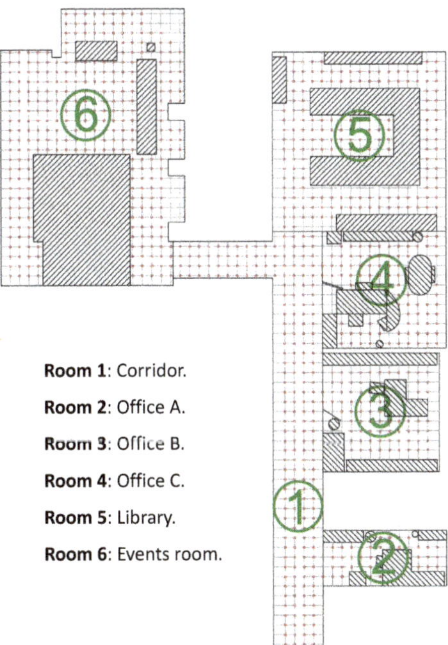

Figure 4. Bird's eye view of the Quorum V database.

Through evaluating these two types of datasets, an analysis of the localization in maps which are completely different is tackled: the first kind of map (Quorum V) is a grid-based map, and the second dataset (COLD) is a trajectory-based map. The Quorum V database presents a distance between images of 40 cm approximately. This distance is considered reasonable for indoor applications. In this case, the expected maximum error (when all the images are used for mapping) is around 28 cm (a case in which the test image is in the middle of four images of the map, which compose a square of a side of

40 cm). This is a reasonable accuracy to solve localization tasks, and additionally, the requirements of memory to store the images of the map are not excessively high in large environments. Regarding the downsampling that is carried out in COLD, this was done with the purpose of obtaining results that can be directly compared with the ones obtained through the Quorum V database (whose minimum available distance is 40 cm). Previous works [6] have shown that the distance between images has a direct relation with the accuracy of localization when global appearance descriptors are used. Lower distances tend to provide more accurate results. Therefore, if a specific application requires a lower error, a more dense initial dataset of images should be used to obtain the map.

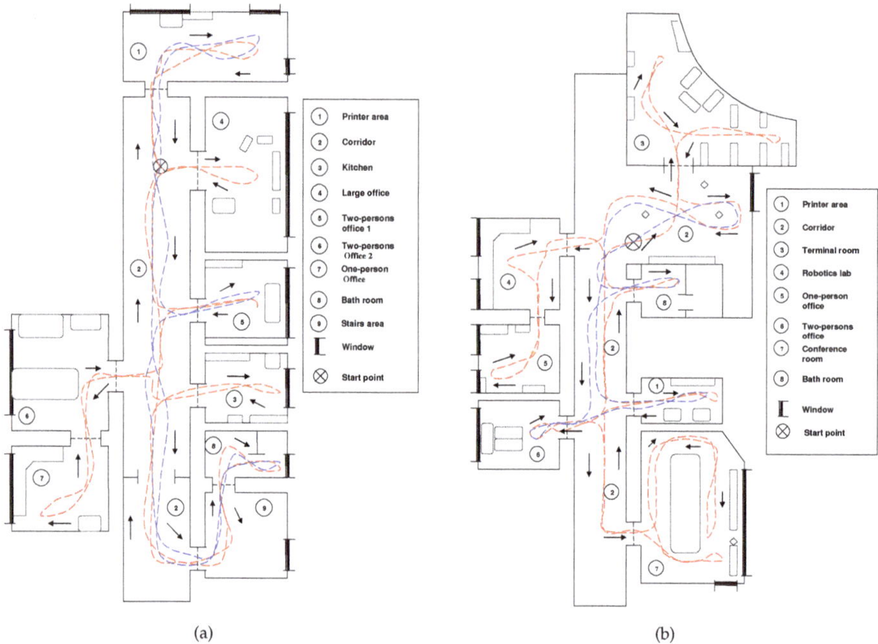

(a)   (b)

Figure 5. Bird's eye view of the COsy Localization Database (COLD). (**a**) Freiburg and (**b**) Saarbrücken environment. Extracted from https://www.nada.kth.se/cas/COLD/.

Table 1. Datasets used to carry out the experiments.

Dataset Name	Number of Images	Number of Rooms
QuorumV_training	872	6
QuorumV_test	77	
Freiburg_training	519	9
Freiburg_test	52	
Saarbrucken_training	566	8
Saarbrucken_test	57	

5.2. Creating Compact Maps through Clustering

This section focuses on the evaluation of clustering methods to compact the information contained in a set of global appearance descriptors. To carry out the experiments, two clustering methods were studied for each environment, and three global appearance descriptors were considered. The first method (Method 1) consists of spectral clustering along with k-means as was explained in Section 3.1. Other configurations were tested, such as to use of SOM instead of k-means to solve Step 5 of the spectral clustering, but the results were quite similar; thus, only the spectral clustering along with k-means to cluster the normalized matrix of the n_c eigenvectors is shown. The second method (Method 2) consists of the use of SOM, which was explained in Section 3.2. Therefore, for the two proposed methods, several experiments were carried out to study the influence of the parameters of the three global appearance descriptors. Table 2 summarizes the experiments developed.

Table 2. Summary of the parameters that have been varied to carry out the clustering experiments. FS, Fourier Signature.

Parameter	Values
Environment	Quorum V Freiburg (COLD) Saarbrücken (COLD)
Descriptor	FS HOG *gist*
Descriptor parameters	FS: k_1 = 4, 8, 16, 32, 64, 128, 256 HOG: k_2 = 2, 4, 16, 32, 64, 128 *gist*: k_3 = 2, 4, 8, 16, 32, 64 *gist*: n_{masks} = 2, 4, 8, 16, 32, 64
Number of clusters	Quorum V: n_c = 15, 25, 40, 60, 80, 100 Freiburg: n_c = 10, 20, 30, 40, 50, 60, 70 Saarbrücken: n_c = 10, 20, 30, 40, 50, 60, 70

The values k_1, k_2, and k_3 define the length of each descriptor, but their meaning is not the same (equal values of k_1, k_2, and k_3 would not lead to the same descriptor size). Therefore, as our aim is to study the correct tuning of these values to use each descriptor as efficiently as possible, we do not apply the same values for all the descriptors in the experiments.

Once the compact map has been produced, it may be interesting to provide some measures that permit quantifying the compactness of the map. In this context, the concept of the silhouette is commonly used. Silhouette values point out the degree of similarity between the instances within the same cluster and at the same time the dissimilarity with the instances that belong to other clusters. The silhouette takes values in the range $[-1, 1]$, and it provides information about how compact the clusters are. Therefore, in order to quantify the goodness of each method, three parameters are considered:

a   The average moment of inertia of the cluster.
b   The average silhouette of the points.
c   The average silhouette of the descriptors.

These values are collected after the clustering process. As for the moment of inertia, it measures the compactness of the clusters (if the clusters group images captured from geometrically-close points) and is calculated as:

$$M = \sum_{i=1}^{n_c} \frac{\sum_{j=1}^{n_i} dist((x,y)_{r_i}, (x_j, y_j))^2}{n_i} \tag{7}$$

where $dist((x,y)_{r_i}, (x_j, y_j))$ is the Euclidean distance between the coordinates of the representative $\vec{r_i}$ and the position of the jth image that belongs to the cluster C_i, and n_i is the number of images within this cluster.

As for the silhouettes values, two types of silhouette are used: the average silhouette of points is defined as:

$$S_{points} = \frac{\sum_{w=1}^{N} S_w}{N} \tag{8}$$

N is the number of instances (images), and s_w is the silhouette of each instance; it is calculated as:

$$s_w = \frac{b_w - a_w}{max(a_w, b_w)} \tag{9}$$

where a_w is the average distance between the capture point of the instance $\vec{d}_w$ and the capture points of the other instances in the same cluster, and b_w is the minimum average distance between the capture point of the instance $\vec{d}_w$ and the capture point of the instances in the other clusters.

Differently, the average silhouette of descriptors is traditionally obtained through:

$$S_{descr} = \frac{\sum_{k=1}^{N} s_k}{N} \tag{10}$$

where N is the total number of instances and s_k is the silhouette of each instance. This value is calculated as:

$$s_k = \frac{b_k - a_k}{max(a_k, b_k)} \tag{11}$$

where a_k is the average distance between the descriptor $\vec{d}_k$ and the descriptor of the rest of the entities contained in the same cluster, and b_k is the minimum average distance between $\vec{d}_k$ and the instances contained in the other clusters.

The silhouette of descriptors has been traditionally used to measure the compactness of the clusters. However, it does not measure the geometrical compactness. This is why we introduce the silhouette of points, which can provide more proper information since we are interested in knowing whether the clusters have grouped images captured nearby.

5.2.1. Clustering in the Quorum V Environment

Figure 6 shows the results of the two clustering methods using FS as the descriptor depending on the parameter k_1. Figure 7 shows the results using HOG depending on the parameter k_2. Figure 8 shows the results using *gist* depending on the parameter k_3 and with n_{masks} = 16. These figures present the graphs that determine the goodness of each configuration to carry out the mapping task through clustering. The three figures show the moment of inertia and average silhouettes vs. the number of clusters. In all cases, the range of the vertical axis is the same, for comparison purposes. Furthermore, Figure 9 shows the computing time necessary to cluster the environment through the two clustering methods.

Regarding the parameters used to measure the compactness of the maps, the lower the moment of inertia and the higher the silhouettes are, the more compact the map is. Generally, Method 1 (spectral clustering) produces the best results. Method 2 (SOM) does not improve these results. As for the use of the global appearance descriptor with the spectral clustering method, FS is not capable of creating reliable clusters. As for HOG, the moment of inertia and silhouettes depend considerably on the value of k_2. When k_2 is low, the results are poor, but when $k_2 > 8$, the moment of inertia, as well as the silhouettes improve significantly. At last, regarding the *gist* descriptor, low values of k_3 produce low silhouettes and high moments of inertia, and high values of this parameter imply better results.

As for the computation time required to carry out the clustering through the two methods, the SOM method presents the highest values. The computing time required for the clustering process through the FS descriptor is the highest, whereas the time through HOG or *gist* is lower, and the fastest one would be determined by the value of either k_2 or k_3.

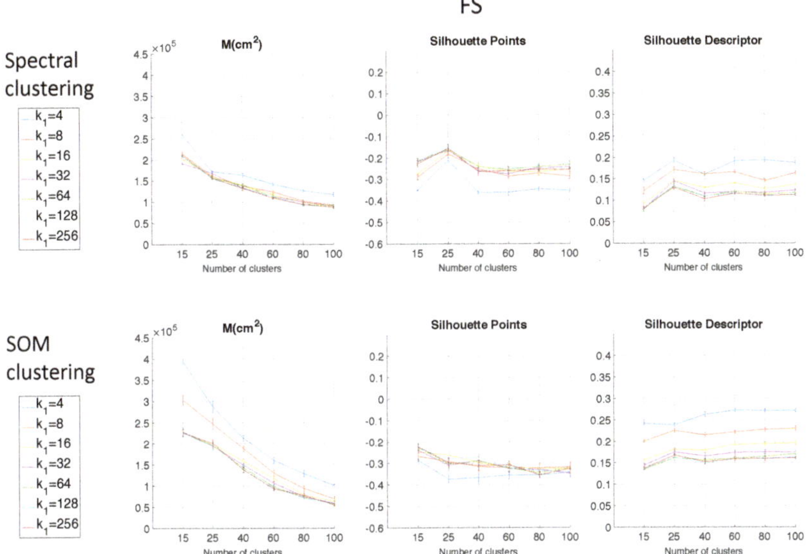

Figure 6. Results of the two clustering methods: average moment of inertia, average silhouette of points, and average silhouette of descriptors vs. number of clusters, when using FS in the Quorum V environment. SOM, Self-Organizing Maps.

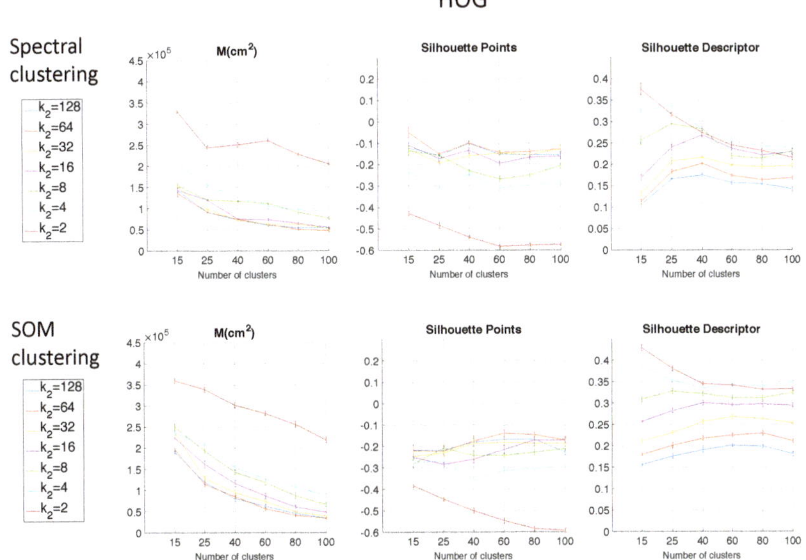

Figure 7. Results of the two clustering methods: average moment of inertia, average silhouette of points, and average silhouette of descriptors vs. number of clusters, when using HOG in the Quorum V environment.

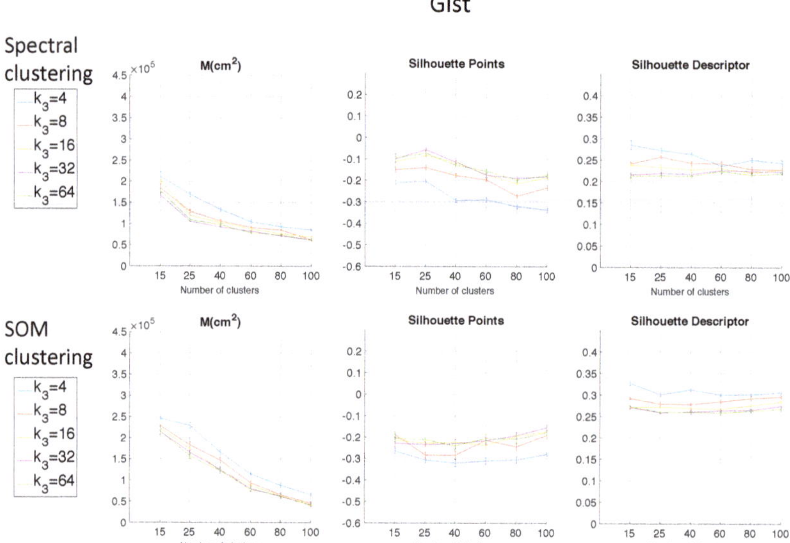

Figure 8. Results of the two clustering methods: average moment of inertia, average silhouette of points, and average silhouette of descriptors vs. number of clusters, when using *gist* in the Quorum V environment.

As expected, the more components the descriptor has, the more time is required. In Section 5.3, the trade-off descriptor size-localization accuracy will be studied.

Therefore, in the case of HOG, a value of $k_2 = 32$ or $k_2 = 64$ could be a good choice to achieve a compromise between compactness and computing time, and in the case of *gist*, an intermediate value of k_3 could be also a good choice for the same purpose. The FS descriptor presents, in general, the worst results: the moment of inertia is higher, and the silhouettes are lower, in general. Hence, the best clustering results are obtained through the use of the spectral clustering method and the use of HOG (for a configuration of $k_2 = [32, 64]$) or *gist* (for a configuration of $k_3 = [16, 32]$ and $n_{masks} = 16$) as the global appearance descriptor. Figure 10 shows a bird's eye view of the clusters obtained with spectral clustering and gistwith $k_3 = 32$ and $n_{masks} = 16$.

5.2.2. Clustering in COLD Environments

The previous results have shown that the use of FS for clustering is less suitable. Considering this, only HOG and *gist* descriptors are analysed in the experiments with the COLD environment. Figure 11 shows the results using HOG depending on the parameter k_2 in the Freiburg environment. Figure 12 shows the results of the clustering methods using *gist* depending on the parameter k_3 and with n_{masks}=16 in the Freiburg environment. In the same way, for the Saarbrücken environment, Figure 13 shows the results using HOG, and Figure 14 shows the results with *gist*. Regarding the use of HOG with the second method (using SOM), it was not able to solve the clustering task for $k_2 = [4, 16]$ when $n_c > 60$.

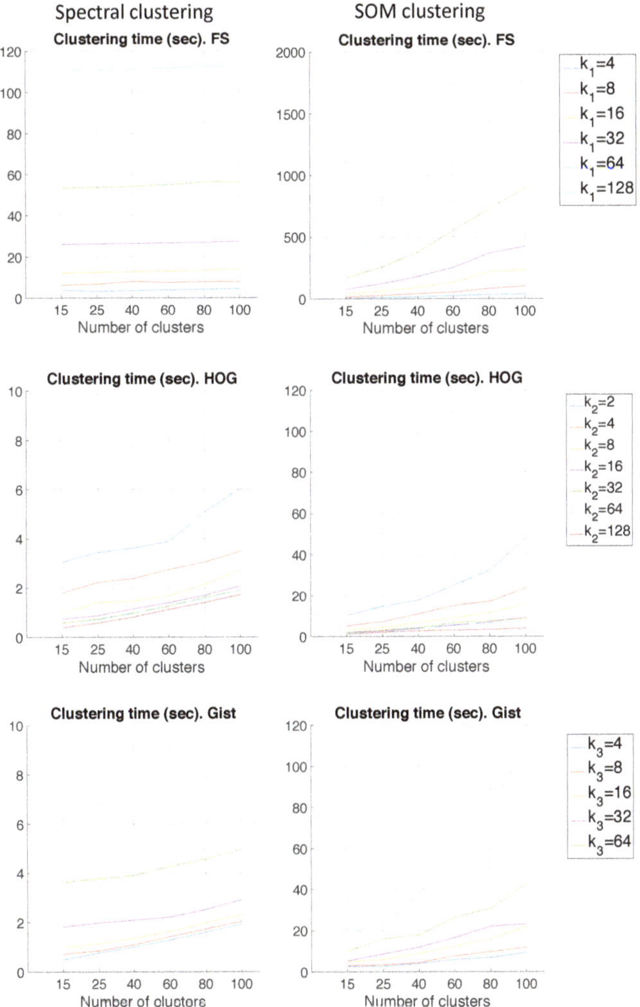

Figure 9. Results of the two clustering methods: computing time vs. number of clusters, when using FS, HOG, and *gist* descriptors in the Quorum V environment.

Again, spectral clustering is the best method, and in this case, *gist* presents better clustering outcomes. Hence, through the experiments carried out in the environments of the COLD database, a confirmation of the results obtained in Quorum V is reached (see Figure 15). Therefore, the proposed method is generalizable despite the use of different types of models (linear or grid). As a conclusion, the best option to carry out the compression of visual maps is reached when spectral clustering with *gist* is applied.

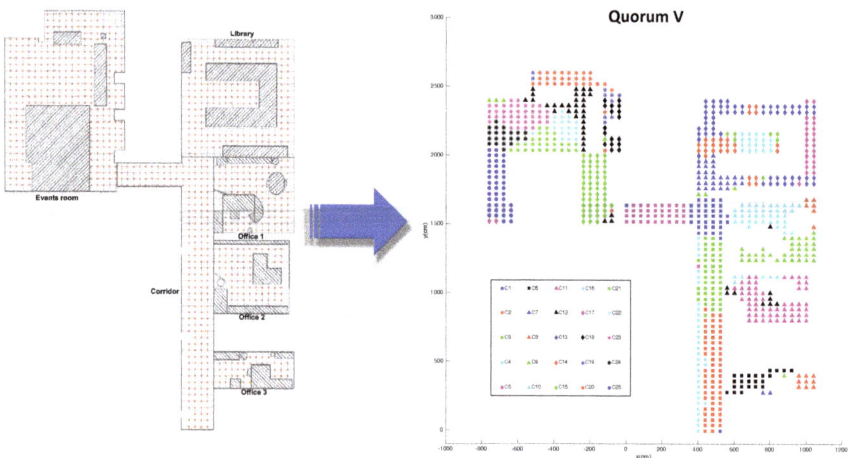

Figure 10. Quorum V environment. Cluster obtained with spectral clustering and *gist* description ($k_3 = 32, n_{masks} = 16$).

HOG. FREIBURG

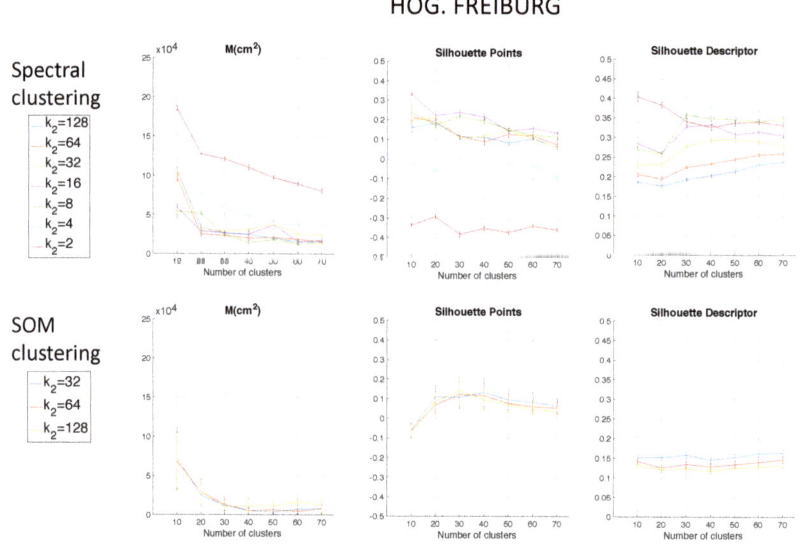

Figure 11. Results of the two clustering methods: average moment of inertia, average silhouette of points, and average silhouette of descriptors vs. number of clusters, when using HOG in the Freiburg environment.

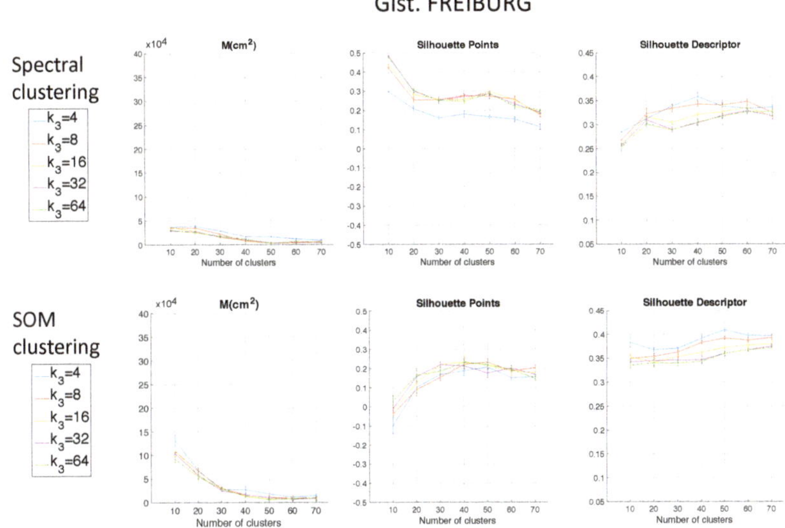

Figure 12. Results of the two clustering methods: average moment of inertia, average silhouette of points, and average silhouette of descriptors vs. number of clusters, when using *gist* in the Freiburg environment.

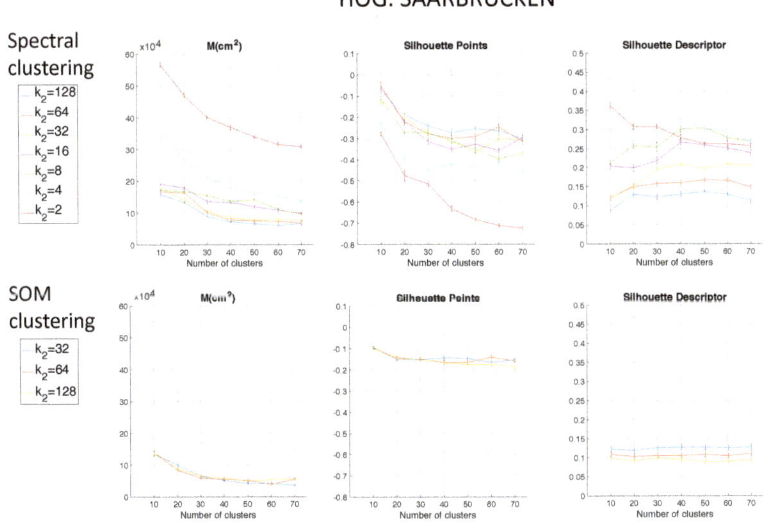

Figure 13. Results of the two clustering methods: average moment of inertia, average silhouette of points, and average silhouette of descriptors vs. number of clusters, when using HOG in the Saarbrücken environment.

Gist. SAARBRÜCKEN

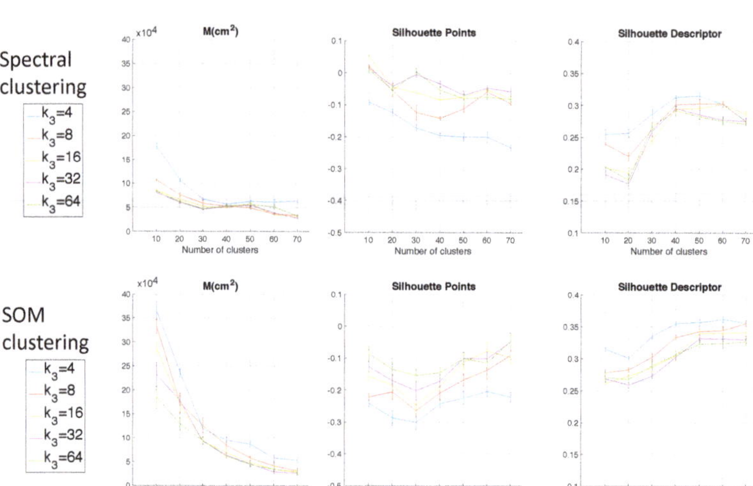

Figure 14. Results of the two clustering methods: average moment of inertia, average silhouette of points, and average silhouette of descriptors vs. number of clusters, when using *gist* in the Saarbrücken environment.

5.3. Localization Using the Compact Maps

This section evaluates the performance of the compact maps to solve the localization problem. The objective is to achieve a compactness that presents a balance between computing time and accuracy of localization. To carry out the evaluation, among the mapping results, the spectral clustering algorithm is selected with the *gist* descriptor ($k_3 = 32$ and $n_{masks} = 16$). With this configuration, a map per environment is built, using the training images. After that, the test images are used to solve the localization problem. The previous subsection proved that the best option to build the compressed map was through the use of the *gist* descriptor. Nevertheless, the three proposed global appearance descriptors are proposed again to solve the localization task (because mapping and localization are two independent processes, and the performance of the descriptors could be different in a localization framework). For each test image, its descriptor is calculated (either by FS, HOG, or *gist*), and then, it is compared with the cluster representatives of the compact map. Afterwards, the most similar cluster is retained. Three distance measures are considered for this comparison: (1) the correlation distance, (2) the cosine distance, and (3) the Euclidean distance. In order to carry out a realistic comparison, despite the real position of the robot being provided by the database, only visual information will be used to estimate the position of the robot. The metric information will be used only as ground truth, for comparison purposes.

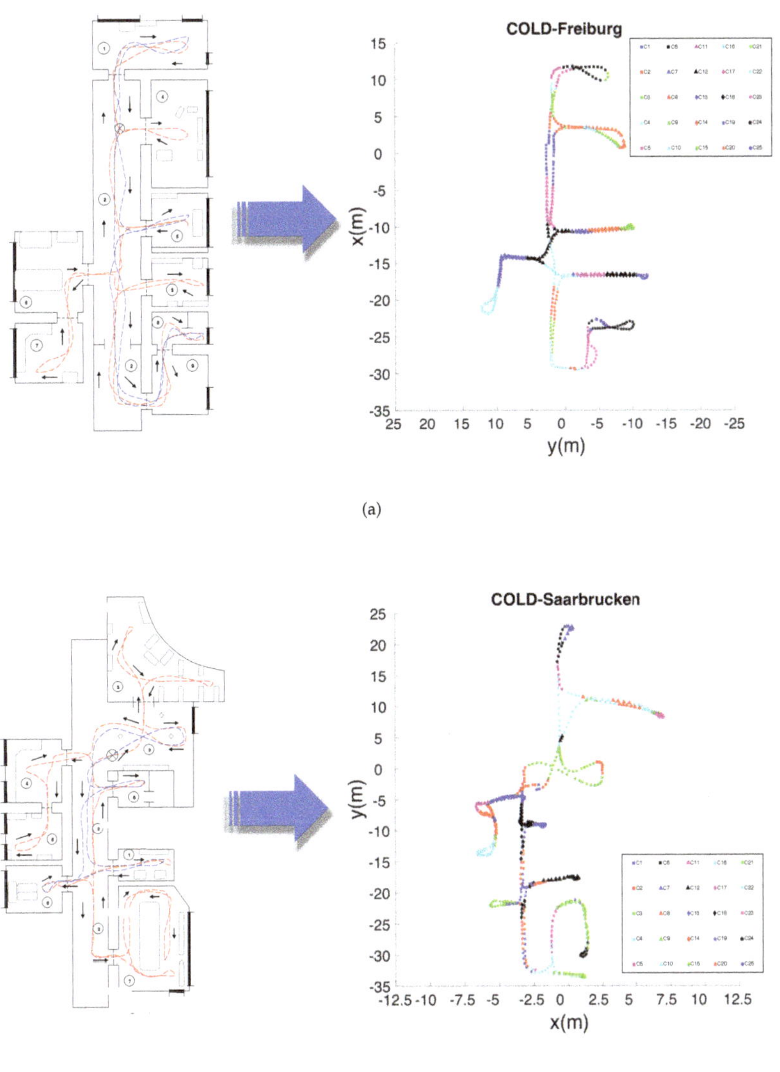

(a)

(b)

Figure 15. Clusters obtained in the COLD environments through the use of Spectral clustering and *gist* description. (**a**) Freiburg and (**b**) Saarbrücken environment.

5.3.1. Localization in the Quorum V Environment

Figure 16 shows the average localization error (cm) obtained when FS (first row), HOG (second row), and *gist* (third row) are used, respectively, as the descriptor. Figure 17 presents the computational time (s). In the case of HOG, the effect of homomorphic filtering adds a constant time of 0.02 s per test image. Regarding the number of clusters, $n_c = 872$ is considered since this value provides the case in which the localization is solved without compacting the map. This value is used as a reference to know the relative utility of the compacted map.

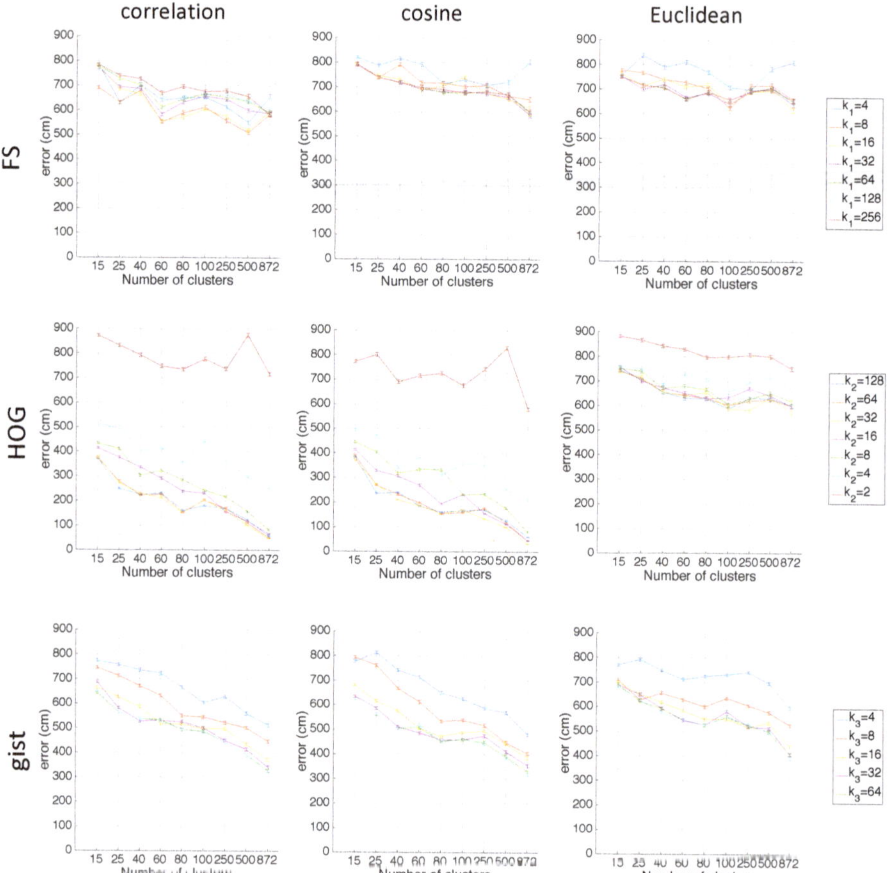

Figure 16. Results of the localization process with FS, HOG, and *gist* used to describe the representatives of the clusters and the test images: average localization error (cm) vs. number of clusters. Quorum V environment.

The FS descriptor is not good for localization since the best choice (correlation distance) presents errors between 650 cm and 800 cm depending on the number of clusters and the size of the descriptor. HOG clearly improves the localization task. Except for the case $k_2 = 2$, the average localization error decreases as the number of clusters increases, and these values go from 500 cm when n_c is low and achieve values under 100 cm (when n_c is high). As for the *gist* descriptor, it also produces relatively good results, but they are not as good as those obtained through the use of HOG. The localization task achieves the best results when the correlation distance is used.

Regarding the computation time, with the FS descriptor, as the number of clusters increases, the computational time required for the localization task increases substantially. With HOG, the time is much lower than FS, and it keeps constant independently of the number of clusters. This means that the time to calculate the descriptor is higher than the time to compare it with the map. The computation time required for *gist* is also worse than HOG. The time required by *gist* is around twice the time with HOG.

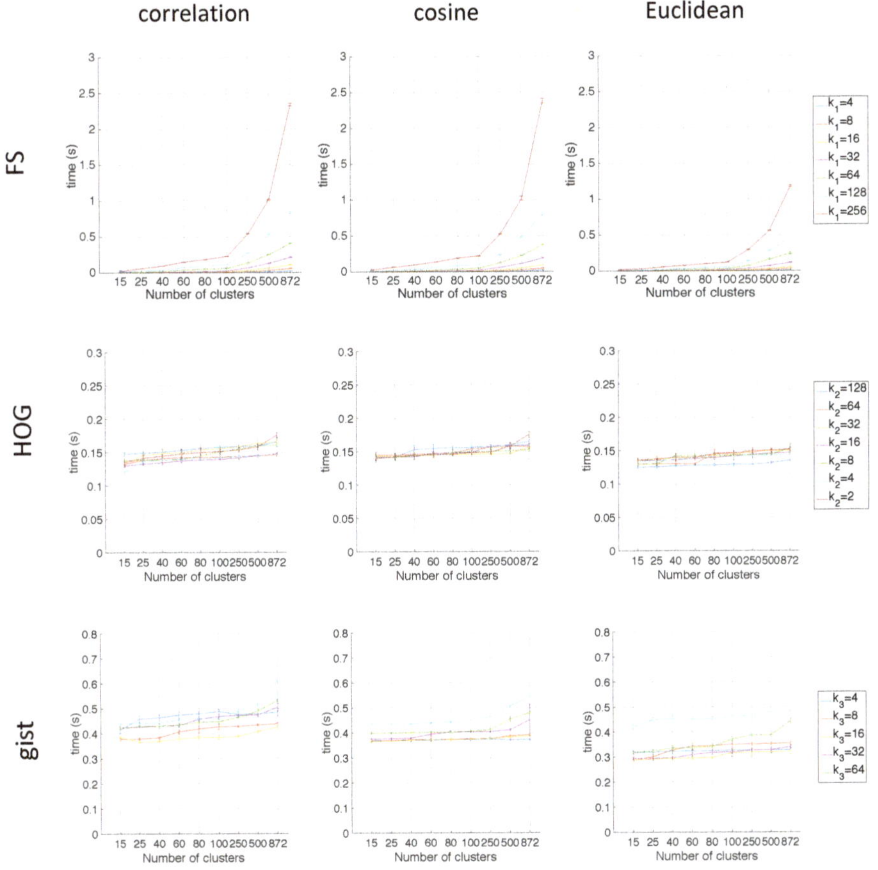

Figure 17. Results of the localization process with FS, HOG, and *gist* used to describe the representatives of the clusters and the test images: average computing time vs. number of clusters. Quorum V environment.

In general, as the number of clusters increases, the computation time required for the localization task also increases, and the average localization error decreases. This is an expected behaviour due to the fact that a high number of clusters means that the map is less compact and the information stays in representatives of the clusters whose distance to the test image is lower. Hence, the more clusters, the more comparisons with representatives must be carried out. This leads to a higher computation time and lower average localization error distance. Thus, a balance between these behaviours must be achieved. Therefore, in order to solve the localization in an environment whose properties are similar to the Quorum V environment (grid-distributed data), the optimal values are reached through the use of a HOG descriptor with $k_2 = [32, 64]$ and correlation distance.

5.3.2. Localization in the Freiburg Environment

As in the previous case (clustering task), with the aim of corroborating the results obtained in Quorum V, an evaluation of the localization task is carried out in the COLD environments. These environments present trajectory maps instead of grid maps. The two COLD environments present a similar configuration and also similar results. This way, only the results obtained in one of them are shown. Freiburg is chosen because this environment presents more rooms and also is more challenging due to the fact that the building presents many glass walls. Moreover, as Figure 16 shows, since the FS descriptor has presented the worst results, this descriptor is discarded in subsequent localization experiments. Furthermore, the Euclidean distance results are omitted in this section because it presented the worst outcomes. Figure 18 shows the average localization error (cm) obtained when HOG (first row) and *gist* (second row) are used respectively as the descriptor. The case of no compaction is also considered ($n_c = 519$).

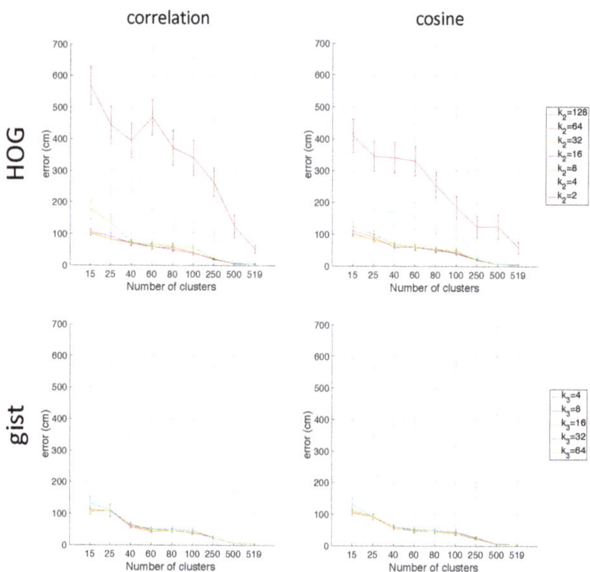

Figure 18. Results of the localization process with HOG and *gist* used to describe the representatives of the clusters and the test images: average localization error (cm) vs number of clusters. Freiburg environment.

In this case, some differences are noticed between the results collected in the Quorum V environment and the results in the Freiburg environment. When the number of clusters is low ($n_c = [15, 25, 40]$), the localization task presents a lower average localization error with *gist*. If this number is higher than 40, the localization error is very similar for HOG and *gist*. Comparing the results obtained with the two evaluated types of distances, no remarkable differences are found. Nevertheless, a slight improvement can be noticed when the cosine distance is used. For instance, the average error value when $n_c = 40$ in HOG is lower with the cosine than with correlation.

Additionally, the value of k_2 in HOG is very important. The average error varies significantly according to it. Therefore, in order to solve the localization in an environment whose properties are similar to Freiburg or Saarbrücken (information along a trajectory), the optimal values are reached through the use of HOG descriptor with $k_2 = [16, 32]$ and cosine distance.

5.3.3. Localization When Several Maps Are Available

In some applications, several maps of some different environments are initially available. If the robot has no information about the environment it is located in, first, it has to use the visual information to select the correct environment. After that, the localization can be solved in the selected environment, as presented in Section 4. Considering this, in this section, the ability to select the right environment is studied. In order to check the goodness of the descriptors for this purpose, the two COLD maps built in Section 5.2.2 are considered. Additionally, a test dataset is created as a combination of images from the Freiburg and Saarbrüken environments. A total of 60 test images compose the test dataset (34 from Freiburg and 26 from Saarbrücken). In this experiment, only HOG and *gist* are tested again. Furthermore, since the cosine distance presented the best solutions for COLD, only this kind of distance is applied. Figure 19 shows the percentage of success in selecting the right environment for the two descriptors.

By and large, the correct environment selection is almost always done. Many cases are given in which 100% success is reached, whereas the worst cases do not present a success rate under 75%. If the environment selection is carried out with HOG, results depend substantially on the chosen k_2 value. For instance, the worst cases are presented for $k_2 = 2, 4$. However, for $k_2 = 32 - 128$, 100% success is reached. Through the use of *gist* descriptor, 100% success is given independently of the number of clusters or the k_3 value.

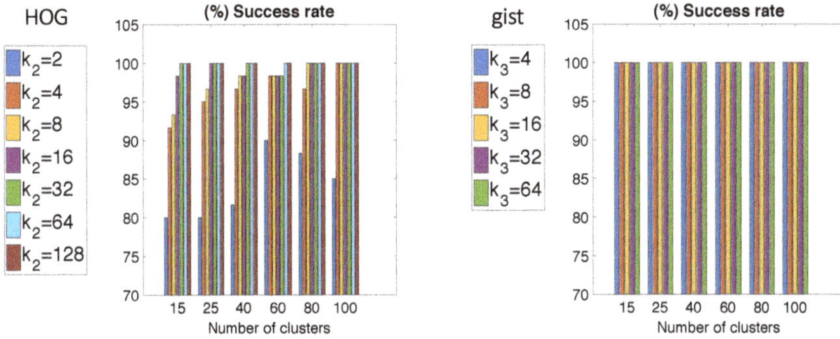

Figure 19. Percentage of success to detect the correct environment between Freiburg and Saarbrücken with FS, HOG, and *gist* used to describe the representatives of the clusters and the test images: percentage of success vs. number of clusters.

5.4. A Comparative Study of Localization with Straightforward and with Compact Maps

Compact maps obtained after clustering present an effective solution to carry out the localization task in a high-level map, as shown in the previous experiments. This process requires capturing a high number of images from the environment to map, prior to the clustering process. At this point, we could ask the following question: is it necessary to capture this high number of images, or could we create a compact model directly, capturing only a limited number of images from the environment? In this section, this issue is studied. Two kinds of models are considered: (a) a compact model obtained after clustering a high number of images and (b) a straightforward model obtained by just capturing a limited number of views from the environment. Both kinds of models will be used to solve the high-level localization task. The straightforward method we propose to retain representatives is downsampling the databases: the COLD databases are downsampled, and only a certain number of images are retained (one of every x images is retained).

The utility of this straightforward model will be compared with the utility of the optimal compact model obtained in Section 5.2.2 with spectral clustering.

Therefore, two models are used as departing points to carry out the localization task: (Model 1) departing from the representative instances obtained through the spectral clustering algorithm and (Model 2) departing from the instances obtained through sampling the databases. Afterwards, the localization task is studied in the Freiburg environment in the same way as was done in Section 5.3.2.

Figure 20 compares the utility of the two models in localization tasks. The cosine distance is selected to show these results, due to the fact that this distance presented good results in previous localization experiments. The two best global appearance descriptors for localization (HOG and *gist*) are shown.

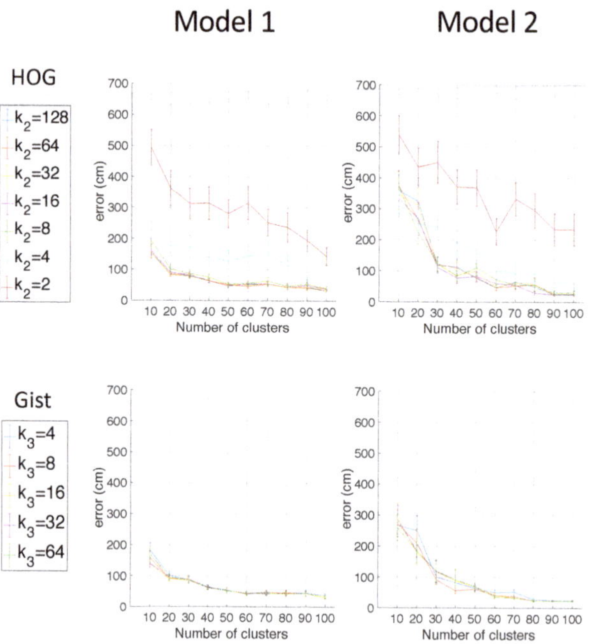

Figure 20. Results of the localization process in the Freiburg environment by using two types of models to retain visual representatives. Average localization error (cm) vs. number of clusters. Model 1 uses representatives obtained through spectral clustering, and Model 2 obtains the representatives through sampling the dataset. The localization task has been carried out with HOG and *gist*, and the distances are calculated through the cosine distance.

As can be seen, the localization error worsens when the straightforward map is used. When the number of clusters is low, the model that has been obtained through spectral clustering presents the best localization results. For example, independent of the descriptor, the average localization error is less than 100 cm when $n_c > 20$ for Model 1 and $n_c > 40$ for Model 2. The average localization error is lower for Model 2 only when the number of clusters is substantially high, $n_c > 80$ (HOG case) and $n_c > 70$ (*gist* case). This outcome means that the proposed alternative to spectral clustering may only be interesting when a low compactness is required. However, if the number of clusters is low (high compactness), spectral clustering provides better results. Therefore, as a conclusion, this experiment has proven that the use of straightforward methods to retain visual representatives is less efficient than using spectral clustering methods. Spectral clustering is able to create compact models that provide accurate localization results.

5.5. Discussion of the Results

This subsection includes a brief discussion related to the results obtained throughout the present work. Regarding the use of methods to compress visual models, spectral clustering has proven to be, in general, more efficient than the SOM clustering. Furthermore, the global appearance descriptor, which presented better behaviour to carry out the clustering task, is *gist*. About the localization task,

HOG presented generally the best outcomes independently of the type of map. The best results are summarized in Figure 21. The best clustering results in Freiburg were obtained with *gist* ($k_3 = [32, 64]$ and $n_{masks} = 16$) and using spectral clustering. Moreover, the best localization outcome in this environment was obtained through the use of HOG ($k_2 = [16, 32]$) with the cosine distance.

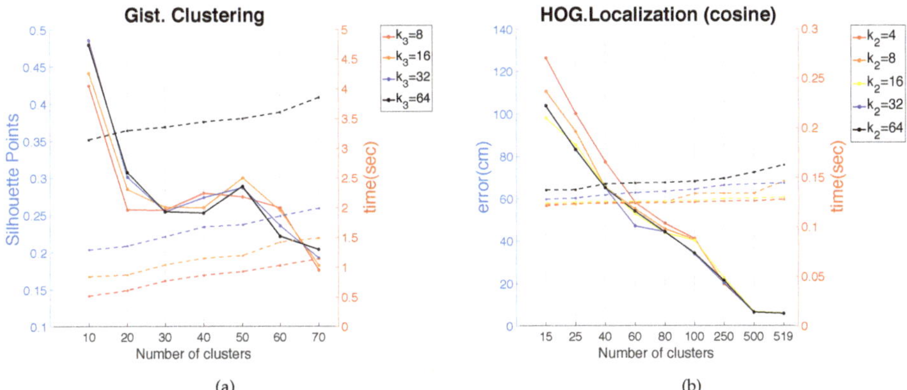

Figure 21. Best results of the clustering and localization processes. (**a**) Clustering with *gist* and spectral clustering: silhouette of points (left axis, solid lines) and computing time (right axis, dashed lines) vs. number of clusters. (**b**) Localization with HOG and cosine distance: average localization error (cm) (left axis, solid lines) and computing time (right axis, dashed lines) vs. the number of clusters. Freiburg environment.

Furthermore, comparing the localization results obtained after compaction and through using raw models, with no compaction ($n_c = 872$, $n_c = 519$, and $n_c = 566$ respectively for Quorum, Freiburg, and Saarbrücken), compact models have proven to be a successful tool to reduce computing time and keep the localization accuracy (see Section 4).

Regarding the use of the global appearance descriptor to select the right map among several options (Section 5.3.3), *gist* has proven to be the most efficient choice. Using this descriptor, 100% of success was reached independently of the number of clusters and the value k_3.

Finally, straightforward methods to compress the information can be discarded since they are not capable of keeping more information about the environment than the proposed spectral clustering method (Section 5.4). Despite that straightforward methods might be faster and easier, the localization outcomes obtained departing from spectral clustering proved to be, in general terms, more accurate.

6. Conclusions and Future Works

This paper proposes two different methods to compact topological maps. With this aim, three datasets from indoor environments were used. These datasets were composed by either panoramic images or omnidirectional images that were transformed to panoramic. During the experiments, with the objective of compacting the information, the number of instances was reduced to a value in the interval from 10–100. That means a reduction of instances up to between 1.1% and 11.5% of the original number. The proposed methods were (1) spectral clustering and (2) self-organizing maps. Moreover, three global appearance descriptors were used since they presented a good solution for environments whose data dimensionality was high. The work shows that it is possible to reduce the visual information drastically from the original model. Among these combinations of method-descriptor, spectral clustering along with the *gist* descriptor was proven to be the best choice to compact the model.

Once the original model is compacted, the resultant map can be used to solve the localization task. Hence, an evaluation is carried out with the aim of measuring the goodness of the localization task through the use of compact maps and global appearance descriptors. In this case, three descriptors and two indoor environments are evaluated. Furthermore, a mixture between indoor environments is created with the aim of evaluating whether it is possible, first, to detect the right environment and, second, estimate the position of the instance. From this study, HOG is the description method whose localization results were the best. Additionally, *gist* presented the most successful results in order to select the correct environment of a test instance from a combined dataset. Finally, the use of clustering methods to tackle the compression step has proven to be more efficient than carrying out a downsampling of the images directly from the database.

The team is now working on how the localization task through compact maps is affected by illumination changes. Additionally, other compacting methods will be studied in order to achieve the Simultaneous Localization And Mapping task (SLAM).

Author Contributions: Conceptualization, L.P. and O.R.; Methodology, L.P. and W.M.; Software, S.C.; Validation, L.P., S.C. and W.M.; Formal Analysis, O.R. and L.P.; Investigation, S.C. and O.R.; Resources, L.P. and W.M.; Data Curation, S.C. and W.M.; Writing—Original Draft Preparation, S.C.; Writing—Review & Editing, L.P. and O.R.; Visualization, S.C. and O.R.; Supervision, L.P.; Project Administration, O.R.; Funding Acquisition, L.P., O.R. and S.C.

Funding: This research was funded by the Generalitat Valenciana through Grant ACIF/2017/146 and by the Spanish government through the project DPI2016-78361-R (AEI/FEDER, UE): "Creación de mapas mediante métodos de apariencia visual para la navegación de robots".

Conflicts of Interest: The authors declare no conflict of interest.

References

1. Okuyama, K.; Kawasaki, T.; Kroumov, V. Localization and position correction for mobile robot using artificial visual landmarks. In Proceedings of the 2011 International Conference on Advanced Mechatronic Systems, Zhengzhou, China, 11–13 August 2011; pp. 414–418.
2. Zhao, Y.; Cheng, W.; Liu, G. The navigation of mobile robot based on stereo vision. In Proceedings of the 2012 Fifth International Conference on Intelligent Computation Technology and Automation, Zhangjiajie, China, 12–14 January 2012; pp. 670–673.
3. Gwinner, K.; Jaumann, R.; Hauber, E.; Hoffmann, H.; Heipke, C.; Oberst, J.; Neukum, G.; Ansan, V.; Bostelmann, J.; Dumke, A.; et al. The High Resolution Stereo Camera (HRSC) of Mars Express and its approach to science analysis and mapping for Mars and its satellites. *Planet. Space Sci.* **2016**, *126*, 93–138. [CrossRef]
4. Jia, Y.; Li, M.; An, L.; Zhang, X. Autonomous navigation of a miniature mobile robot using real-time trinocular stereo machine. In Proceedings of the IEEE International Conference on Robotics, Intelligent Systems and Signal, Changsha, China, 8–13 Octorber 2003.
5. Valiente, D.; Gil, A.; Reinoso, Ó.; Juliá, M.; Holloway, M. Improved Omnidirectional Odometry for a View-Based Mapping Approach. *Sensors* **2017**, *17*, 325. [CrossRef] [PubMed]
6. Berenguer, Y.; Payá, L.; Ballesta, M.; Reinoso, O. Position Estimation and Local Mapping Using Omnidirectional Images and Global Appearance Descriptors. *Sensors* **2015**, *15*, 26368–26395. [CrossRef] [PubMed]
7. Tardif, J.P.; Pavlidis, Y.; Daniilidis, K. Monocular visual odometry in urban environments using an omnidirectional camera. In Proceedings of the 2008 IEEE/RSJ International Conference on Intelligent Robots and Systems, Nice, France, 22–26 Septtmber 2008; pp. 2531–2538.
8. Murillo, A.; Guerrero, J.; Sagues, C. SURF features for efficient robot localization with omnidirectional images. In Proceedings of the 2007 IEEE International Conference on Robotics and Automation, Roma, Italy, 10–14 April 2007; pp. 3901–3907.
9. Menegatti, E.; Pretto, A.; Scarpa, A.; Pagello, E. Omnidirectional vision scan matching for robot localization in dynamic environments. *IEEE Trans. Robot.* **2006**, *22*, 523–535. [CrossRef]
10. Payá, L.; Gil, A.; Reinoso, O. A State-of-the-Art Review on Mapping and Localization of Mobile Robots Using Omnidirectional Vision Sensors. *J. Sens.* **2017**, *2017*, 3497650. [CrossRef]

11. Pantazi, X.E.; Tamouridou, A.A.; Alexandridis, T.; Lagopodi, A.L.; Kashefi, J.; Moshou, D. Evaluation of hierarchical self-organising maps for weed mapping using uas multispectral imagery. *Comput. Electron. Agric.* **2017**, *139*, 224–230. [CrossRef]

12. Hagiwara, Y.; Inoue, M.; Kobayashi, H.; Taniguchi, T. Hierarchical Spatial Concept Formation Based on Multimodal Information for Human Support Robots. *Front. Neurorobot.* **2018**, *12*, 11. [CrossRef] [PubMed]

13. Hwang, Y.; Choi, B. Hierarchical System Mapping for Large-Scale Fault-Tolerant Quantum Computing. *arXiv* **2018**, arXiv:1809.07998.

14. Lowe, D.G. Object recognition from local scale-invariant features. In Proceedings of the Seventh IEEE International Conference on Computer Vision, Kerkyra, Greece, 20–27 September 1999; pp. 1150–1157.

15. Bay, H.; Tuytelaars, T.; Gool, L. SURF: Speeded Up Robust Features. In *Computer Vision at ECCV 2006*; Springer: Berlin/Heidelberg, Germany, 2006; pp. 404–417.

16. Calonder, M.; Lepetit, V.; Strecha, C.; Fua, P. Brief: Binary robust independent elementary features. In Proceedings of the 11th European Conference on Computer Vision, Crete, Greece, 5–11 September 2010.

17. Rublee, E.; Rabaud, V.; Konolige, K.; Bradski, G. ORB: An efficient alternative to SIFT or SURF. In Proceedings of the 2011 International Conference on Computer Vision, Barcelona, Spain, 6–13 November 2011; pp. 2564–2571.

18. Angeli, A.; Doncieux, S.; Meyer, J.; Filliat, D. Visual topological SLAM and global localization. In Proceedings of the 2009 IEEE International Conference on Robotics and Automation, Kobe, Japan, 12–17 May 2009; pp. 2029–2034.

19. Menegatti, E.; Maeda, T.; Ishiguro, H. Image-based memory for robot navigation using properties of omnidirectional images. *Robot. Autom. Syst.* **2004**, *47*, 251–267. [CrossRef]

20. Liu, M.; Scaramuzza, D.; Pradalier, C.; Siegwart, R.; Chen, Q. Scene recognition with omnidirectional vision for topological map using lightweight adaptive descriptors. In Proceedings of the 2009 IEEE/RSJ International Conference on Intelligent Robots and Systems, St. Louis, MO, USA, 10–15 Octorber 2009; pp. 116–121.

21. Payá, L.; Fernández, L.; Gil, A.; Reinoso, O. Map Building and Monte Carlo Localization Using Global Appearance of Omnidirectional Images. *Sensors* **2010**, *10*, 11468–11497. [CrossRef]

22. Rituerto, A.; Murillo, A.C.; Guerrero, J. Semantic labeling for indoor topological mapping using a wearable catadioptric system. *Robot. Autom. Syst.* **2014**, *62*, 685–695. [CrossRef]

23. Leonardis, A.; Bischof, H. Robust recognition using eigenimages. *Comput. Vis. Image Understand.* **2000**, *78*, 99–118. [CrossRef]

24. Oliva, A.; Torralba, A. Modeling the shape of the scene: A holistic representation of the spatial envelope. *Int. J. Comput. Vis.* **2001**, *42*, 145–175. [CrossRef]

25. Radon, J. Uber die bestimmug von funktionen durch ihre integralwerte laengs geweisser mannigfaltigkeiten. *Ber. Saechsishe Acad. Wiss. Math. Phys.* **1917**, *69*, 262.

26. Zivkovic, Z.; Bakker, B.; Krose, B. Hierarchical map building and planning based on graph partitioning. In Proceedings of the 2006 IEEE International Conference on Robotics and Automation, Orlando, FL, USA, 15–19 May 2006; pp. 803–809.

27. Grudic, G.Z.; Mulligan, J. Topological Mapping with Multiple Visual Manifolds. In Proceedings of the Robotics Science and Systems 2005 Workshop, Cambridge, MA, USA, 8–11 June 2005.

28. Valgren, C.; Lilienthal, A. SIFT, SURF & seasons: Appearance-based long-term localization in outdoor environments. *Robot. Autom. Syst.* **2010**, *58*, 149–156.

29. Stimec, A.; Jogan, M.; Leonardis, A. Unsupervised learning of a hierarchy of topological maps using omnidirectional images. *Int. J. Pattern Recognit. Artif. Intell.* **2007**, *22*, 639–665. [CrossRef]

30. Shi, X.; Shen, Y.; Wang, Y.; Bai, L. Differential-Clustering Compression Algorithm for Real-Time Aerospace Telemetry Data. *IEEE Access* **2018**, *6*, 57425–57433. [CrossRef]

31. Payá, L.; Mayol, W.; Cebollada, S.; Reinoso, O. Compression of topological models and localization using the global appearance of visual information. In Proceedings of the 2017 IEEE International Conference on Robotics and Automation (ICRA), Singapore, 29 May–3 June 2017.

32. Mekonnen, A.A.; Briand, C.; Lerasle, F.; Herbulot, A. Fast HOG based person detection devoted to a mobile robot with a spherical camera. In Proceedings of the 2013 IEEE/RSJ International Conference on Intelligent Robots and Systems, Tokyo, Japan, 3–7 November 2013; pp. 631–637.

33. Dong, L.; Yu, X.; Li, L.; Hoe, J.K.E. HOG based multi-stage object detection and pose recognition for service robot. In Proceedings of the 2010 11th International Conference on Control Automation Robotics & Vision, Singapore, 7–10 December 2010; pp. 2495–2500.

34. Dalal, N.; Triggs, B. Histograms of Oriented Gradients fot Human Detection. In Proceedings of the IEEE Conference on Computer Vision and Pattern Recognition, San Diego, CA, USA, 20–25 June 2005; pp. 886–893.

35. Zhu, Q.; Avidan, S.; Yeh, M.; Cheng, K. Fast Human Detection Using a Cascade of Histograms of Oriented Gradients. In Proceedings of the 2006 IEEE Computer Society Conference on Computer Vision and Pattern Recognition (CVPR'06), New York, NY, USA, 17–22 June 2006; pp. 1491–1498.

36. Payá, L.; Amorós, F.; Fernández, L.; Reinoso, O. Performance of global-appearance descriptors in map building and localization using omnidirectional vision. *Sensors* **2014**, *14*, 3033–3064. [CrossRef] [PubMed]

37. Oliva, A.; Torralba, A. Building the gist of ascene: The role of global image features in recognition. *Prog. Brain Res.* **2006**, *155*, 23–36.

38. Siagian, C.; Itti, L. Biologically Inspired Mobile Robot Vision Localization. *IEEE Trans. Robot.* **2009**, *25*, 861–873. [CrossRef]

39. Chang, C.; Siagian, C.; Itti, L. Mobile robot vision navigation and localization using Gist and Saliency. In Proceedings of the 2010 IEEE/RSJ International Conference on Intelligent Robots and Systems, Taipei, Taiwan, 18–22 Octorber 2010; pp. 4147–4154.

40. Murillo, A.C.; Singh, G.; Kosecka, J.; Guerrero, J.J. Localization in Urban Environments Using a Panoramic Gist Descriptor. *IEEE Trans. Robot.* **2013**, *29*, 146–160. [CrossRef]

41. Fernández, L.; Payá, L.; Reinoso, Ó.; Gil, A.; Juliá, M. Robust Methods for Robot Localization under Changing Illumination Conditions-Comparison of Different Filtering Techniques. *ICAART* **2010**, *1*, 223–228.

42. Gonzalez, R.C.; Woods, R.E. *Digital Image Processing*; Prentice Hall: Upper Saddle River, NJ, USA, 2002.

43. Payá, L.; Reinoso, O.; Berenguer, Y.; Úbeda, D. Using Omnidirectional Vision to Create a Model of the Environment: A Comparative Evaluation of Global-Appearance Descriptors. *J. Sens.* **2016**, *2016*, 1–21. [CrossRef] [PubMed]

44. Fernández, L.; Paya, L.; Amoros, F.; Reinoso, O. Using Global Appearance Descriptors to Solve Topological Visual SLAM. In *Encyclopedia of Information Science and Technology*, 4th ed.; IGI Global: Philadelphia, PA, USA, 2018; pp. 6894–6905.

45. Luxburg, U. A tutorial on spectral clustering. *Stat. Comput.* **2007**, *17*, 395–416. [CrossRef]

46. Ng, A.Y.; Jordan, M.I.; Weiss, Y. On Spectral Clustering: Analysis and an algorithm. In *Advances in Neural Information Processing Systems*; MIT Press: Cambridge, MA, USA, 2001; pp. 849–856.

47. Valgren, C.; Duckett, T.; Lilienthal, A. Incremental spectral clustering and its application to topological mapping. In Proceedings of the IEEE International conference on Robotics and Automation, Roma, Italy, 10–14 April 2007; pp. 4283–4288.

48. Sorensen, D.C. Implicitly restarted Arnoldi/Lanczos methods for large scale eigenvalue calculations. In *Parallel Numerical Algorithms*; Springer: Berlin/Heidelberg, Germany, 1997; pp. 119–165.

49. Kohonen, T. The self-organizing map. *Neurocomputing* **1998**, *21*, 1–6. [CrossRef]

50. Van Gassen, S.; Callebaut, B.; Van Helden, M.J.; Lambrecht, B.N.; Demeester, P.; Dhaene, T.; Saeys, Y. FlowSOM: Using self-organizing maps for visualization and interpretation of cytometry data. *Cytometry Part A* **2015**, *87*, 636–645. [CrossRef] [PubMed]

51. Thrun, S.; Fox, D.; Burgard, W.; Dellaert, F. Robust Monte Carlo localization for mobile robots. *Artif. Intell.* **2001**, *128*, 99–141. [CrossRef]

52. Pérez, J.; Caballero, F.; Merino, L. Enhanced Monte Carlo localization with visual place recognition for robust robot localization. *J. Intell. Robot. Syst.* **2015**, *80*, 641–656. [CrossRef]

53. Rui, Y.; Huang, T.S.; Chang, S.F. Image retrieval: Current techniques, promising directions, and open issues. *J. Vis. Commun. Image Represent.* **1999**, *10*, 39–62. [CrossRef]

54. Automation, Robotics and Computer Vision Research Group. Quorum 5 Set of Images. Available online: http://arvc.umh.es/db/images/quorumv/ (accessed on 1 June 2018).

55. Pronobis, A.; Caputo, B. COLD: COsy Localization Database. *IJRR* **2009**, *28*, 588–594. [CrossRef]

 *applied sciences*

Article

Mixed-Degree Cubature H_∞ Information Filter-Based Visual-Inertial Odometry

Chunlin Song, Xiaogang Wang * and Naigang Cui

School of Astronautics, Harbin Institute of Technology, Harbin 150001, China; 14b918036@hit.edu.cn (C.S.); cui_naigang@163.com (N.C.)
* Correspondence: wangxiaogang@hit.edu.cn; Tel.: +86-0451-8641-3459

Received: 3 December 2018; Accepted: 17 December 2018; Published: 24 December 2018

Abstract: Visual–inertial odometry is an effective system for mobile robot navigation. This article presents an egomotion estimation method for a dual-sensor system consisting of a camera and an inertial measurement unit (IMU) based on the cubature information filter and H_∞ filter. The intensity of the image was used as the measurement directly. The measurements from the two sensors were fused with a hybrid information filter in a tightly coupled way. The hybrid filter used the third-degree spherical-radial cubature rule in the time-update phase and the fifth-degree spherical simplex-radial cubature rule in the measurement-update phase for numerical stability. The robust H_∞ filter was combined into the measurement-update phase of the cubature information filter framework for robustness toward non-Gaussian noises in the intensity measurements. The algorithm was evaluated on a common public dataset and compared to other visual navigation systems in terms of absolute and relative accuracy.

Keywords: visual-inertial odometry; cubature information filter; navigation; IMU; RGBD camera

---

1. Introduction

Visual odometry (VO) is the process of estimating the egomotion of an agent with a single or multiple visual sensors in mobile robot navigation. The aim of VO is to estimate the pose incrementally. Compared with the simultaneous localization and mapping (SLAM) method seeking globally consistent estimation, VO is concerned with the local consistency of the pose estimation [1]. As an inexpensive method independent of external reference systems such as global positioning systems (GPS) and motion capture systems, VO is widely applied in domains including robotics, automotive, and wearable computing [2]. In robotic applications, VO is the basis of studies on visual control [3], obstacle avoidance in robot navigation [4,5], and unmanned aerial vehicle navigation [6].

In practical applications, VO is commonly combined with the inertial measurement units (IMUs) with the aid of Kalman filters. As the visual–inertial integrated navigation system serves as an egomotion estimator and is similar to VO functionally, we used the term of visual–inertial odometry (VIO) in this paper. The IMU generally operates independently of the environmental conditions at a higher rate than cameras. Therefore, the robustness and accuracy of the visual odometry can be enhanced by fusing the IMU measurements.

The VO methods are grouped into indirect methods and direct methods, depending on the usage of the images. The indirect methods preprocess the images of the camera to extract features that are used for motion estimation with a matching process. Direct methods use the raw measurements of the visual sensors without feature extraction and matching processes.

Indirect methods have dominated the research field for a long time, and many effective descriptors of features have been developed as shown in [7]. Guang [8] applied a line-feature extractor in the pre-processing of images and estimated the attitude error with line features. The IMU measurements were fused with the visual measurements loosely with the conventional Kalman filter. Mostafa [9] constructed a multi-sensor fusion system based on the extended Kalman filter (EKF). An indirect VO algorithm with speeded up robust features (SURF) was applied to process the visual measurements. A tightly coupled nonlinear optimization-based indirect VO was applied on the micro aerial vehicle (MAV) in [10,11]. Forster developed a hybrid approach with a direct method for initial alignment and an indirect method for joint optimization. Aladem [12] built a light-weight VO using the adaptive and generic accelerated segment test (AGAST) corner detector and the binary robust independent elementary features (BRIEF) descriptor.

Recently, direct methods have gained more attention due to their independence of extra feature extraction and matching processes. Bloesch [13] used the intensity errors of image patches as measurements and fused the measurements of IMU with the direct VO with EKF. The resulting VIO method was applied in the navigation of a MAV in [14]. Furthermore, Bloesch used the iterated EKF method that operates in a recursive form like the Gauss–Newton optimization to reduce the linearization errors in the regular EKF. A binocular vision system was used to construct a direct VIO in [15]. Engel designed a direct sparse odometry (DSO) using a monocular camera [16]. The even sampling and keyframe management methods used in Engle's DSO contributed to improving the accuracy and robustness of the photometric-error optimization.

The Jacobian-based EKF method suffers from linearization errors in nonlinear systems. To improve its accuracy, Arasaratnam designed the cubature Kalman filter (CKF) by applying the third-degree spherical-radial cubature rule on the nonlinear Bayes filter [17]. The resulting CKF outperformed the EKF and unscented Kalman filter (UKF) on nonlinear systems. Based on the third-degree CKF, Jia designed a fifth-degree CKF for higher accuracy [18]. Wang developed another CKF method with the regular simplex and moment matching method [19]. The developed fifth-degree spherical simplex-radial cubature Kalman filter (SSRCKF) required fewer cubature points than the fifth-degree CKF in [18] in high-dimensional applications. Zhang developed an arbitrary degree interpolatory cubature Kalman filter (ICKF) with additional free parameters [20]. By adjusting the free parameters, CKF and UKF are special cases of ICKF. An estimator with an online measuring of non-linearity was developed in [21]. The method adaptively switches between cubature rules with different degrees. The CKF was used in [22] for a loosely coupled nonlinear attitude estimator. Tseng [23] embedded the Huber M-estimation method with CKF in [17] to deal with the outliers and non-Gaussian noises in the measurements.

As an equivalent description of the Kalman filter, the information filter has also been extended with the cubature rules for nonlinear systems. Pakki directly applied the technique used in [17] on the information filter and obtained a third-degree spherical-radial cubature rule-based cubature information filter (CIF). Jia applied the fifth-degree spherical-radial cubature rule on the information filter for multi-sensor fusion [24,25]. Yin developed the third-degree spherical simplex-radial cubature information filter. CIFs with different cubature rules have been used in the state estimation for the biased nonlinear system [26], bearing-only tracking with multi sensors [27], and the trajectory estimation for the ballistic missile [28].

The H_∞ filter was developed for minimizing the estimation errors in the presence of unsatisfactory noises. Yang developed a solution of the H_∞ filter in the form of the regular Kalman filter [29]. Chandra directly embedded Yang's H_∞ filter in the CKF and CIF according to the computational procedures of the filters [30,31] to improve the CKF and CIF in the presence of non-Gaussian noises.

In this paper, motivated by the recently developed cubature information filters for nonlinear systems, a mixed-degree cubature H_∞ information filter-based VIO (MCH$_\infty$IF-VIO) was designed by fusing the measurements of IMU and intensity errors in a tightly coupled manner. Unlike previous studies that have focused on refining the feature descriptions, we developed the MCH$_\infty$IF-VIO system according to the system characteristics and used a simple frame-to-frame alignment model. The proposed VIO method mainly focuses on the nonlinearity and non-Gaussian noises in the VIO system, and is suitable for the camera-IMU system consisting of an IMU and an RGBD camera. The main contributions of this paper are as follows.

1. A mixed-degree nonlinear filter framework was proposed for fusing the IMU measurements and the intensity measurements of the camera. The proposed filter applied two cubature rules with different degrees to guarantee numerical stability with high accuracy.
2. The H_∞ filter was used in the measurement update phase for robustness toward the non-Gaussian noises in intensity errors.

The remainder of this paper is organized as follows. The model of the visual–inertial odometry is presented in Section 2. The VIO system based on a mixed-order cubature H_∞ information filter is detailed in Section 3. The test results with a public dataset are presented in Section 4. The conclusions are presented in Section 5.

2. Model of Visual–Inertial Odometry

In this section, we present the model of a keyframe-based VIO system consisting of the motion model, the keyframe management, and the measurement model. The motion model was formulated in form of classical rigid-body kinematics with the modified Rodrigues parameter representing the attitude. The keyframe management was mainly inspired by Engle's DSO system with some simplifications. The measurement model was formulated in the form of the simple frame-to-frame alignment with the intensity as the measurement.

2.1. Coordinate Systems

Two moving coordinate systems connected with the sensors were used in this paper: (1) The IMU coordinate system (ICS) with its origin at the center of the IMU, where the linear accelerations and angular rates are measured; and (2) The camera coordinate system (CCS) with its origin at the optical center of the camera and the z-axis aligned with the optical axis of the lens.

Another two reference coordinate systems were defined correspondingly: The static IMU coordinate system (SICS) with the same origin and axes as the ICS at the time of constructing the current keyframe; and the static camera coordinate system (SCCS) with the same origin and axes as the CCS at the time of constructing the current keyframe. The SICS and SCCS were static relative to the inertial space in the egomotion estimation.

In this paper, the superscript at the top-left corner of a vector represents the coordinate system the vector is expressed in. Specifically, ^Ip represents the vector p in the ICS, ^Cp represents the vector p in the CCS, ^{SI}p represents the vector p in the SICS, and ^{SC}p represents the vector p in the SCCS.

2.2. Motion Model of Camera-IMU System

A 9-dimension system was built to represent the motion of the camera–IMU system. The state vector is defined as follows:

$$x = \begin{bmatrix} ^I_{SI}\rho^T & ^{SI}p^T & ^{SI}v^T \end{bmatrix}^T \tag{1}$$

where $^I_{SI}\rho$ is the modified Rodrigues parameter (MRP) representing the rotation of the ICS relative to the SICS; ^{SI}p is the position of the origin of the ICS in the SICS; and ^{SI}v is the velocity of the origin of the ICS relative to the SICS. A discrete motion model was constructed according to the six degrees of the freedom kinematic model of a rigid body as follows:

$$x_k = Fx_{k-1} + G(x_{k-1})u_{k-1} + C \tag{2}$$

where $u_{k-1} = \begin{bmatrix} ^I\omega_{k-1}^T & ^Ia_{k-1}^T \end{bmatrix}^T$ is the input vector of the navigation system; $^I\omega_{k-1}$ and $^Ia_{k-1}$ are the measurements of the gyroscope and the accelerator at time $k-1$, respectively; $C = \begin{bmatrix} 0_{1\times6} & \Delta t^{SI}g^T \end{bmatrix}^T$

is a constant vector; Δt is the sampling period; $^{SI}g = {}^{SI}_E R {}^E g$ is the gravity acceleration vector in the SICS; $^{SI}_E R$ is the transformation matrix from the earth-fixed coordinate system to the SICS, of which initial value determined by a simple attitude and heading reference system (AHRS) [32]; and F and $G(x_{k-1})$ are the state-transition matrix and input matrix, respectively, formulated as follows:

$$
F = \begin{bmatrix} \mathcal{I}_3 & 0_3 & 0_3 \\ 0_3 & \mathcal{I}_3 & \Delta t \mathcal{I}_3 \\ 0_3 & 0_3 & \mathcal{I}_3 \end{bmatrix}
$$
$$
G(x_{k-1}) = \begin{bmatrix} A\left({}^I_{SI}\rho_{k-1}\right) & 0_3 \\ 0_3 & 0_3 \\ 0_3 & {}^I_{SI}R\left({}^I_{SI}\rho_{k-1}\right)^T \end{bmatrix}
\tag{3}
$$

where $\mathcal{I}_3$ is the three-dimensional identity matrix; and 0_3 is the three-dimensional zero matrix. $A\left({}^I_{SI}\rho_{k-1}\right)$ and $^I_{SI}R\left({}^I_{SI}\rho_{k-1}\right)$ are formulated as follows. The right subscript $k-1$, left subscript 'SI' and left prescript 'I' in the right side of the equations are neglected to simplify the description

$$
A\left({}^I_{SI}\rho_{k-1}\right) = \frac{1}{4}\begin{bmatrix} \rho_1^2 - \rho_2^2 - \rho_3^2 + 1 & 2(\rho_1\rho_2 - \rho_3) & 2(\rho_1\rho_3 + \rho_2) \\ 2(\rho_1\rho_2 + \rho_3) & \rho_2^2 - \rho_1^2 - \rho_3^2 + 1 & 2(\rho_2\rho_3 - \rho_1) \\ 2(\rho_1\rho_3 - \rho_2) & 2(\rho_2\rho_3 + \rho_1) & \rho_3^2 - \rho_1^2 - \rho_2^2 + 1 \end{bmatrix}
\tag{4}
$$

$$
{}^I_{SI}R\left({}^I_{SI}\rho_{k-1}\right) = \begin{bmatrix} 1 - \dfrac{8\left(\rho_2^2 + \rho_3^2\right)}{\left(|\rho|^2 + 1\right)^2} & -\dfrac{4\left(|\rho|^2\rho_3 - 2\rho_1\rho_2 - \rho_3\right)}{\left(|\rho|^2 + 1\right)^2} & \dfrac{4\left(|\rho|^2\rho_2 + 2\rho_1\rho_3 - \rho_2\right)}{\left(|\rho|^2 + 1\right)^2} \\ \dfrac{4\left(|\rho|^2\rho_3 + 2\rho_1\rho_2 - \rho_3\right)}{\left(|\rho|^2 + 1\right)^2} & 1 - \dfrac{8\left(\rho_1^2 + \rho_3^2\right)}{\left(|\rho|^2 + 1\right)^2} & -\dfrac{4\left(|\rho|^2\rho_1 - 2\rho_2\rho_3 - \rho_1\right)}{\left(|\rho|^2 + 1\right)^2} \\ -\dfrac{4\left(|\rho|^2\rho_2 - 2\rho_1\rho_3 - \rho_2\right)}{\left(|\rho|^2 + 1\right)^2} & \dfrac{4\left(|\rho|^2\rho_1 + 2\rho_2\rho_3 - \rho_1\right)}{\left(|\rho|^2 + 1\right)^2} & 1 - \dfrac{8\left(\rho_1^2 + \rho_2^2\right)}{\left(|\rho|^2 + 1\right)^2} \end{bmatrix}
\tag{5}
$$

In fact, $^I_{SI}R\left({}^I_{SI}\rho_{k-1}\right)$ is the direct cosine matrix representing the rotation of the IMU coordinate system relative to the static IMU coordinate system.

2.3. Keyframe Management

A keyframe consisting of a RGB image and a depth image represents the reference for the egomotion estimation. The very first RGB image and depth image are used to set up the first keyframe. A point set is sampled from the images of the keyframe together with the intensity. Unlike the indirect VO, no extraction or matching process of the distinct features is needed in the direct VO. As a result, the direct VO is generally faster than the indirect VO. However, the accuracy and robustness are hard to ensure without carefully designed features. To ensure the accurate estimation, the point sampling in the direct VO should follow two basic criteria: (1) The sampled points are distributed as evenly as possible in the image; and (2) The absolute intensity gradient at the sampled point is significantly higher than its neighbors. The absolute intensity gradient g at point j with the pixel coordinate of $\left[u_{P_j}, v_{P_j}\right]$ can be evaluated with the intensity differences as follows:

$$
g = \sqrt{\left(\frac{I\left(u_{P_j} + 1, v_{P_j}\right) - I\left(u_{P_j} - 1, v_{P_j}\right)}{2}\right)^2 + \left(\frac{I\left(u_{P_j}, v_{P_j} + 1\right) - I\left(u_{P_j}, v_{P_j} - 1\right)}{2}\right)^2}
\tag{6}
$$

where $I(\cdot)$ is the intensity at the corresponding pixel coordinate. For even sampling, the RGB image in the keyframe is divided into patches with the size of $b \times b$. The pixel with the highest absolute intensity gradient among the pixels with valid depth measurements in each patch is chosen as a candidate. If the absolute intensity gradient of a candidate is no less than a predesigned threshold g_{thres}, the 3D coordinate of the candidate pixel is constructed. We used a simple pin-hole camera model for the 3D construction.

$$^{SC}p_{P_j} = \begin{bmatrix} ^{SC}x_{P_j} \\ ^{SC}y_{P_j} \\ ^{SC}z_{P_j} \end{bmatrix} = d\left(u_{P_j}, v_{P_j}\right) \begin{bmatrix} \frac{u_{P_j}-c_x}{f_x} \\ \frac{v_{P_j}-c_y}{f_y} \\ 1 \end{bmatrix} \tag{7}$$

where $d\left(u_{P_j}, v_{P_j}\right)$ is the depth measurement of corresponding point; and f_x, f_y, c_x, and c_y are the intrinsic parameters of the RGB camera. To simplify the measurement function, the constructed points were transformed into static IMU coordinate system. The transformed 3-dimensional coordinate was added into the point set of the keyframe.

$$^{SC}p_{P_j} = {}^{C}_{I}R^{T}\left(^{SC}p_{P_j} - {}^{C}p\right) \tag{8}$$

where ${}^{C}_{I}R$ and ${}^{C}p$ represent the transformation between the ICS and the CCS. We assumed that the two sensors were mounted on a rigid body so that the transformation did not change over time. The camera–IMU transformation and the intrinsic parameters were calibrated in advance and the calibration methods are not included in this paper.

The absolute intensity gradient threshold g_{thres} was designed according to the intensity characteristic of each patch and evaluated adaptively online. We treated g_{thres} as the sum of two parts:

$$g_{\text{thres}} = \bar{g} + g_b \tag{9}$$

where $\bar{g}$ is the average intensity of the image patch and g_b is a positive value to ensure that the absolute intensity gradient of the candidate is sufficiently high. It is easy to understand that the absolute intensity gradient of the candidate should exceed more pixels in a larger patch, which means that g_{thres} is higher with a larger patch and g_b has a positive correlation with the patch size b. We used a simple linear function to represent the positive correlation and formulated the lower-bound constraint of the absolute intensity gradient as follows:

$$g \geq \bar{g} + \lambda(b-1), \lambda > 0 \tag{10}$$

where λ is a positive constant. Particularly in the case of $b = 1$, i.e., only one pixel exits in each patch, the above lower bound constraint degrades into $g \geq g$, which means that all points with valid depth measurements are constructed. Figure 1 shows an example of point sampling on a 640×480 image. In the example, the above method was able to sample points not only from the edges of the objects, but also from regions with weak intensity variations such as the computer screen and the floor. The points were distributed more densely with a smaller patch. The numbers of sampled points with different patch sizes in this example are shown in Table 1.

Table 1. The number of sampled points with different patch sizes.

b	Number of Sampled Points
2	55752
4	14445
8	3707
16	964
32	242

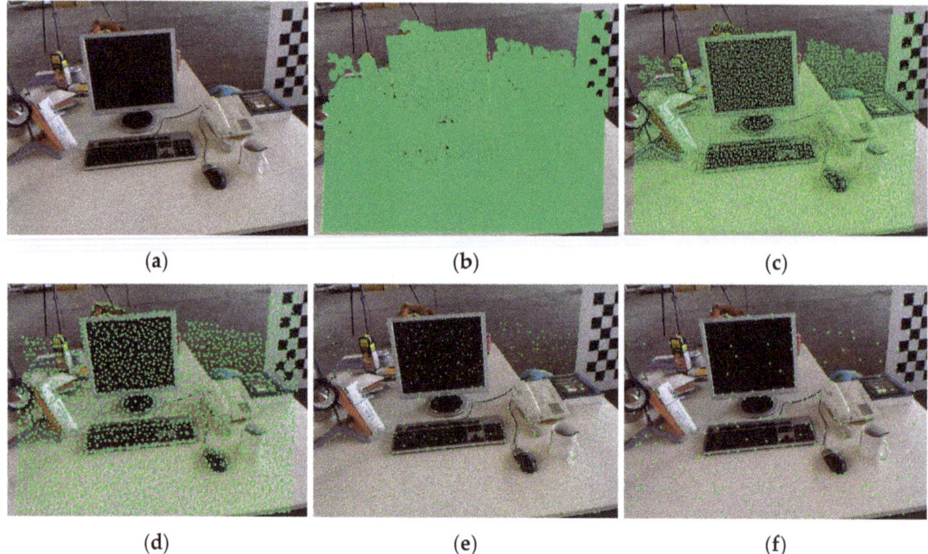

Figure 1. Example of point sampling on a 640 × 480 image with different patch sizes. The green points represent the sampled points. (**a**) The original image; (**b**) The result of point sampling with $b = 2$; (**c**) The result of point sampling with $b = 4$; (**d**) The result of point sampling with $b = 8$; (**e**) The result of point sampling with $b = 16$; and (**f**) The result of point sampling with $b = 32$.

The above point sampling method adjusts the number of the points by tuning b and guarantees evenly distributed points. The usage of images approximated the dense visual odometry with a small b. The usage of images approximated the sparse visual odometry with a large b. The keyframe may not sufficient to estimate the egomotion as the camera moves, so we used two judging criteria to change the keyframe:

1. A new keyframe is needed if there are not enough points in the current view. This is measured via the ratio of points in the current view.
2. Observing the same area from different locations far from each other is likely to lead to large differences in occlusion and illumination. In such cases, new keyframes are needed even if enough points are in the current view. This criterion is quantified with the mean square optical flow ignoring rotational motion $f = \sqrt{\frac{1}{m}\sum_{i=1}^{m}\left((u_{i,0} - u_{i,t})^2 + (v_{i,0} - v_{i,t})^2\right)}$. $[u_{i,0}, v_{i,0}]$ is the pixel coordinate of point i in the keyframe, and $[u_{i,t}, v_{i,t}]$ is the pixel coordinate transformed only by the translational motion.

A new keyframe is constructed with the current RGBD images if any of the two criteria are satisfied. After a new keyframe is constructed, the previous keyframe will be replaced by the new one, and the reference coordinate systems are changed in turn. The estimations of ^{SI}v and $^{SI}_E R$ are transformed with $^I_{SI}\rho$. Then, the estimations of $^I_{SI}\rho$ and ^{SI}p are set to 0.

The above keyframe and point sampling method mainly refers to Engel's DSO method, except for several simplifications as follows:

- Only one keyframe is maintained for the sake of complexity.
- The depth measurements are generated with an RGBD camera instead of being estimated in the optimization scheme.
- The gradient constraint in the sampling points is evaluated according to the patch size.

2.4. Measurement Model

A direct VO was used as the measurement model where the intensity was used as the measurement directly without feature extraction and matching processes. The position of each point j was transformed from the SICS into the CCS by the motion state as follows:

$$^C\boldsymbol{p}_{\mathrm{P}_j} = {}^C_I R\, {}^I_{\mathrm{SI}} R\left({}^I_{\mathrm{SI}}\rho\right)\left({}^{\mathrm{SI}}\boldsymbol{p}_{\mathrm{P}_j} - {}^{\mathrm{SI}}\boldsymbol{p}\right) + {}^C\boldsymbol{p} \tag{11}$$

The pixel coordinate of point j in the current view was evaluated according to the pin-hole camera model.

$$\begin{bmatrix} u_{\mathrm{P}_j} \\ v_{\mathrm{P}_j} \end{bmatrix} = \pi\left({}^C\boldsymbol{p}_{\mathrm{P}_j}\right) = \begin{bmatrix} {}^Cx_{\mathrm{P}_j}\frac{f_x}{{}^Cz_{\mathrm{P}_j}} + c_x \\ {}^Cy_{\mathrm{P}_j}\frac{f_y}{{}^Cz_{\mathrm{P}_j}} + c_y \end{bmatrix} \tag{12}$$

The intensity $I_k\left(u_{\mathrm{P}_j}, v_{\mathrm{P}_j}\right)$ of point j in the RGB image at time k was used as the measurement

$$z_{j,k} = I_k\left(u_{\mathrm{P}_j}, v_{\mathrm{P}_j}\right) + n_{j,k} \tag{13}$$

where $n_{j,k}$ is the measurement noise and assumed to obey a Gaussian distribution with covariance $R_{j,k}$. Combining Equations (11)–(13), one can obtain the following measurement equation.

$$z_k = h\left({}^I_{\mathrm{SI}}\rho_k, {}^{\mathrm{SI}}\boldsymbol{p}_k\right) \tag{14}$$

3. Mixed-Degree Cubature H_∞ Information Filter-Based VIO

The above keyframe management method and measurement model of intensity led to a simple frame-to-frame alignment-based egomotion estimator. However, the simple estimator with linearization-based solvers such as the Gaussian–Newton optimization and EKF performed poorly as the models were highly nonlinear and the measurement noises vary dramatically. The linear approximations and Gaussian assumptions used in the solvers were not valid practically. For a robust estimator, the previous studies on VO generally maintained a slide window of multiple keyframes or a local map. In this section, we present a nonlinear filter-based VIO method for the frame-to-frame alignment problem. The designed hybrid cubature H_∞ information filter used two cubature rules with different degrees to reduce the linearization error in a numerically stable way and the H_∞ filter to estimate the states in the presence of non-Gaussian noises.

3.1. Bayes Filter for Nonlinear Visual–Inertial Navigation System

Consider the following nonlinear system

$$\begin{aligned} x_k &= f(x_{k-1}) + w_{k-1} \\ z_k &= h(x_k) + n_k \end{aligned} \tag{15}$$

where $f(\cdot)$ and $h(\cdot)$ are arbitrary nonlinear functions and w_{k-1} and n_k are the process noise and the measurement noise, respectively. Under the Gaussian assumption of noises, the Bayes filter for nonlinear system described by the information vector and information matrix was formulated as an iterated process of a time-update phase and a measurement-update phase.

(1) Time-update phase

$$\hat{x}_{k|k-1} = \int_{\mathbb{R}^n} f(x)\mathcal{N}\left(x; \hat{x}_{k-1|k-1}, P_{k-1|k-1}\right)dx$$
$$P_{k|k-1} = \int_{\mathbb{R}^n} f(x)f^{\mathrm{T}}(x)\mathcal{N}\left(x; \hat{x}_{k-1|k-1}, P_{k-1|k-1}\right)dx - \hat{x}_{k|k-1}\hat{x}_{k|k-1}^{\mathrm{T}}$$
$$Y_{k|k-1} = P_{k|k-1}^{-1}$$
$$y_{k|k-1} = Y_{k|k-1}\hat{x}_{k|k-1} \tag{16}$$

(2) Measurement-update phase

$$\hat{z}_{k|k-1} = \int_{\mathbb{R}^n} h(x)\mathcal{N}\left(x; \hat{x}_{k|k-1}, P_{k|k-1}\right)dx$$
$$P_{xz} = \int_{\mathbb{R}^n} xh^{\mathrm{T}}(x)\mathcal{N}\left(x; \hat{x}_{k|k-1}, P_{k|k-1}\right)dx - \hat{x}_{k|k-1}\hat{z}_{k|k-1}^{\mathrm{T}}$$
$$H_k = \left(Y_{k|k-1}P_{xz}\right)^{\mathrm{T}}$$
$$\hat{y}_{k|k} = \hat{y}_{k|k-1} + H_k^{\mathrm{T}}R_k^{-1}\left(z_k - \hat{x}_{k|k-1}\right) + H_k\hat{x}_{k|k-1}$$
$$Y_{k|k} = Y_{k|k-1} + H_k^{\mathrm{T}}R_k^{-1}H_k \tag{17}$$

The state estimation could be recovered from the information state

$$P_{k|k} = Y_{k|k}^{-1}$$
$$\hat{x}_{k|k} = P_{k|k}\hat{y}_{k|k} \tag{18}$$

According to the measurement model in Equation (14), only a part of the state vector appeared explicitly in the measurement equation. We defined the state vector explicitly shown in the measurement model as follows:

$$x_1 = \begin{bmatrix} {}_{SI}^{I}\rho \\ {}^{SI}p \end{bmatrix} \in \mathbb{R}^{n_1} \tag{19}$$

where $n_1 = 6$. The rest state vector is

$$x_2 = {}^{SI}v \in \mathbb{R}^{n_2}. \tag{20}$$

Correspondingly, the covariance matrix P of the state vector is composed of four blocks as follows:

$$P = \begin{bmatrix} P_{11} & P_{12} \\ P_{21} & P_{22} \end{bmatrix} \tag{21}$$

Applying the above partition of the state vector to the evaluation of the predicted measurements (the first equation in Equation (17)), one can obtain.

$$\begin{aligned} \hat{z}_{k|k-1} &= \int_{\mathbb{R}^n} h(x)\mathcal{N}\left(x; \hat{x}_{k|k-1}, P_{k-1|k-1}\right)dx \\ &= \int_{\mathbb{R}^{n_1}} h(x)\mathcal{N}\left(x_1; \hat{x}_{1,k|k-1}, P_{11,k|k-1}\right)dx_1 \end{aligned} \tag{22}$$

As shown by Equation (22), the original 9-dimensional integral is simplified to a 6-dimensional integral. The evaluation of the cross-covariance matrix (the second equation in Equation (17)) can also be simplified in a similar way.

$$\begin{aligned} P_{xz} &= \int_{\mathbb{R}^n} xh^{\mathrm{T}}(x)\mathcal{N}\left(x; \hat{x}_{k-1|k-1}, P_{k-1|k-1}\right)dx - \hat{x}_{k|k-1}\hat{z}_{k|k-1}^{\mathrm{T}} \\ &= \int_{\mathbb{R}^{n_1}} \int_{\mathbb{R}^{n_2}} \begin{bmatrix} x_1 \\ x_2 \end{bmatrix} h^{\mathrm{T}}(x_1)p(x_1x_2)dx_2dx_1 - \hat{x}_{k|k-1}\hat{z}_{k|k-1}^{\mathrm{T}} \end{aligned} \tag{23}$$

where $p(x_1, x_2)$ is the joint probability density function of two random vectors x_1 and x_2 i.e., the probability density function of the Gaussian distribution. Based on the formula of conditional probability, one can obtain

$$p(x_1, x_2) = p(x_2|x_1)p(x_1) \tag{24}$$

Substituting Equation (24) into Equation (23) yields

$$P_{xz} = \int_{\mathbb{R}^{n_1}} \int_{\mathbb{R}^{n_2}} \begin{bmatrix} x_1 \\ x_2 \end{bmatrix} h^{\mathrm{T}}(x_1)p(x_2|x_1)dx_2 p(x_1)dx_1 - \hat{x}_{k|k-1}\hat{z}_{k|k-1}^{\mathrm{T}} \tag{25}$$

With the Gaussian assumption of the state vector, the distribution of x_2 conditional on x_1 is a multivariate Gaussian with the mean value of $\hat{x}_2 + P_{21}P_{11}^{-1}(x_1 - \hat{x}_1)$ and the covariance matrix of $P_{22} - P_{21}P_{11}^{-1}P_{12}$. Thus, the inner integral in Equation (25) is solved as

$$\int_{\mathbb{R}^{n_2}} \begin{bmatrix} x_1 \\ x_2 \end{bmatrix} h^{\mathrm{T}}(x_1)p(x_2|x_1)dx_2 = \begin{bmatrix} x_1 \\ \hat{x}_2 + P_{21}P_{11}^{-1}(x_1 - \hat{x}_1) \end{bmatrix} h^{\mathrm{T}}(x_1). \tag{26}$$

The cross-covariance matrix is evaluated as follows:

$$
\begin{aligned}
P_{xz} &= \int_{\mathbb{R}^{n_1}} \begin{bmatrix} x_1 \\ \hat{x}_{2,k|k-1} + P_{21,k|k-1}P_{11,k|k-1}^{-1}\left(x_1 - \hat{x}_{1,k|k-1}\right) \end{bmatrix} h^{\mathrm{T}}(x_1)p(x_1)dx_1 - \hat{x}_{k|k-1}\hat{z}_{k|k-1}^{\mathrm{T}} \\
&= \begin{bmatrix} \int_{\mathbb{R}^{n_1}} x_1 h^{\mathrm{T}}(x_1)p(x_1)dx_1 \\ \int_{\mathbb{R}^{n_1}} \left(\hat{x}_{2,k|k-1} + P_{21,k|k-1}P_{11,k|k-1}^{-1}\left(x_1 - \hat{x}_{1,k|k-1}\right)\right) h^{\mathrm{T}}(x_1)p(x_1)dx_1 \end{bmatrix} - \hat{x}_{k|k-1}\hat{z}_{k|k-1}^{\mathrm{T}} \\
&= \begin{bmatrix} \int_{\mathbb{R}^{n_1}} x_1 h^{\mathrm{T}}(x_1)p(x_1)dx_1 - \hat{x}_{1,k|k-1}\hat{z}_{k|k-1}^{\mathrm{T}} \\ P_{21,k|k-1}P_{11,k|k-1}^{-1}\left(\int_{\mathbb{R}^{n_1}} x_1 h^{\mathrm{T}}(x_1)p(x_1)dx_1 - \hat{x}_{1,k|k-1}\hat{z}_{k|k-1}^{\mathrm{T}}\right) \end{bmatrix}
\end{aligned}
\tag{27}
$$

In this way, the integrals in 9-dimensional space are simplified into integrals in 6-dimensional space. This is practically useful when applying high-degree cubature rules and will be discussed later.

3.2. Mixed-Degree Cubature Information Filter

In general, the four integrals in Equations (16) and (17) cannot be solved directly. A group of numerical integration-based approximation methods named by cubature rules are used for the filter process and the resulting filters are called cubature filters. Using a cubature rule, a Gaussian weighted integral of a nonlinear function $g(x)$ is approximated with a weighted sum as follows:

$$\int_{\mathbb{R}^n} g(x)\mathcal{N}\left(x; \hat{x}_{k-1|k-1}, P_{k-1|k-1}\right)dx \approx \sum_{j=1}^{N_p} w_j g(X_j) \tag{28}$$

where X_j represents the cubature points; w_j is the corresponding weight that satisfies $\sum_{j=1}^{N_p} w_j = 1$; and N_p is the number of the cubature points. The formulations of the cubature points and weights are different in different cubature rules. The spherical-radial (SR) cubature rules use the spherical-radial coordinates to transform the original integral into a double integral and select the cubature points on the spherical coordinate and the radial coordinate respectively. Based on the SR rules, the spherical simplex-radial (SSR) cubature rules use the n-simplex to evaluate the cubature points on the spherical coordinate.

The degree of a cubature rule is defined based on the degree of used monomials in the Tayler polynomial of the nonlinear function. A p-degree cubature rule integrates all of the monomials in the Tayler polynomial of $g(x)$ up to degree p exactly but not exactly for some monomials of degree $p + 1$. In general, the cubature rule with a higher degree achieves higher accuracy, but suffers from more computation with more cubature points and worse numerical stability with potential negative weights. Practically, third-degree cubature rules and fifth-degree cubature rules are commonly used in the Bayes filter. The numbers of cubature points of the four cubature rules are listed in Table 2.

Table 2. Number of cubature points of different rules.

Cubature rules	N_p
3rd-degree spherical-radial cubature rule (3-SR)	$2n$
5th-degree spherical-radial cubature rule (5-SR)	$2n^2 + 1$
3rd-degree spherical simplex-radial cubature rule (3-SSR)	$2n + 2$
5th-degree spherical simplex-radial cubature rule (5-SSR)	$n^2 + 3n + 3$

A cubature rule with all weights being positive is more desirable than one with both positive and negative weights for the sake of numerical stability. A stability parameter was defined in [33] as the sum of absolute values of the weights:

$$\Theta = \sum_{i=0}^{N_p} |w_i| \tag{29}$$

When Θ is larger than 1, the truncation error will decrease the accuracy of the numerical integration. The curves of Θ with respect to n for different cubature rules are shown in Figure 2. The stability parameters of 3-SR and 3-SSR were maintained at 1, while the stability parameter of 5-SR increased since $n > 4$, and the stability parameter of 5-SSR increased since $n > 7$.

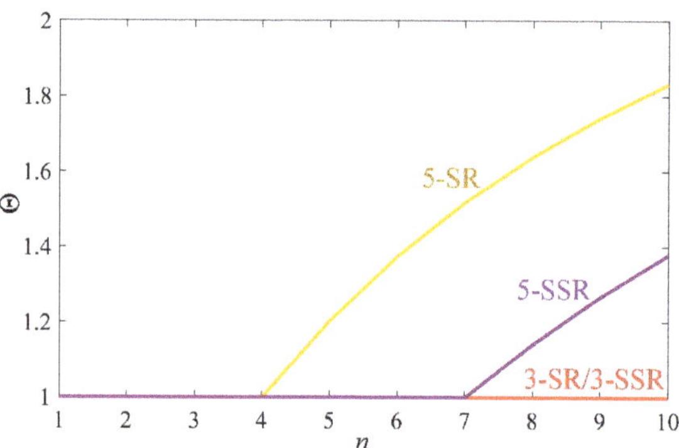

Figure 2. The curves of Θ with respect to n for different cubature rules.

According to the previous motion and measurement models, the time-update phase involves Gaussian-weighted integrals in 9-dimensional space while the measurement-update phase involves Gaussian-weighted integrals in 6-dimensional space. To obtain high precision and numerical stability with as few cubature points as possible, we applied the 3-SR rule in the time-update phase and the 5-SSR rule in the measurement-update phase. Substituting the chosen cubature rules, we formulated the VIO process based on the mixed-degree cubature information filter as follows.

(1) Time-update phase

Evaluate the cubature points based on the 3-SR rule:

$$\xi_{k-1|k-1,j} = S_{k-1|k-1}\gamma_j^{SR3} + \hat{x}_{k-1|k-1}, 1 \leq j \leq 2n \tag{30}$$

where $S_{k-1|k-1}$ is the square root of the covariance matrix $P_{k-1|k-1}$ so that $P_{k-1|k-1} = S_{k-1|k-1}S_{k-1|k-1}^T$ and could be evaluated using the Cholesky decomposition; and γ_j^{SR3} is formulated as follows:

$$\gamma_j^{SR3} = \begin{cases} a_j & , & 1 \le j \le n \\ -a_{j-n} & , & n+1 \le j \le 2n \end{cases} \tag{31}$$

$$a_j = [a_{j,1}, a_{j,2}, \cdots, a_{j,n}]^T$$
$$a_{j,i} = \begin{cases} \sqrt{9} & , & i = j \\ 0 & , & i \ne j \end{cases} \quad 1 \le i \le n, 1 \le j \le n \tag{32}$$

Evaluate the predicted state and the covariance matrix:

$$\hat{x}_{k|k-1} = \frac{1}{2n} \sum_{j=1}^{2n} \left(F\xi_j + G(\xi_j) u_{k-1} + c \right)$$
$$P_{k|k-1} = \frac{1}{2n} \sum_{j=1}^{2n} \left(F\xi_j + G(\xi_j) u_{k-1} + c - \hat{x}_{k|k-1} \right) \left(F\xi_j + G(\xi_j) u_{k-1} + c - \hat{x}_{k|k-1} \right)^T + Q \tag{33}$$

Evaluate the information matrix and the information vector:

$$Y_{k|k-1} = P_{k|k-1}^{-1}$$
$$\hat{y}_{k|k-1} = Y_{k|k-1} \hat{x}_{k|k-1} \tag{34}$$

(2) Measurement-update phase

Evaluate the cubature points based on the 5-SSR rule

$$\zeta_{k|k-1,j}^{m} = S_{1,k|k-1} \gamma_j^{SSR5} + \hat{x}_{1,k|k-1} \tag{35}$$

where γ_j^{SSR5} is formulated as follows:

$$\gamma_j^{SSR5} = \begin{cases} \begin{bmatrix} 0 & \cdots & 0 \end{bmatrix}^T & , & j = 0 \\ \sqrt{n_1 + 2} c_j & , & j = 1, 2, \cdots, n_1 + 1 \\ -\sqrt{n_1 + 2} c_{j-(n_1+1)} & , & j = n_1 + 2, \cdots, 2(n_1 + 1) \\ \sqrt{n_1 + 2} b_{j-2(n_1+1)} & , & j = 2n_1 + 3, \cdots, (n_1^2 + 5n_1 + 4)/2 \\ -\sqrt{n_1 + 2} b_{j-(n_1^2+5n_1+4)/2} & , & j = (n_1^2 + 5n_1 + 6)/2, \cdots, n_1^2 + 3n_1 + 2 \end{cases} \tag{36}$$

where $\{c_j\}$ are the vertexes of the n_1-simplex, and

$$c_j = [c_{j,1}, c_{j,2}, \cdots, c_{j,n_1}]^T$$
$$c_{j,i} = \begin{cases} -\sqrt{\frac{n_1+1}{n_1(n_1-i+2)(n_1-i+1)}} & , & i < j \\ \sqrt{\frac{(n_1+1)(n_1-j+1)}{n_1(n_1-j+2)}} & , & i = j \quad 1 \le i \le n_1, 1 \le j \le n_1 + 1 \\ 0 & , & i > j \end{cases} \tag{37}$$

$\{b_i\}$ are the projections of the midpoints of the edges constructed by $\{c_j\}$.

$$\{b_i\} = \left\{ \sqrt{\frac{n_1}{2(n_1-1)}} (c_l + c_m) : l < m, 1 \le l \le n_1 + 1, 1 \le m \le n_1 + 1 \right\} \tag{38}$$

The weights for the 5-SSR rule are as follows:

$$w_j = \begin{cases} \frac{2}{n_1+2} & , & j = 0 \\ \frac{(7-n_1)n_1^2}{2(n_1+1)^2(n_1+2)^2} & , & j = 1, \cdots, 2(n_1 + 1) \\ \frac{2(n_1-1)^2}{(n_1+1)^2(n_1+2)^2} & , & j = 2n_1 + 3, \cdots, n_1^2 + 3n_1 + 2 \end{cases} \tag{39}$$

Evaluate the predicted measurement $\hat{z}_{k|k-1}$ and the cross-covariance matrix P_{xz}:

$$\hat{z}_{k|k-1} = \sum_{j=1}^{n_1^2+3n_1+3} w_j h\left(\zeta_{k|k-1,j}^{\mathrm{m}}\right)$$

$$P_{xz} = \begin{bmatrix} \sum_{j=1}^{n_1^2+3n_1+3} w_j \zeta_{k|k-1,j}^{\mathrm{m}} h\left(\zeta_{k|k-1,j}^{\mathrm{m}}\right)^{\mathrm{T}} - \hat{x}_{1,k|k-1}\hat{z}_{k|k-1}^{\mathrm{T}} \\ P_{21,k|k-1}P_{11,k|k-1}^{-1}\left(\sum_{j=1}^{n_1^2+3n_1+3} w_j \zeta_{k|k-1,j}^{\mathrm{m}} h\left(\zeta_{k|k-1,j}^{\mathrm{m}}\right)^{\mathrm{T}} - \hat{x}_{1,k|k-1}\hat{z}_{k|k-1}^{\mathrm{T}}\right) \end{bmatrix} \tag{40}$$

As the dimension of the measurement vector is generally large, the inversion of the matrix in Equation (17) is very time-consuming. To omit the inversion, we assumed that the m measurements were independent of each other, and the covariance matrix was a diagonal matrix with diagonal elements of $R_{j,k}(j=1,\ldots,m)$. Further assuming that $R_{j,k}$ was invariant with time and equal to the other diagonal elements, $R_{j,k}$ was treated as a constant R. Thus, the inverse of the $m \times m$ matrix R was replaced by the division of the scalar R, and the measurement update of the information vector and matrix was obtained as follows.

$$\begin{aligned} \hat{y}_{k|k} &= \hat{y}_{k|k-1} + R^{-1}H_k^{\mathrm{T}}\left(z_k - h\left(\hat{x}_{1,k|k-1}\right) + H_k\hat{x}_{k|k-1}\right) \\ Y_{k|k} &= Y_{k|k-1} + R^{-1}H_k^{\mathrm{T}}H_k \end{aligned} \tag{41}$$

Finally, the estimated state vector and covariance matrix were recovered based on Equation (18).

3.3. Robust H_∞ Filter

The aim of the H_∞ filter is to obtain the state estimation minimizing the following cost function [34].

$$J_\infty = \frac{\sum_{k=0}^{N-1}\|x_k - \hat{x}_k\|_{M_k}^2}{\|x_0 - \hat{x}_0\|_{Y_0}^2 + \sum_{k=0}^{N-1}\left(\|w_k\|_{Q_k}^2 + \|n_k\|_{R_k}^2\right)} \tag{42}$$

Based on game theory, the design of the H_∞ filter is a game between the filter designer and nature. Nature's goal is to maximize the estimation error by properly choosing x_0, w_k, and n_k. The cost function is defined in a fractional form to create a fair game by preventing nature using extremely large noises.

The above optimization problem is difficult to solve directly. An alternative strategy is to choose an upper bound and estimate the state satisfying the following constraint.

$$J_\infty < \gamma^2 \tag{43}$$

where γ is the performance bound to be designed. According to the above equation, a new cost function is defined as follows:

$$J = -\frac{1}{\theta}\|x_0 - \hat{x}_0\|_{Y_0}^2 + \sum_{k=0}^{N-1}\left(\|x_k - \hat{x}_k\|_{M_k}^2 - \frac{1}{\theta}\left(\|w_k\|_{Q_k}^2 + \|n_k\|_{R_k}^2\right)\right) < 1 \tag{44}$$

Therefore, the original optimization problem is transformed into the following minimax problem.

$$J^* = \min_{\hat{x}_k} \max_{x_0,w_k,n_k} J \tag{45}$$

Yang [29] solved the minimax problem and proposed an extended H_∞ filter (EH$_\infty$F) for the nonlinear system, which was similar to EKF in form.

$$x_{k|k-1} = f\left(x_{k-1|k-1}\right)$$
$$P_{k|k-1} = \nabla f P_{k-1|k-1} \nabla f^{\mathrm{T}} + Q \tag{46}$$

$$K_k = P_{k|k-1} \nabla h^{\mathrm{T}} \left(\nabla h P_{k|k-1} \nabla h^{\mathrm{T}} + R_k\right)^{-1}$$
$$x_{k|k} = x_{k|k-1} + K_k \left(z_k - h\left(x_{k-1|k-1}\right)\right)$$
$$P_{k|k}^{-1} = P_{k|k-1}^{-1} + \nabla h^{\mathrm{T}} R_k^{-1} \nabla h - \gamma^{-2} \mathcal{I}_n \tag{47}$$

where $\mathcal{I}_n$ is the n-dimension identity matrix. The $EH_\infty F$ is a Jacobian-based filter, and can be extended with cubature rules for higher precision in nonlinear systems as Chandra did in [30]. The time-update phase of the cubature H_∞ filter ($CH_\infty F$) is the same as the CKF and expressed as the equation. Therefore, we only provided the measurement-update equations as follows:

$$P_{zz} = H_k P_{k|k-1} H_k^{\mathrm{T}} + R_k$$
$$K_k = P_{xz} P_{zz}^{-1}$$
$$\hat{x}_{k|k} = \hat{x}_{k|k-1} + K_k \left(z_k - h\left(\hat{x}_{k|k-1}\right)\right)$$
$$P_{k|k}^{-1} = P_{k|k-1}^{-1} + H_k^{\mathrm{T}} R_k^{-1} H_k - \gamma^{-2} \mathcal{I}_n \tag{48}$$

where the cross-covariance matrix P_{xz} and the measurement matrix H_k are evaluated using a cubature rule. It can be seen from Equations (47) and (48) that the main difference between the $EH_\infty F$ and the Kalman filter was the formula to update the covariance matrix $P_{k|k}$. In this way, the $EH_\infty F$ acts to reduce the confidence of the state estimation in case the unexpected noises negatively impact the performance. Applying the information-description of the Kalman filter on the above $CH_\infty F$, a cubature H_∞ information filter ($CH_\infty IF$) can be formulated as follows:

$$P_{zz} = H_k P_{k|k-1} H_k^{\mathrm{T}} + R_k$$
$$i_k = H_k^{\mathrm{T}} R_k^{-1} \left(z_k - h\left(\hat{x}_{k|k-1}\right) + H_k \hat{x}_{k|k-1}\right) - \gamma^{-2} \left(\hat{x}_{k|k-1} + P_{xz} P_{zz}^{-1} \left(z_k - h\left(\hat{x}_{k|k-1}\right)\right)\right)$$
$$\hat{y}_{k|k} = \hat{y}_{k|k-1} + i_k$$
$$Y_{k|k} = Y_{k|k-1} + H_k^{\mathrm{T}} R_k^{-1} H_k - \gamma^{-2} \mathcal{I}_n \tag{49}$$

where the covariance matrix of measurement error vector P_{zz} is an m-dimension matrix. As previously stated, m is generally large in direct VO, which means that the inversion of P_{zz} in the above equation is not practical. We used the expressions of the gain matrix in the Kalman filter to simplify the inversion. In this way, the measurement-update phase in $CH_\infty IF$ was divided into two steps as follows:

(1) Evaluate the measurement matrix H_k with a cubature rule and update the information matrix using the information filter.

$$Y_k' = Y_{k|k-1} + H_k^{\mathrm{T}} R_k^{-1} H_k \tag{50}$$

(2) Evaluate the gain matrix K_∞ and execute the measurement update of the H_∞ filter.

$$K_\infty = Y_k'^{-1} H_k^{\mathrm{T}} R_k^{-1}$$
$$i_k = H_k^{\mathrm{T}} R_k^{-1} \left(z_k - h\left(\hat{x}_{k|k-1}\right) + H_k \hat{x}_{k|k-1}\right) - \gamma^{-2} \left(\hat{x}_{k|k-1} + K_\infty \left(z_k - h\left(\hat{x}_{k|k-1}\right)\right)\right)$$
$$\hat{y}_{k|k} = \hat{y}_{k|k-1} + i_k$$
$$Y_{k|k} = Y_k' - \gamma^{-2} \mathcal{I}_n \tag{51}$$

It is easy to prove that the above two-step measurement-update process is equal to Equation (49). The inversion of the m-dimension matrix is replaced by the inversion of a 9-dimensional matrix Y_k'.

It is worth noting that the measurement-update phase of the proposed $CH_\infty IF$ was different to the one with the same abbreviation that Chandra derived in [31]. Compared with Equation (49), Chandra's method omitted the second term of i_k when using the technique to analyze the estimation error in [35,36]. The error transition function of the measurement-update phase of our method is as follows:

$$\tilde{x}_{k|k} = \tilde{x}_{k|k-1} - K_k\tilde{z}_k \tag{52}$$

$$K_k = P_{k|k}\left(H_k^{\mathrm{T}}R_k^{-1} - \gamma^{-2}K_\infty\right) \tag{53}$$

where $\tilde{z}_k$ is the predicted error of the measurement z_k, and formulated as follows:

$$\tilde{z}_k = z_k - h\left(\hat{x}_{k|k-1}\right) \tag{54}$$

While the error transition function of the measurement-update phase of Chandra's method is related to the predicted mean value of the state vector $\hat{x}_{k|k-1}$ as follows:

$$\tilde{x}_{k|k} = \tilde{x}_{k|k-1} - P_{k|k}H_k^{\mathrm{T}}R_k^{-1}\tilde{z}_k - \gamma^{-2}P_{k|k}\hat{x}_{k|k-1} \tag{55}$$

This goes against the convergence of the filter. In contrast, the error transition function of our method is in the form of a weighted error function as used in [35,36], which makes a convergent filter.

3.4. Summary of MCH$_\infty$IF-Based VIO

Combining the above methods, a mixed-order cubature H_∞ information filter (MOCH$_\infty$IF) was designed for the nonlinear non-Gaussian VIO system. The time-update phase was the same as Equations (30)–(34). The measurement-update phase was performed in the following steps.

- Evaluate cubature points based on the 5-SSR rule with the time complexity of $O\left(N_p\right)$

$$\zeta_{k|k-1,j}^{\mathrm{m}} = S_{1,k|k-1}\gamma_j^{\mathrm{SSR5}} + \hat{x}_{1,k|k-1} \tag{56}$$

- Evaluate the predicted measurements $\hat{z}_{k|k-1}$ and the cross-covariance matrix P_{xz} with the time complexity of $O\left(mN_p\right)$

$$\hat{z}_{k|k-1} = \sum_{j=1}^{n_1^2+3n_1+3} w_j h\left(\zeta_{k|k-1,j}^{\mathrm{m}}\right)$$

$$P_{xz} = \begin{bmatrix} \sum\limits_{j=1}^{n_1^2+3n_1+3} w_j\zeta_{k|k-1,j}^{\mathrm{m}}h\left(\zeta_{k|k-1,j}^{\mathrm{m}}\right)^{\mathrm{T}} - \hat{x}_{1,k|k-1}\hat{z}_{k|k-1}^{\mathrm{T}} \\ P_{21,k|k-1}P_{11,k|k-1}^{-1}\left(\sum\limits_{j=1}^{n_1^2+3n_1+3} w_j\zeta_{k|k-1,j}^{\mathrm{m}}h\left(\zeta_{k|k-1,j}^{\mathrm{m}}\right)^{\mathrm{T}} - \hat{x}_{1,k|k-1}\hat{z}_{k|k-1}^{\mathrm{T}}\right) \end{bmatrix} \tag{57}$$

- Evaluate the measurement matrix H_k with the third equation in Equation (17) and update the information matrix using the information filter with the time complexity of $O(m)$

$$Y_k' = Y_{k|k-1} + R^{-1}H_k^{\mathrm{T}}H_k \tag{58}$$

- Evaluate the gain matrix K_∞ and execute the measurement update of the H_∞ filter with the time complexity of $O(m)$

$$\begin{aligned} K_\infty &= R^{-1}Y_k'^{-1}H_k^{\mathrm{T}} \\ i_k &= R^{-1}H_k^{\mathrm{T}}\left(z_k - h\left(\hat{x}_{k|k-1}\right) + H_k\hat{x}_{k|k-1}\right) - \gamma^{-2}\left(\hat{x}_{k|k-1} + K_\infty\left(z_k - h\left(\hat{x}_{k|k-1}\right)\right)\right) \\ \hat{y}_{k|k} &= \hat{y}_{k|k-1} + i_k \\ Y_{k|k} &= Y_k' - \gamma^{-2}I_n \end{aligned} \tag{59}$$

In conclusion, the time complexity of the measurement-update phase of MOCH$_\infty$IF is $O(mN_p)$. In the case of $n_1 = 6$, we have $N_p = 57$. Generally, the number of measurements is much larger than that i.e., $m \gg N_p$. Thus, the above measurement-update phase is a linear algorithm. In addition, the process of the measurement-update phase is highly parallelizable except for the Cholesky decomposition and the inverse of Y_k' as it mainly consists of matrix additions and multiplications.

4. Evaluation Results

As analyzed in the previous section, the measurement-update phase with operations of large matrixes is the bottleneck for real-time applications. However, the measurement-update phase of an information-based filter is parallelizable and can be implemented on a parallel processor. We used a laptop with the Geforce 940M GPU (NVIDIA, Santa Clara, CA, USA) and the i7-5600U CPU (Intel, Santa Clara, CA, USA) to test the proposed MOCH$_\infty$IF-VIO system. We implemented our method with CUDA (in Supplementary Materials) (Version 8.0, NVIDIA, Santa Clara, CA, USA, 2016) language. The time-update phase was executed on the CPU and the measurement-update phase was executed on the GPU. We used the Technical University of Munich (TUM) RGBD dataset [37] for evaluation. The TUM RGBD dataset contains multiple sequences of videos collected using an RGBD camera under different environmental conditions. The RGB images and depth images are collected at the rate of 30 Hz with a resolution of 640×480. The TUM dataset does not contain IMU measurements, we simply used the differential and quadratic differential of the ground truth with the addition of random noises as the simulated IMU measurements.

We compared the proposed MCH$_\infty$IF-VIO method with RGBDSLAM v2.0 in [38] and the lightweight visual tracking (LVT) in [12]. The results of RGBDSLAM were calculated with the corresponding ROS package, and the results of the LVT were directly obtained from the original paper. We conducted the evaluation for MCH$_\infty$IF-VIO with five different patch sizes: 2, 4, 8, 16, and 32. The numbers were selected as powers of two for the convenience of CUDA implementation. The absolute trajectory (ATE) and relative pose error (RPE) defined in [37] were used as error metrics. The ATE measures the global consistency of the estimated trajectory and is suitable for visual SLAM methods. On the other hand, the RPE measures the local accuracy of the trajectory over a fixed time interval and is suited for visual odometry methods. We used five sequences in the TUM dataset for testing the three methods. The basic parameters of the five sequences are shown in Table 3. The test results are shown in Figures 3–7. In these figures, MCH$_\infty$IF-VIO(n) represents the MCH$_\infty$IF-VIO with a patch size of n.

Table 3. Parameters of the used sequences.

Sequence	Duration (s)	Trajectory length (m)
fr1_desk	23.40	9.263
fr1_desk2	24.86	10.161
fr1_room	48.90	15.989
fr1_xyz	30.09	7.112
f3_office	87.09	21.455

The RPE of MCH$_\infty$IF-VIO increased along with the patch size in all of the tests. The change in the ATE of MCH$_\infty$IF-VIO was similar to RPE with some exceptions. This is in line with the common knowledge that more measurements lead to better accuracy in visual navigation. The MCH$_\infty$IF-VIO in this paper and LVT are dead-reckoning algorithms, while RGBDSLAM is a compete SLAM algorithm with joint optimization and loop-closure. Therefore, RGBDSLAM is more consistent globally than MCH$_\infty$IF-VIO and LVT according to the ATE results. In our tests, MCH$_\infty$IF-VIO achieved a smaller ATE than LVT with only one exception, where the ATE of MCH$_\infty$IF-VIO(32) was larger than LVT in the test on *fr1_room*. MCH$_\infty$IF-VIO(2) had the smallest RPE in all of the tests. The estimation errors caused by the IMU measurement noises increased over time. Moreover, the simple frame-to-frame alignment used in the measurement-update phase was less robust than the optimization with the global and local map used in RGBDSLAM and LVT. Therefore, the performance of $MCH_\infty IF\text{-}VIO$ in the test with long duration such as *fr3_office* was less distinctive.

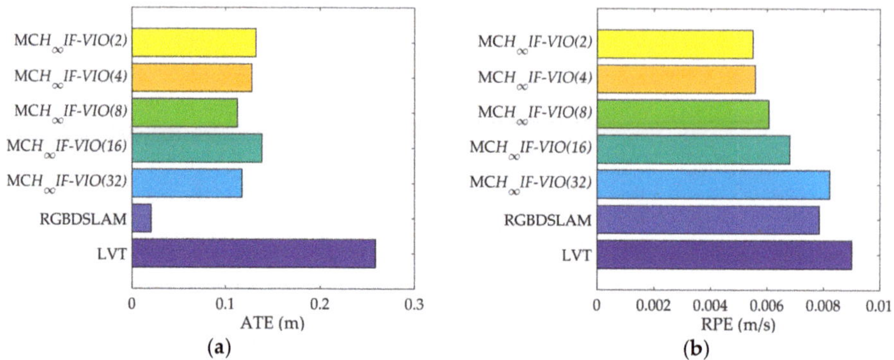

Figure 3. Test results on the sequence *fr1_desk*: (**a**) ATE; (**b**) RPE.

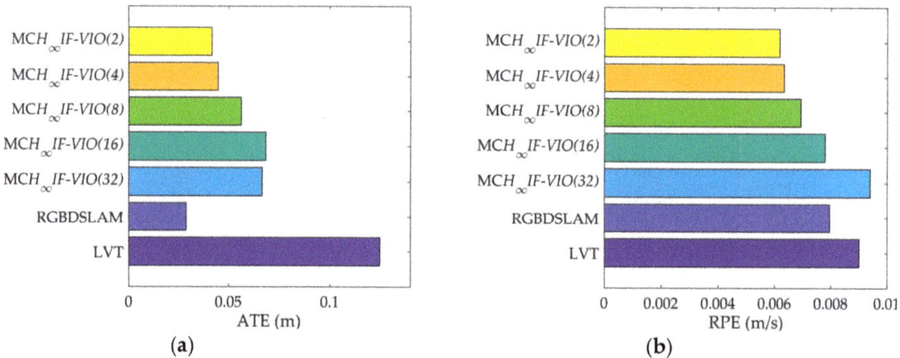

Figure 4. Test results on the sequence *fr1_desk2*: (**a**) ATE; (**b**) RPE.

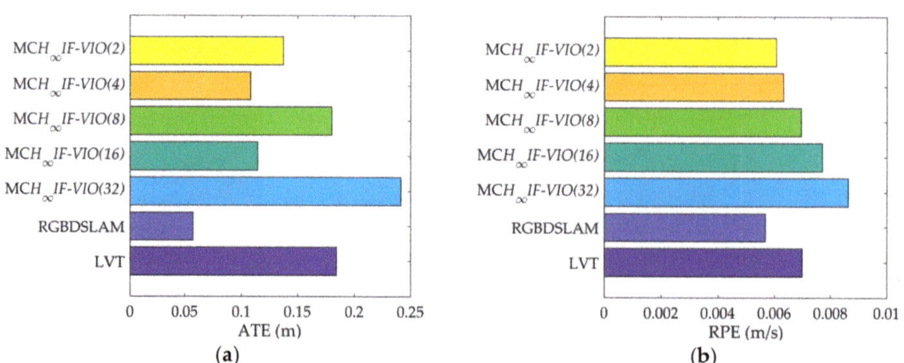

Figure 5. Test results on the sequence *fr1_room*: (**a**) ATE; (**b**) RPE.

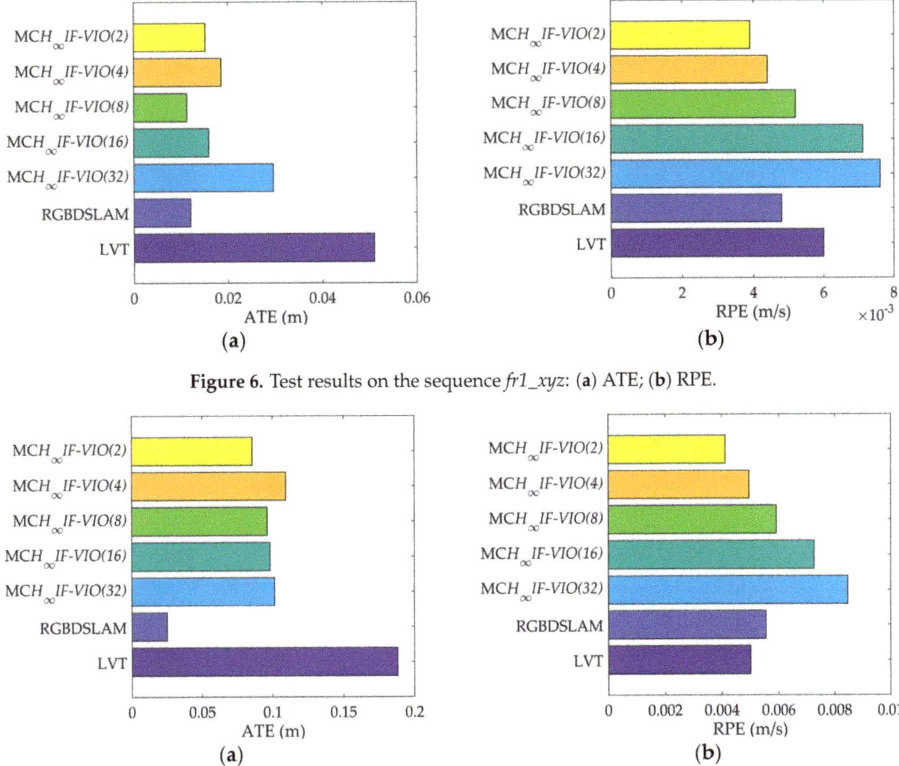

Figure 6. Test results on the sequence *fr1_xyz*: (**a**) ATE; (**b**) RPE.

Figure 7. Test results on the sequence *fr3_office*: (**a**) ATE; (**b**) RPE.

5. Conclusions

In this paper, we presented our visual–inertial navigation system called MCH_∞IF-VIO, which uses a raw intensity-based measurement model. By analyzing the integrated navigation system, we constructed a numerically stable mixed-degree cubature information filter scheme for the state estimation problem. The H_∞ filter was combined for the non-Gaussian noises in the intensity measurements. The proposed VIO system was suitable for the RGBD camera–IMU system. The system was evaluated on the TUM RGBD dataset and compared with the RGBDSLAM and LVT systems.

Given the raw visual measurement model with intensities directly used as measurements in frame-to-frame alignment, the proposed MCH_∞IF-VIO system with consideration of the system nonlinearity and non-Gaussian measurement noises was shown to achieve good performance in our tests. Though the method prefers short durations, the implementation with a patch size of two outperformed the RGBDSLAM and LVT in the long-duration test. With the aid of adjustable patch size, the proposed method could be tuned from an accurate-dense algorithm to a fast-sparse algorithm.

This paper aimed to build a robust state-estimation framework according to the fundamental characteristics of the visual–inertial navigation system. The camera measurements were used in a raw and direct way without heuristics in selecting the feature descriptors or tracking a sliding window. Other well-designed intensity-based and feature-based measurement models can be applied in this hybrid filter-based framework.

Supplementary Materials: The CUDA code used in the evaluations is available online at https://1drv.ms/f/s!ApzqIuEnpaPHiDMOyneS8UV2KO6N.

Author Contributions: Investigation, C.S. and X.W.; Methodology, C.S.; Project Administration, X.W.; Resources, N.C.; Software, C.S.; Supervision, N.C.; Writing – Original Draft, C.S.

Funding: This research received no external funding.

Conflicts of Interest: The authors declare no conflict of interest.

References

1. Scaramuzza, D.; Fraundorfer, F. Visual odometry part i: The first 30 years and fundamentals. *IEEE Robot. Autom. Mag.* **2011**, *18*, 80–92. [CrossRef]
2. Aqel, M.O.A.; Marhaban, M.H.; Saripan, M.I.; Ismail, N.B. Review of visual odometry: Types, approaches, challenges, and applications. *SpringerPlus* **2016**, *5*, 26. [CrossRef] [PubMed]
3. Papoutsidakis, M.; Kalovrektis, K.; Drosos, C.; Stamoulis, G. Intelligent design and algorithms to control a stereoscopic camera on a robotic workspace. *Int. J. Comput. Appl.* **2017**, *167*. [CrossRef]
4. Basaca-Preciado, L.C.; Sergiyenko, O.Y.; Rodriguez-Quinonez, J.C.; Garcia, X.; Tyrsa, V.V.; Rivas-Lopez, M.; Hernandez-Balbuena, D.; Mercorelli, P.; Podrygalo, M.; Gurko, A.; et al. Optical 3D laser measurement system for navigation of autonomous mobile robot. *Opt. Lasers Eng.* **2014**, *54*, 159–169. [CrossRef]
5. Garcia-Cruz, X.M.; Sergiyenko, O.Y.; Tyrsa, V.; Rivas-Lopez, M.; Hernandez-Balbuena, D.; Rodriguez-Quinonez, J.C.; Basaca-Preciado, L.C.; Mercorelli, P. Optimization of 3D laser scanning speed by use of combined variable step. *Opt. Lasers Eng.* **2014**, *54*, 141–151. [CrossRef]
6. Lindner, L.; Sergiyenko, O.; Rivas-Lopez, M.; Ivanov, M.; Rodriguez-Quinonez, J.C.; Hernandez-Balbuena, D.; Flores-Fuentes, W.; Tyrsa, V.; Muerrieta-Rico, F.N.; Mercorelli, P.; et al. Machine vision system errors for unmanned aerial vehicle navigation. In Proceedings of the 2017 IEEE 26th International Symposium on Industrial Electronics, Edinburgh, UK, 19–21 June 2017; IEEE: New York, NY, USA, 2017; pp. 1615–1620.
7. Fraundorfer, F.; Scaramuzza, D. Visual odometry part ii: Matching, robustness, optimization, and applications. *IEEE Robot. Autom. Mag.* **2012**, *19*, 78–90. [CrossRef]
8. Guang, X.X.; Gao, Y.B.; Leung, H.; Liu, P.; Li, G.C. An autonomous vehicle navigation system based on inertial and visual sensors. *Sensors* **2018**, *18*, 2952. [CrossRef]
9. Mostafa, M.; Zahran, S.; Moussa, A.; El-Sheimy, N.; Sesay, A. Radar and visual odometry integrated system aided navigation for UAVS in GNSS denied environment. *Sensors* **2018**, *18*, 2776. [CrossRef]
10. Qin, T.; Li, P.L.; Shen, S.J. Vins-mono: A robust and versatile monocular visual-inertial state estimator. *IEEE Trans. Robot.* **2018**, *34*, 1004–1020. [CrossRef]
11. Lin, Y.; Gao, F.; Qin, T.; Gao, W.L.; Liu, T.B.; Wu, W.; Yang, Z.F.; Shen, S.J. Autonomous aerial navigation using monocular visual-inertial fusion. *J. Field Robot.* **2018**, *35*, 23–51. [CrossRef]
12. Aladem, M.; Rawashdeh, S.A. Lightweight visual odometry for autonomous mobile robots. *Sensors* **2018**, *18*, 2837. [CrossRef] [PubMed]
13. Bloesch, M.; Omani, S.; Hutter, M.; Siegwart, R. Robust visual inertial odometry using a direct EKF-based approach. In Proceedings of the 2015 IEEE/RSJ International Conference on Intelligent Robots and Systems, Hamburg, Germany, 28 September–2 October 2015; IEEE: New York, NY, USA, 2015; pp. 298–304.
14. Sa, I.; Kamel, M.; Burri, M.; Bloesch, M.; Khanna, R.; Popovic, M.; Nieto, J.; Siegwart, R. Build your own visual-inertial drone a cost-effective and open-source autonomous drone. *IEEE Robot. Autom. Mag.* **2018**, *25*, 89–103. [CrossRef]
15. Usenko, V.; Engel, J.; Stuckler, J.; Cremers, D. Direct visual-inertial odometry with stereo cameras. In Proceedings of the 2016 IEEE International Conference on Robotics and Automation, Stockholm, Sweden, 16–21 May 2016; Okamura, A., Menciassi, A., Ude, A., Burschka, D., Lee, D., Arrichiello, F., Liu, H., Moon, H., Neira, J., Sycara, K., et al., Eds.; IEEE: New York, NY, USA, 2016; pp. 1885–1892.
16. Engel, J.; Koltun, V.; Cremers, D. Direct sparse odometry. *IEEE Trans. Pattern Anal. Mach. Intell.* **2018**, *40*, 611–625. [CrossRef] [PubMed]
17. Arasaratnam, I.; Haykin, S. Cubature kalman filters. *IEEE Trans. Autom. Control* **2009**, *54*, 1254–1269. [CrossRef]
18. Jia, B.; Xin, M.; Cheng, Y. High-degree cubature kalman filter. *Automatica* **2013**, *49*, 510–518. [CrossRef]
19. Wang, S.Y.; Feng, J.C.; Tse, C.K. Spherical simplex-radial cubature kalman filter. *IEEE Signal Process. Lett.* **2014**, *21*, 43–46. [CrossRef]

20. Zhang, Y.G.; Huang, Y.L.; Li, N.; Zhao, L. Interpolatory cubature kalman filters. *IET Control Theory Appl.* **2015**, *9*, 1731–1739. [CrossRef]

21. Zhang, L.; Li, S.; Zhang, E.Z.; Chen, Q.W. Robust measure of non-linearity-based cubature kalman filter. *IET Sci. Meas. Technol.* **2017**, *11*, 929–938. [CrossRef]

22. Guo, X.T.; Sun, C.K.; Wang, P. Multi-rate cubature kalman filter based data fusion method with residual compensation to adapt to sampling rate discrepancy in attitude measurement system. *Rev. Sci. Instrum.* **2017**, *88*, 11. [CrossRef]

23. Tseng, C.H.; Lin, S.F.; Jwo, D.J. Robust huber-based cubature kalman filter for gps navigation processing. *J. Navig.* **2017**, *70*, 527–546. [CrossRef]

24. Bin, J.; Xin, M.; Pham, K.; Blasch, E.; Chen, G.S. Multiple sensor estimation using a high-degree cubature information filter. In *Sensors and Systems for Space Applications VI*; Pham, K.D., Cox, J.L., Howard, R.T., Chen, G., Eds.; Spie-Int Soc Optical Engineering: Bellingham, WA, USA, 2013; Volume 8739.

25. Jia, B.; Xin, M. Multiple sensor estimation using a new fifth-degree cubature information filter. *Trans. Inst. Meas. Control* **2015**, *37*, 15–24. [CrossRef]

26. Zhang, L.; Rao, W.B.; Xu, D.X.; Wang, H.L. Two-stage high-degree cubature information filter. *J. Intell. Fuzzy Syst.* **2017**, *33*, 2823–2835. [CrossRef]

27. Jiang, H.; Cai, Y. Adaptive fifth-degree cubature information filter for multi-sensor bearings-only tracking. *Sensors* **2018**, *18*, 3241. [CrossRef] [PubMed]

28. Wang, X.G.; Qin, W.T.; Cui, N.G.; Wang, Y. Robust high-degree cubature information filter and its application to trajectory estimation for ballistic missile. *Proc. Inst. Mech. Eng. Part G-J. Aerosp. Eng.* **2018**, *232*, 2364–2377. [CrossRef]

29. Yang, F.W.; Wang, Z.D.; Lauria, S.; Hu, X.H. Mobile robot localization using robust extended h-infinity filtering. *Proc. Inst. Mech. Eng. Part I-J Syst Control Eng.* **2009**, *223*, 1067–1080. [CrossRef]

30. Chandra, K.P.B.; Gu, D.W.; Postlethwaite, I. A cubature h-infinity filter and its square-root version. *Int. J. Control* **2014**, *87*, 764–776. [CrossRef]

31. Chandra, K.P.B.; Gu, D.W.; Postlethwaite, I. Cubature h-infinity information filter and its extensions. *Eur. J. Control* **2016**, *29*, 17–32. [CrossRef]

32. Madgwick, S. An efficient orientation filter for inertial and inertial/magnetic sensor arrays. *Report x-io* **2010**, *25*, 113–118.

33. Wu, Y.X.; Hu, D.W.; Wu, M.P.; Hu, X.P. A numerical-integration perspective on gaussian filters. *IEEE Trans. Signal Process.* **2006**, *54*, 2910–2921. [CrossRef]

34. Simon, D. *Optimal State Estimation: Kalman, H Infinity, and Nonlinear Approaches*; Wiley: New York, NY, USA, 2006.

35. Xiong, K.; Zhang, H.Y.; Chan, C.W. Performance evaluation of ukf-based nonlinear filtering. *Automatica* **2006**, *42*, 261–270. [CrossRef]

36. Xiong, K.; Zhang, H.Y.; Chan, C.W. Author's reply to "comments on 'performance evaluation of ukf-based nonlinear filtering'". *Automatica* **2007**, *43*, 569–570. [CrossRef]

37. Sturm, J.; Engelhard, N.; Endres, F.; Burgard, W.; Cremers, D. A benchmark for the evaluation of RGB-D slam systems. In Proceedings of the 2012 IEEE/RSJ International Conference on Intelligent Robots and Systems, Vilamoura, Portugal, 7–12 October 2012; IEEE: New York, NY, USA, 2012; pp. 573–580.

38. Endres, F.; Hess, J.; Sturm, J.; Cremers, D.; Burgard, W. 3-D mapping with an RGB-D camera. *IEEE Trans. Robot.* **2014**, *30*, 177–187. [CrossRef]

 *applied sciences*

Article

Framework for Fast Experimental Testing of Autonomous Navigation Algorithms

Miguel Á. Muñoz–Bañón [*,†], **Iván del Pino** [†], **Francisco A. Candelas and Fernando Torres**

Group of Automation, Robotics and Computer Vision (AUROVA), University of Alicante,
San Vicente del Raspeig S/N, 03690 Alicante, Spain; ivan.delpino@ua.es (I.d.P.);
francisco.candelas@ua.es (F.A.C.); fernando.torres@ua.es (F.T.)
* Correspondence: miguelangel.munoz@ua.es
† These authors contributed equally to this work.

Received: 12 April 2019; Accepted: 10 May 2019; Published: 15 May 2019

Abstract: Research in mobile robotics requires fully operative autonomous systems to test and compare algorithms in real-world conditions. However, the implementation of such systems remains to be a highly time-consuming process. In this work, we present an robot operating system (ROS)-based navigation framework that allows the generation of new autonomous navigation applications in a fast and simple way. Our framework provides a powerful basic structure based on abstraction levels that ease the implementation of minimal solutions with all the functionalities required to implement a whole autonomous system. This approach helps to keep the focus in any sub-problem of interest (i.g. localization or control) while permitting to carry out experimental tests in the context of a complete application. To show the validity of the proposed framework we implement an autonomous navigation system for a ground robot using a localization module that fuses global navigation satellite system (GNSS) positioning and Monte Carlo localization by means of a Kalman filter. Experimental tests are performed in two different outdoor environments, over more than twenty kilometers. All the developed software is available in a GitHub repository.

Keywords: autonomous navigation; mobile robots; Monte Carlo localization; SLAM; GNSS; planning; control; Kalman filter

1. Introduction

Autonomous navigation is currently one of the more important topics in robotics since robots capable of moving freely in their environments can produce a large number of new applications in many fields, like logistics [1], agriculture [2] or passenger transport [3]. There are researches covering this topic since the nineteen-seventies [4], so it is a mature field with lots of published algorithms and available tools. In addition, recent advances like the emergence of deep learning are providing researchers with very advanced scene understanding algorithms [5,6]. However, real applications in real conditions—especially in outdoor environments—remain to be a challenge [7].

The usual architecture of an autonomous navigation system—whether it be terrestrial, marine or aerial—relies on the use of several dedicated subsystems, that solve different required tasks like localization [8,9], mapping, path-planning, and control, among others [10–12]. Each one of these tasks belongs to different research topics making mobile robotics an extremely interdisciplinary field [13]. One approach to cope with this complexity is to study the different subsystems separately, what simplifies—and reduces the costs of—the experimental processes. For instance, localization algorithms can be developed making use of public datasets [14], and researchers in control can take advantage of robotic simulators to develop their algorithms [15]. However, these approaches are not sufficient to make a comprehensive evaluation of the developed algorithms, because real systems present interactions—and even feedbacks—between their different modules whose effects can only be observed when the whole system is implemented and tested in real-world conditions [16]. Only extensive experimental tests spanning different conditions and environments can bring enough information for a comprehensive system evaluation. Moreover, these kinds of tests are extremely useful to guide the development processes in a robust and productive way. In the present paper, we introduce a robotic operating system (ROS)-based navigation framework that provides a powerful basic structure based on abstraction levels. This framework is designed to generate minimal but complete autonomous navigation solutions, thus speeding-up the implementation processes required to obtain a full system suitable for experimental testing. This approach permits to save efforts and to keep the focus in concrete research problems, without the drawbacks of simulators and datasets. To show that our framework can be an excellent tool to test and compare different algorithms, we first implement a fully operative autonomous navigation system and then we show how easy results to modify it. Concretely, we change its initial 2D simultaneous localization and mapping (SLAM) localization module by a new one that implements a loosely coupled architecture integrating global navigation satellite system GNSS information and 2D SLAM by means of a Kalman filter. Thanks to this GNSS fusion approach our ground vehicle is able to navigate autonomously in the University of Alicante campus, going in and out from the mapped area. This mixed on-map/off-map navigation is straightforward using our framework, but it would have been very hard to implement using the existing alternatives, as explained later. The system has been evaluated through three experimental sessions, in two different environments, with two different localization algorithms, accumulating more than twenty kilometers of navigation in real-world conditions. Summarizing, our contributions are the following (All the software described in this paper is publicly available and can be found in a GitHub repository https://github.com/AUROVA-LAB/aurova_framework.):

- A generic navigation framework. We propose a conceptual structure for navigation problems that permits to implement complete autonomous navigation systems in a fast and easy way (although demonstrated in a ground vehicle, these implementations can be tailored to different kind of robots, whether it be terrestrial, marine or aerial). This framework permits us to easily arrange the system complexity, enabling researchers to focus on their topics of interest while generating minimal but complete applications suitable for real-world experimental testing. This feature improves the research productivity and is a direct consequence of the proposed architecture.
- A Kalman filter (KF)-based 2D SLAM and GNSS fusion module. To demonstrate how easy is to replace any module using the proposed framework, we developed a new localization module that became a contribution itself. This module is based on a Kalman filter that fuses the poses generated by two complementary localization sources as 2D SLAM and GNSS are. This module permits to recover the SLAM localization after exploring unmapped areas, so mixed navigation on-map/off-map can be performed.
- A set of tools for basic system implementation. In addition to the conceptual framework, we provide a set of tools that brings the basic functionalities required to implement a fully operative terrestrial autonomous navigation system. It comprises planning, car-like control, and reactive safety modules.

The rest of this paper is organized as follows: in Section 2, we describe some related works focusing on general frameworks for developing autonomous navigation systems. In Section 3, we explain the requirements and the design decisions to fulfill them, giving a conceptual architecture for our framework. Then, in Section 4, we explain the basic implementation and its features, from perception to control. Section 5 is devoted to explaining the second particularization of our framework, which implements a fusion of GNSS and Monte Carlo localization by means of a Kalman filter. In Section 6, the different experimental sessions are described and discussed, including real autonomous navigation in two different challenging outdoor environments. Finally, in Section 7, we give conclusions and future works.

2. Related Work

Due to the complexity of autonomous navigation systems, each work presenting a complete application has—in one way or another—implemented its own system architecture and navigation framework. Early examples of papers describing a complete system able to navigate outdoors in large environments are found in the eighties, like in [17,18] where Carnegie Mellon researchers developed a system to navigate autonomously through a network of sidewalks and intersections in the CMU campus. More recently, the two editions of the DARPA grand challenge [19] and the DARPA urban challenge [20] boosted the development of autonomous cars, producing a great number of contributions to the field and several papers describing the architectures and designs developed by the participant teams [11,21,22]. Observing these works, one can find that the different architectures share some features—like parallel processes, communications, tasks, etc.—which could be reusable between them, and that was one of the major reasons that motivated the creation of ROS [23]. The ROS navigation stack is the most widely spread and well-known framework to develop autonomous navigation applications. The navigation stack provides a great number of useful tools but has some limitations [24]: it is designed to operate only with differential drive and holonomic robots, it assumes that the robot can be controlled by means of a twist message indicating x, y and theta velocities, it needs a planar laser for localization and mapping and it performs best with nearly square or circular robots. There exist some workarounds like plug-ins to use it with car-like robots [25] and data conversions that can help to incorporate other sensors like the well-known Kinect depth camera [24], however the tightly coupled architecture of the navigation stack results rigid in operation [26]. In contrast, we adopt—as it will be later explained with detail—a clear architecture based on abstraction levels that makes easy to keep the system modular and scalable, so any module can be replaced without affecting the rest of the system. Looking at other recent literature, one can find some complete applications like [27–29], but these works aim to solve specific problems and are not designed for general purpose. Some efforts in producing more general frameworks for different levels of autonomous system development can also be found, ranging from very high-level project management and agile software development like in [30] to intermediate and low levels, like local trajectory planning and obstacle avoidance in car-like robots [31–33]. However, frameworks aiming to produce a fully operative "template system" to develop and test autonomous navigation algorithms are less common. In [12] a framework for fast designing and prototyping of autonomous multi-robot systems for unmanned aerial vehicles (UAV) is presented. This work shares several aspects with our approach, but it is focused exclusively on aerial multi-agent systems. A whole system architecture for autonomous surface vehicles (ASV) can be found in [34], while in [35] a generic framework is presented as an alternative to the navigation stack, but it focuses in planning and control of wheeled robots with different kinematic constraints rather than covering the whole navigation problem.

3. Framework Design

Our research group is currently working on developing localization algorithms for mobile robots. More specifically we are interested in the integration of GNSS information in graph SLAM algorithms. On the other hand, we would like to test these algorithms in the context of a complete autonomous application to study how the localization influences the system performance and to be able of conducting such experiments in the widest possible variety of environments. With this purpose, we first tried to implement an autonomous navigation system using the well-known ROS navigation stack. Nevertheless, we found that the tools provided by this stack were too much geared towards the use of two-dimensional grid maps, which makes difficult the integration of alternative localization algorithms as can be seen in [36]. For this reason, we designed an alternative framework that—among other interesting features—provides enough flexibility to integrate localization algorithms of different nature in an easy and convenient way. In advance of what will be discussed in this section, we summarize here the main advantages that our final framework design brings with respect to navigation stack:

1. Our planning is independent of the environment representation. In the navigation stack, global planning depends on a grid map, making it difficult to use alternative representations of the environment. On the contrary, following our approach any planning module must be independent of the environment representation, as can be seen in Figure 1. This favors modularity and eliminates conversions and other undesired extra processes required to integrate alternative environment representations with the navigation stack. An example of this can be found in [36], where the authors explain the integration of a graph-based visual SLAM system with the navigation stack planning and control modules. The authors report that they had to create a grid map from their native graph representation and that this extra process introduced additional problems that even forced to discard two of the three scenarios for the path-planning experiments carried out for the paper.

2. Every ROS node belongs to a single module. The navigation stack does not follow this rule, which makes difficult the substitution of certain components. For example, the move_base node fuses planning and control, which makes it rigid in its operation [26]. In contrast, replacing modules in our framework is as easy as changing a single line in the ROS launch file. This makes the produced systems flexible and easy to adapt to different specifications (e.g., different robot kinematics, highlighted as a ROS navigation stack limitation in [24,35]) because only the related nodes need to be modified or substituted.

3. Our framework follows a clear conceptual structure. In [37] we see an example of how developing a complex system using the navigation stack can lead to an intricate architecture. On the contrary, we follow a conceptual structure based on abstraction levels to make the applications clear, organized and scalable. Moreover, a neat division in conceptually different sub-problems makes easier to keep the research focus on the topics of interest without giving up the advantages that a complete system in real-world operation provides for experimental testing, as happens when using datasets or simulators.

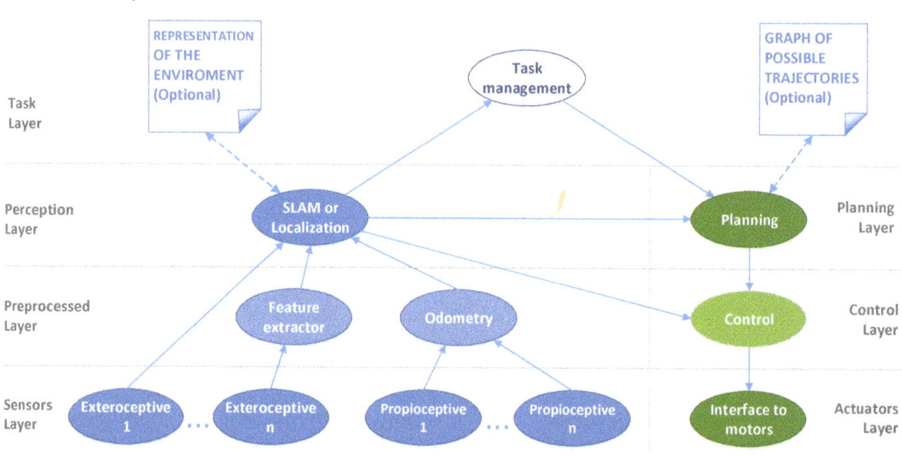

Figure 1. Proposed framework design. Each node represents a software module, and each solid line represents a communication channel between modules. The dashed lines represent optional channels that might connect the localization module with a map of the environment and the planning module with a map of trajectories. The existence of these optional channels depends on the concrete localization and planning algorithms used. The different modules are organized in abstraction layers (sensors, preprocessed...) as well as in conceptual blocks denoted by colours: perception (blue nodes), motion (green nodes) and high-level (white nodes).

Next, we describe the design process, starting with the requirement analysis focused on obtaining a framework flexible enough to allow fast and easy experimentation in different environments with different robots. Then, we describe the proposed design solution taking into account the specified requirements.

3.1. Framework Requirements

The key to developing a generic framework for mobile robotics is to identify precisely which parts of the software are common to different applications and design neat interfaces that permit sufficient modularity without adding unnecessary complexity to the system design. Therefore our framework requires the formalization of different software modules with standardized inputs and outputs to ease the test and comparison of different algorithms. Any algorithm change in any of its modules must be carried out in a simple way, requiring minimal changes in the rest of the system. Moreover, in order to be useful in the long term, we need to keep the framework as independent of the robot and the environment as possible.

3.2. Proposed Approach

As mentioned above, in mobile robotics applications we can distinguish different abstraction levels common to all of them. One of the requirements to develop the proposed framework is to define these abstraction levels and the software modules that are part of each one, as well as their communication channels. As shown in Figure 1, layers below task level are split in two halves to separate perception from control, resulting in three areas: perception (blue nodes), motion (green nodes) and high-level (white nodes). The links shown in the Figure 1 can be modified depending on the inputs/outputs of each particular node.

3.2.1. Perception

The lowest level we define in perception is the sensor layer. Software modules belonging to this abstraction layer are drivers to decode the information captured by the sensors. We divide the sensors into two types: exteroceptive and proprioceptive. Exteroceptive sensors are those that provide environment perception, while proprioceptive ones are those that give the information to generate the odometry. In the next perception level, we define the preprocessing layer in which we place the necessary modules to extract the information from the different sensors, implementing data fusion algorithms if required. Then, we define the highest abstraction level in the perception area to place the localization or SLAM module. Depending on the localization system we are using, the output pose of this module can be expressed with respect to different frames (local, tracked object, map or global coordinates). This localization module is designed to receive their inputs in a ROS standard format making transparent the lower layers, so there is no difference if the data comes from the sensor layer or preprocessing one. In this level, there might also be modules for scene understanding or to detect obstacles and moving objects.

3.2.2. Motion

At the highest level of motion, we define the planning layer. In this level we can integrate different software modules, that will be selected depending on the task we want to perform. Some task examples could be reaching a global goal within an environment or following a moving target, such as a vehicle or a pedestrian. For these two examples, we would need two different planners. Planner selection comes from the high-level area of the system. We also receive the robot location from the perception layer and the information obtained through the scene understanding. The intermediate level in motion is the local control. Controllers integrated into this level receive local goals from the upper layer, as well as the robot pose. The lower layer in the motion area is the actuators layer, where the software modules are drivers for communication with the robot motors.

3.2.3. High Level

At the highest level, we define the task management module. A task gets completely defined by a specific combination of modules. Therefore the task manager is in charge of deciding which modules of the lower layers will be executed. This module can also serve as a high-level interface with the agent (the user) or other robots, allowing to develop collaborative robotics applications, which is a relevant topic [38]. This last feature covers one of the limitations of the navigation stack described in [26].

4. Initial Framework Implementation

In this section we describe the basic implementation of the proposed architecture using the abstraction levels described above (Figures 2 and 3). We can use this basic structure of nodes and topics as a starting point to create more sophisticated applications easily and quickly, simply by replacing the desired modules. The creation of a real-world autonomous navigation application using our framework consists of two phases: pre-execution and execution. In the first one, we gather the data to generate the environment information that we will need during the execution phase.

Of course, an equivalent system could have been generated using just the navigation stack since in this first example we based the localization in a grid map. But our framework adds conceptual structuring and better modularity. That is to say, in this implementation the only module depending on the grid map is the localization. Therefore replacing this module would not affect the rest of the system (as can be seen in Section 5) contrary to what happens with the navigation stack where the planning needs a grid map to generate trajectories.

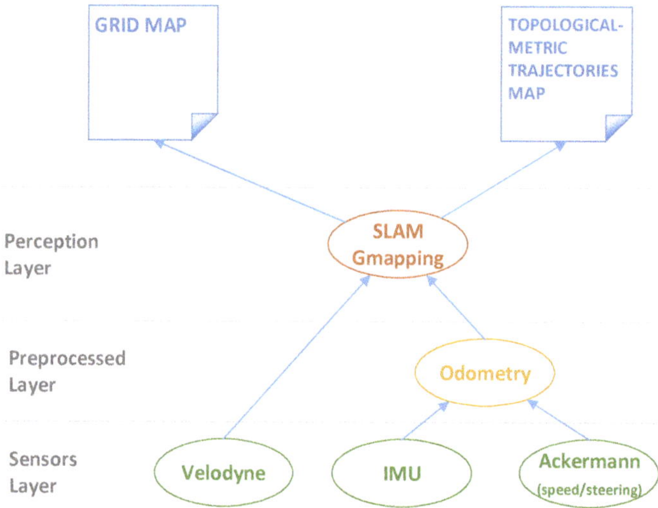

Figure 2. Robot operating system (ROS) nodes scheme that shows the system architecture in the pre-execution phase for our first framework implementation. Notice that each node is placed at its corresponding abstraction level. In this phase we record data from the sensors layer and process them offline using the architecture shown to obtain the grid map and trajectories that will be used in the execution phase.

4.1. Pre-Execution Phase

In Figure 2 we show the pre-execution phase where we generate a grid map of the area in which we want to run the application. Using the localization estimated through the SLAM node we can save the trajectories traveled by the vehicle in order to reproduce them during the execution phase. In this case, we use a GMapping node [39] but it could be replaced by any other SLAM algorithm—if we want to have an environment representation—or even by any localization algorithm (including GNSS) if we do not need a map.

4.2. Execution Phase

Figure 3 shows the architecture proposed as a closed-loop application for testing software modules in autonomous navigation. In the left part of the diagram, we show the different abstraction levels in perception, from the sensor layer to the localization module. This localization module could be replaced for any another, either SLAM-based or GNSS-based. This feature is useful to compare, in the same environment, the localization algorithms under development with well-known state-of-the-art algorithms. In the basic configuration, our localization module consists of a Monte Carlo localization (AMCL) algorithm. On the right side of the diagram, we show the different abstraction levels in the action domain, from high-level planning to the actuators layer. This is the closed-loop part where we can observe how localization affects planning and control. In the upper part of the scheme, we show a node that will serve as a high-level interface with the external agent. Using this interface we can visualize the position of the robot in the map and we can act on the system by sending global goals.

Below we describe the most important modules of our application in perception and motion. We also dedicate the last subsection to describe the reactive safety system implemented to avoid collisions.

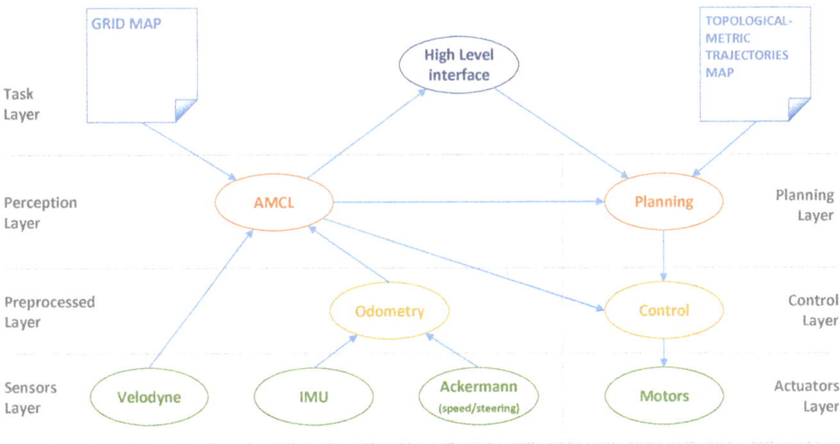

Figure 3. ROS nodes scheme that shows the system architecture in the execution phase for our first framework implementation. Notice that each node is placed at its corresponding abstraction level. The execution is now online because it is a closed loop application that acts on the environment through the control node. For both localization and planning, we will use the files generated in the previous phase shown in Figure 2.

4.2.1. Localization Module: AMCL

In the basic configuration of our system, once we have a grid map obtained using GMapping (Figure 2), we need an algorithm to locate the robot inside it. We rely on the well-known AMCL. This algorithm is based on the method described in [40] and is part of the ROS navigation stack [24]. AMCL is based on the Monte Carlo approximation applied sequentially, which in the literature is called a particle filter. This method uses the sensor model described in the section beam models of range finders in [40]. It also uses the action model described in the odometry motion model section in [40]. In order to adjust the AMCL parameters, we must follow the processes described in the given references. In Section 6.2 we show the parameters obtained through these processes. This module accepts as inputs the odometry of the robot, and laser scans from a light detection and ranging (LiDAR) sensor, in our case a Velodyne VLP-16. These inputs are standard ROS messages, which eases the process of using different LiDAR sensors or even robots with different odometers. The output of this module is also a standard ROS pose message containing pose and covariance information. These covariances will play an important role in the new localization module described in Section 5, where the AMCL estimation will be fused with GNSS information by using a Kalman filter. Thanks to this structure of inputs and outputs we can replace this localization node without affecting the rest of the architecture.

4.2.2. Planning

At the top level of the motion part of the system, we defined the planning module. At first, we assessed the possibility of implementing the planner available in the navigation stack of ROS. Finally, we discarded this possibility due to the lack of versatility of this system in terms of the localization algorithms that can be used, since the planners of this stack are based solely on grid maps to obtain the trajectories. The interest of our research group is focused on the localization of robots, and we plan to use other ways to represent the environment apart from grid maps, or even transit through unmapped areas (Section 5). For this reason, we decided to develop our own system: a simple and functional planner that is more easily adaptable to any localization algorithm.

For the developed planning module, we use a topological-metric map that contains all the possible trajectories required by the target application that we want to generate using our framework. This trajectory map is generated in the pre-execution phase as shown in Figure 2. We represent these trajectories by means of a graph in which each node represents an intersection, and each link represents the path between the intersections. Each node contains information of its location in map coordinates, and each link contains equidistant points also expressed in map coordinates. This metric information is obtained in pre-execution by recording the trajectories during manual driving, next sub-sampling them, and finally assigning each point to the node or link to which it belongs. In Figure 4 we can see an example of a graph that expresses in a topological way all the possible trajectories for a certain environment and application. The inputs to this module are the graph described above, the position obtained through AMCL (or an alternative localization module), and the global goal received from the high-level interface. With this data, the module generates as output the combination of nodes and links that describes the shortest path to the global goal within the graph. This path is composed of the points expressed in map coordinates that are sent sequentially to the low-level control as local goals. This planner assures that the robot circulates always through traversable areas.

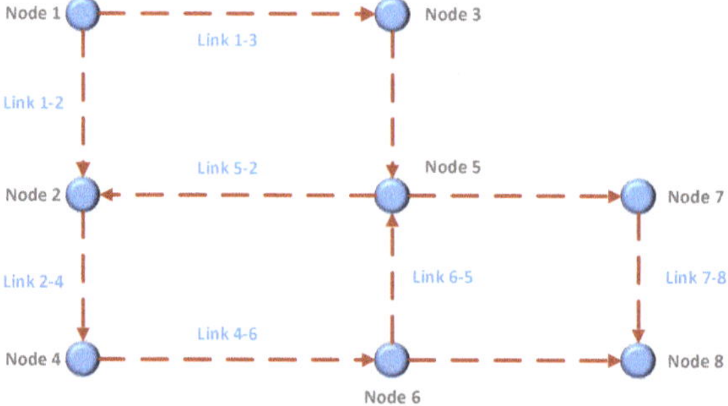

Figure 4. Example of a trajectories graph expressed abstractly as a topological map. This map also includes the metric information. Each node represents the location of roads intersections, and each link contains a series of points in 2D Cartesian coordinates that will serve as local goals to the controller. This graph is obtained as shown in Figure 2, and used in execution as shown in Figure 3.

4.2.3. Control

At the level below the planning, we find the local control level of the vehicle. In this case, we also developed a basic and functional controller designed for low-speed vehicles—in the case of our robot roughly 1.3 m/s of maximum speed. The developed controller is designed for Ackermann vehicles, but given the versatility of the framework, it would be easy to integrate a new controller for vehicles with different kinematic constraints.

The aim of local control is to reach a local goal. Once it is reached, this module warns the upper level (planning) requesting a new local goal. The controller tracks the straight line that joins two consecutive goals (Figure 5, left). These lines can be generated with only one point since the orientation is specified in the ROS pose message used to communicate the goals. The controller is proportional to the error signals in distance and angle with respect to the line. Error signal in distance e_d is obtained as the distance between the vehicle base link and the nearest point in the goal straight line (green in Figure 5, right). Error signal in angle e_α is obtained as the angle between the goal line (green in Figure 5, right) and the orientation vehicle line (red in Figure 5, right). The control action consists of two variables: steering angle (u), and speed (v). On the one hand, we compute the desired steering angle u as (Figure 6):

$$u = e_d k_d + e_\alpha k_\alpha, \tag{1}$$

where e_d and e_α are the errors in distance and orientation respectively, and where k_d and k_α are constants adjusted experimentally for each vehicle in which the framework is implemented. On the other hand, the e_d error is saturated so that $|u|$ does not exceed the maximum possible value u_{max}. This saturation is required because the linear speed applied as a part of a control action is calculated as a function of the steering as:

$$v = v_{max} - |u| k_v, \tag{2}$$

where $k_v = v_{max} - v_{min} / u_{max}$, v_{max}, and v_{min}, are the maximum and minimum velocities configurable as parameters. Note that if $|u| > u_{max}$, the resulting speed would be less than v_{min}. For this reason we saturated the e_d error signal shown in (1). Figure 6 shows the block diagram of the described controller for steering variable.

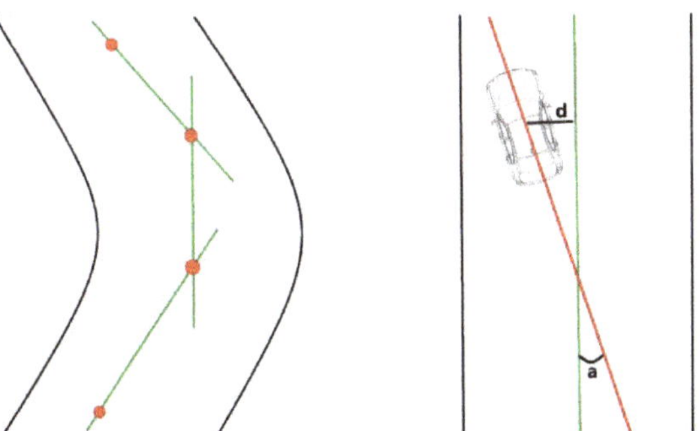

Figure 5. (left) Representation of local goals on the road. Each local goal is expressed as a point (x, y, θ). With this information, we can generate a line (in green) with an orientation θ that crosses the point (x, y) (in red). **(right)** Error signal in distance d is obtained as the distance between the vehicle base link and the nearest point in the goal straight line (green). The error signal in angle a is obtained as the angle between the goal line (green) and the orientation vehicle line (red).

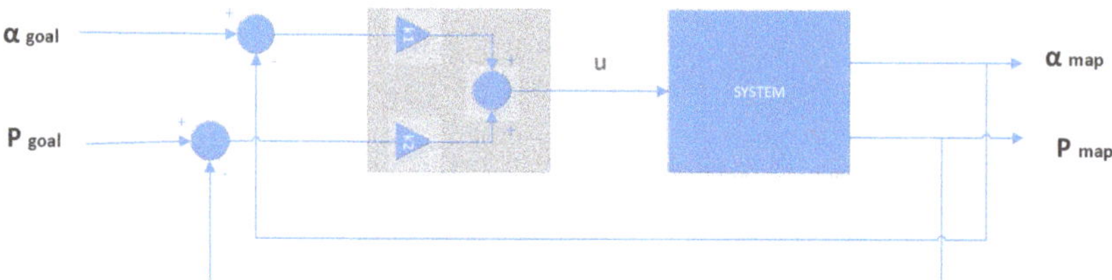

Figure 6. Block diagram that shows the proportional controller described in (1). The set-point is the position and orientation of the point closest to the line shown in Figure 5, left. The error signals are obtained depending on the position and orientation of the vehicle (Figure 5, right). Both, the set-point and the vehicle pose are expressed in map coordinates. The control signal u is the steering value expressed in degrees, that we send to the low-level control module of our vehicle.

4.2.4. Safety

Between the levels of sensors and preprocessing in perception, and between the levels of control and actuators in motion, we implement a reactive hidden layer for safety reasons. In the perception part, we designed a module that given the LiDAR point-clouds and the steering angle returns the distance to the closest point of the closest reachable obstacle. Based on this distance and a safety margin, we calculate the maximum safe speed that is defined as the one required to reach the minimum distance allowed to the obstacle in a safety period of seconds (user configurable). As the vehicle gets progressively closer to the object the speed is reduced due to the fact that the safety period is constant. Finally, when the minimum distance is reached the safety speed is just zero. In the motion part, we implement a speed recommender that reads the speed contained in the control action and the maximum speed calculated by the obstacle detector sending the lowest of them to the actuators. Thus, if the vehicle finds an obstacle along its trajectory it will stop at a defined safety distance and it will wait until the obstacle gets away from the robot. If there are no close obstacles the vehicle will recover its original speed.

5. New Localization Module: GNSS/SLAM Fusion

One of the research lines that our group wants to develop in the future is the fusion of SLAM algorithms with GNSS-based systems. The motivation for this fusion is the complementarity of both localization systems since SLAM algorithms need relatively close landmarks to extract positional information—which can cause occlusions and multi-path problems to GNSS positioning—while GNSS systems work better in clear areas that are challenging for SLAM algorithms because of landmark scarcity. As a starting point for our future research and in order to demonstrate the versatility of our framework, we developed a new localization system based on GNSS/SLAM fusion. Figure 7 shows the ROS nodes scheme of our new localization module, and Figure 8 details its connection with the rest of the system. In our solution, we fuse the positions obtained by the two systems using a Kalman filter in which these positions will be considered asynchronous observations and the odometry is used as a control signal in the prediction phase of the filter. This module allows off-map localization, so it would be difficult to integrate it with the navigation stack but its integration is straightforward using our framework. We next describe each subsystem shown in Figure 7.

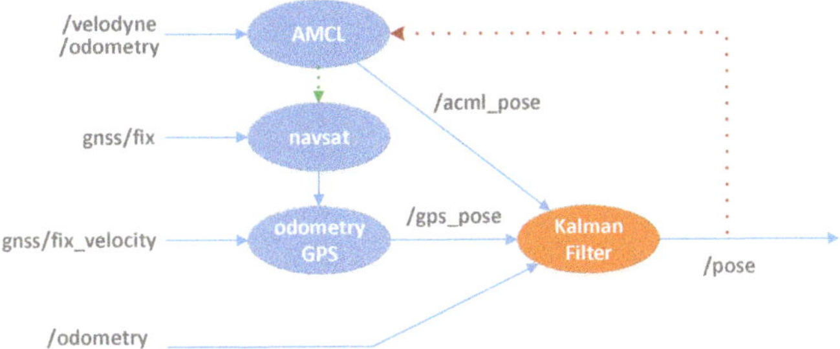

Figure 7. ROS nodes scheme that shows the combination of subsystems that make up the new localization module that will be tested using our framework. This scheme corresponds to the "global navigation satellite system (GNSS)-a Monte Carlo localization (AMCL) fusion" node shown in Figure 8. The "Kalman filter" node uses odometry as a control action, and odometry-GNSS and pose AMCL as observations.

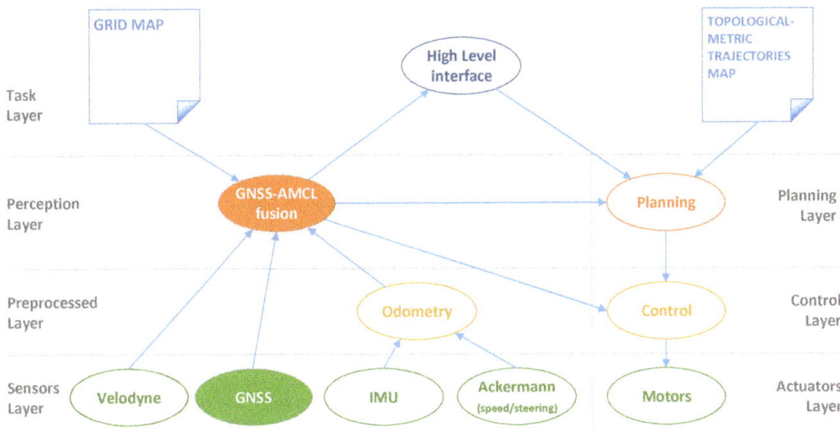

Figure 8. *ROS nodes* scheme that shows the second implementation of our framework which incorporates our novel GNSS fusion localization algorithm. This new implementation has been completed just substituting the previous localization module and adding the GNSS sensor driver. We show these modifications with solid fill in the nodes added or replaced with respect to Figure 3.

5.1. AMCL Subsystem

One of these subsystems is AMCL described in the previous section. To properly implement the fusion, we need to pay attention to the variances that this subsystem provides, associated to each estimated pose (Section 4.2.1) because they provide information about the confidence of the solution. This information will be used by the Kalman filter in the correction step, and it also will be useful for the integrity monitoring system.

5.2. GNSS Subsystem

The GNSS system provides geolocalization in world coordinates (world frame) but to perform the fusion with the subsystem described above we need to convert the global coordinates to map coordinates. For this we use the navsat node available in the ROS robot localization package [41]. In the initialization step, this node receives a position expressed in the frame we want to use as reference—in our case the position calculated through AMCL depicted as a green dotted line in Figure 7—and generates a T transformation between the pose in world frame to the map frame. Then T is applied to all the measurements made by the GNSS. By using only the navsat node, we could only read the GNSS fix and generate a 2D position in the map frame, but without orientation. To overcome this issue we generated a new ROS node called odometry GPS (Figure 7) in which we receive the velocity fix provided by the GNSS and the position and the T calculated by navsat. The output of this node is the orientation in the map frame, obtained using the velocity vector (provided by GNSS velocity fix) and the transform between the global frame and the map frame given by the navsat node.

5.3. Odometry

In a previous work [42], we described a generic module for low-level control of Ackermann type vehicles, called control logic for easy Ackermann robotization (CLEAR). From this low-level module, we obtain the linear speed and steering angle of the vehicle. For this first implementation of the framework, we developed a module to extract the positions and orientations by fusing the information obtained through CLEAR and an inertial measurement unit (IMU). This module provides as output the standard ROS message for odometry.

5.4. Kalman Filter

As mentioned above, we decided to use a Kalman filter for the fusion of the two high-level localization subsystems. This choice is motivated by the fact that we have found a simple linear model with Gaussian noise that is sufficient to represent our fusion problem, and the Kalman filter is the optimal solution under these assumptions. It should be noted that the aim of this filter is not the localization itself, but to fuse sources from complex localization algorithms that provide filtered information. For this reason, we can simplify the model and avoid the Extended Kalman filter and its problems derived from linearization. The state vector in this model is a pose vector: $\mathbf{x} = (x\ y\ \theta)^T$. The state transition equations $\mathbf{x} \leftarrow f(\mathbf{x}_{k-1}, \mathbf{u}, \mathbf{w})$ are defined as:

$$x_k = x_{k-1} + \Delta u_x + w_x \tag{3}$$

$$y_k = y_{k-1} + \Delta u_y + w_y \tag{4}$$

$$\theta_k = \theta_{k-1} + \Delta u_\theta + w_\theta, \tag{5}$$

where $\mathbf{u} = (u_x\ u_y\ u_\theta)^T$ is the control signal, which in this case is the odometry generated by our low-level system, being $\Delta\mathbf{u} = \mathbf{u}_k - \mathbf{u}_{k-1} = (\Delta u_x\ \Delta u_y\ \Delta u_\theta)^T$, where $\mathbf{w} = (w_x\ w_y\ w_\theta)^T$ is the system perturbation noise. As the state is fully observable by the two subsystems described, the observation equation is defined as:

$$\mathbf{y} = H\mathbf{x}_k + v, \tag{6}$$

where $\mathbf{y} = (y_x\ y_y\ y_\theta)^T$ is the observation vector from the localization subsystems, and where $\mathbf{v} = (v_x\ v_y\ v_\theta)^T$ is the observations noise. As the observations are of the same nature as the state vector, the matrix that relates them is an identity matrix $H = I_{3x3}$. Using this Kalman filter implementation, we can obtain a high-frequency prediction step which rate is given by the odometry (see (3)–(5)), while applying the filter correction step only when a new high-level observation becomes available. In this way, the positioning rate is sufficient for the control loop. We follow the notation used in [40] where a thorough explanation of Kalman filter equations can be found.

5.5. Integrity Monitoring

The entire information flow between the localization subsystems and the Kalman filter passes through an integrity monitoring module to avoid possible undesired behaviors. Each time an observation is generated from any high-level subsystem, we compute the Mahalanobis distance between the prediction and the observation to reject outliers. The rejected observations are excluded from the Kalman filter correction step.

The integrity monitoring module also permits to correct the AMCL subsystem using the Kalman filter output, as we can see in Figure 7. This recovery procedure is especially useful to leave and return from/to the mapped area. The monitoring module will allow this correction step in cases where the AMCL variances exceed a certain threshold. To ensure the localization stability, this correction is applied only to the state vector of the AMCL, leaving the covariances as they are. This procedure helps to the convergence of the system when re-entering to a previously mapped area.

6. Experiments

In this section, we give a brief description of the robotic research platform used for the experimental sessions (Figure 9) and a thorough explanation of the experiments performed to test both the autonomous navigation application proposed in Section 4 and the localization module proposed in Section 5. In addition, we justify the choice of the adjustable parameters of the system.

Figure 9. Experimental platform BLUE. This professional research platform—derived from a conventional electric cart—has been developed by our research group, in the University of Alicante. It is integrated with ROS and provides all the features required for developing state-of-the-art research in autonomous outdoor navigation.

6.1. Experimental Platform

To carry out the experimental part of this work, we implemented the solution described in Section 4 using the robotic research platform BLUE. This robot is derived from a conventional electric cart and has been developed entirely by our research group, AUROVA. It has been designed to ease the experimental processes in mobile robotics research [42] and incorporates a powerful low-level module—also developed by our group—that is called control logic for easy Ackermann robotization (CLEAR). This module permits to turn conventional Ackermann vehicles into research robots integrated into ROS [42], and provides—besides other interesting features—a very accurate estimation of the steering angle and the platform speed even using extremely-low resolution sensors [43]. The robot main hardware components are the following: 3D LiDAR sensor Velodyne VPL16, 2D LiDAR sensor Hokuyo UBG-04LX-F01, RGBD camera Intel Realsense D435, GNSS real-time kinematic (RTK) capable Ublox M8P, and IMU sensor CHR-UM7, among others. It also has an onboard computer in which sensors and actuators are integrated via ROS. This computer is also used to run the system implemented following the architecture shown in Figures 2 and 3.

6.2. Settings

In Sections 4 and 5, we commented about different configurable parameters, some that depend on the robot used to implement the application, and others depending on the selected environment. In Table 1 we specify the parameters used for the experimental sessions described in this section. In [44], we can see the large number of default parameters that GMapping uses. We choose to limit the experimental tunning process to just the three parameters shown in Table 1. However, our tunning process ended up replicating the default values, so our GMapping configuration is the same as the standard one. In the case of AMCL we obtained the α parameters from the motion model as described in [40], leaving the rest of default AMCL parameters [45] unchanged. The planning module has only one adjustable parameter, which is the distance between local goals (Section 4.2.2). For the control subsystem, we show the value of the control adjustment constants described in (1) which are obtained empirically, as well as the maximum and minimum speeds and the maximum steering angle, that were selected by design, taking into account the physical limitations of the robot. For the Kalman filter based fusion system, we adjusted the model noise variances empirically and used the variances provided by the AMCL and the GNSS modules as observation noises. In Section 5.5 we discussed the outlier rejection process—that depends on a Mahalanobis distance threshold—and the maximum AMCL variance threshold that activates the recovery process of the Monte Carlo estimator. The adjustment of these thresholds has been done experimentally, looking to obtain the best qualitative results.

In the rest of this section, we describe and discuss the experiments performed using the configuration described in Table 1.

Table 1. Adjustable parameters grouped in subsystems. The table shows the values used in the experimental sessions. Some parameters depend on the robot, as is the case with the parameters α of AMCL, while others depend on the environment, as the map resolution in GMapping.

Module	Parameter	Value	Units
GMapping	Grid resolution	0.05	m
	Maximum map dimensions	100×100	m
	Number of particles	30	none
AMCL	α_1	1.12	none
	α_2	0.1	none
	α_3	1.05	none
	α_4	0.1	none
Planning	Subsampling distance	2.0	m
Control	Constant k_d	8.0	none
	Constant k_α	0.5	none
	Maximum speed	1.0	m/s
	Minimum speed	0.6	m/s
	Maximun steering	25.0	deg
Kalman filter	Noise model for xy components	0.05	m
	Noise model for θ component	0.58	deg
Integrity monitoring	Mahalanobis distance threshold	3.0	none
	AMCL correction threshold for xy components	3.0	m
	AMCL correction threshold for θ component	10	deg

6.3. Initial Framework Experiments

To test the initial framework implementation described in Section 4, we carried out an experimental session in the surroundings of the University Institute for Computing Research building, where the secondary laboratory of our research group is located (Figure 10, right). For this experiment, we used our framework to create a basic but complete autonomous navigation application able to navigate indefinitely repeating the same trajectory in the described environment. Following the procedure detailed in Section 4, the experimental session was divided into two phases: pre-execution, and execution.

In the pre-execution phase, we recorded a dataset in manual driving making a closed loop around the building shown in Figure 10, right. Then, offline we ran this dataset in order to generate the inputs for the architecture shown in Figure 2. As result of this process we obtained a grid map (shown in Figure 10, left) and a subsampled trajectory as described in Section 4.2.2. Thanks to the use of standard ROS launch files we can run the whole pre-execution system with just one click. In the online execution phase, we used the architecture shown in Figure 3 to replicate autonomously the trajectory obtained in the previous phase, cyclically and indefinitely.

The use of a roslaunch file made the starting process fast and easy. During the experiment, the system behavior was quite good. The localization was very stable, allowing the robot to complete the experiment—consisting in five consecutive laps around the building—without interruption. The system demonstrated a good repetitiveness with very small variations in the trajectories from lap to lap. The control was smooth and stable along the entire path adapting perfectly to the given trajectories without suffering any oscillations. This first experiment proved that our framework is able to produce a real autonomous navigation application quickly and easily, requiring just the execution of two launch files. As we can see in [24], it would be also possible to perform an equivalent experiment using the tools provided by the navigation stack and some plug-ins. However, in the following section, we demonstrate the versatility of our framework by changing the localization system to navigate on and off-map, which would have not been possible using the navigation stack.

Figure 10. (left) Grid map of the secondary laboratory building, obtained using the architecture shown in Figure 2. **(right)** The actual building and the environment where the initial framework experiments have been carried out.

6.4. GNSS/SLAM Fusion Framework Experiments

To demonstrate our framework versatility, we generated a second application in a different environment. In this case, we replaced the original localization module with the fusion module described in Section 5. Our framework permitted to make this replacement by just changing the corresponding extensible markup language (XML) tag in the launch file of the execution phase without worrying about other nodes and component interactions. The rest of the process to create the application was exactly the same as the one described in the previous section. In this way, we verified the easiness that our framework provides to test new algorithms within a fully operative autonomous navigation application. Concretely, in this case, it would have been very complicated to generate an equivalent application using the navigation stack since it was not prepared for off-map navigation.

In the rest of this section, we describe and comment on the experiments carried out to test the second application generated using our framework. We tested the new localization module observing how it affected the whole autonomous navigation system. We performed these experiments on the University of Alicante campus (in the area whose grid map is shown in Figure 11) because it was an environment where we can find different scenarios that may affect the localization of our vehicle. For example, there were areas with high trees and buildings—that could cause satellite occlusions affecting the GNSS—and lots of dynamic obstacles, such as pedestrians, electric maintenance cars, or even road vehicles, that can affect the Monte Carlo localization.

Figure 11. Grid map obtained in the pre-execution phase (Figure 2), and latter used in the execution phase (Figure 3). We generated this map using datasets previously recorded during the experimental sessions described in [42].

6.4.1. Localization

To test our new localization module (described in Section 5) we followed different paths through the campus. These routes were recorded in manual driving using a remote control. To test the fusion of the different localization subsystems these paths pass both within a grid map, previously obtained using the architecture shown in Figure 2, and outside it. We obtained this map through the datasets recorded for the experiments of our previous work [42]. Using this map, we are able to locate our vehicle in the environment using AMCL. However, it has been six months since the data was captured. This means that in the map may appear dynamic objects that were in the environment at the time the map was made and that might not be in the environment when the AMCL runs. We can see this map in Figure 11. The paths we followed in the experiments are those shown in Figures 12 and 13. In the path *R1* (Figure 12 *left*) we circulated entirely within the map. In the path *R2* (Figure 12 *right*) we circulated in part out of the map, in this way we tested how the system behaves when using only GNSS. In the path *R3* (Figure 13 *left*) we circulated approximately the half of the route out of map. The difference with respect to the path *R2* is that in this case we went out and entered the map describing a rectilinear trajectory, which can be more favorable to the relocation because AMCL suffers in the turns. The last route was the *R4* (Figure 13, right), in which we circulated most of the journey off-map to test the non-divergence of the system. In this route, we found very unfavorable areas for the GNSS because of high trees and nearby buildings. All the paths were done twice, once with GNSS standalone, and another using the GNSS-RTK in floating mode (details about the single-frequency RTK system used can be found in [42]).

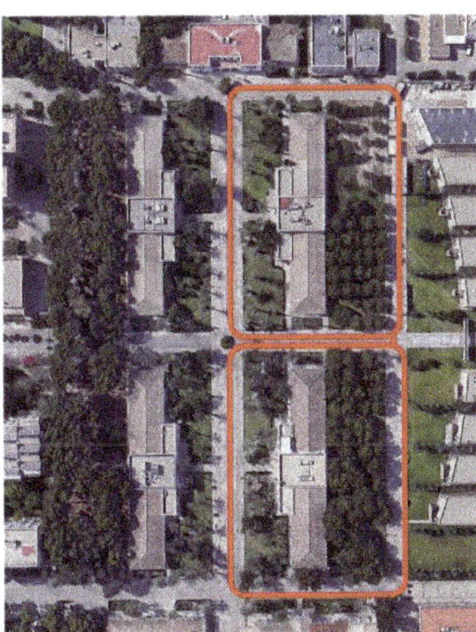

Figure 12. (**left**) Path *R1* around the building located in the upper right quadrant. This path runs entirely through the map shown in Figure 11. (**right**) path *R2* around the two buildings on the right side of the image. This path leaves the map in the area of the building in the lower right quadrant of the image.

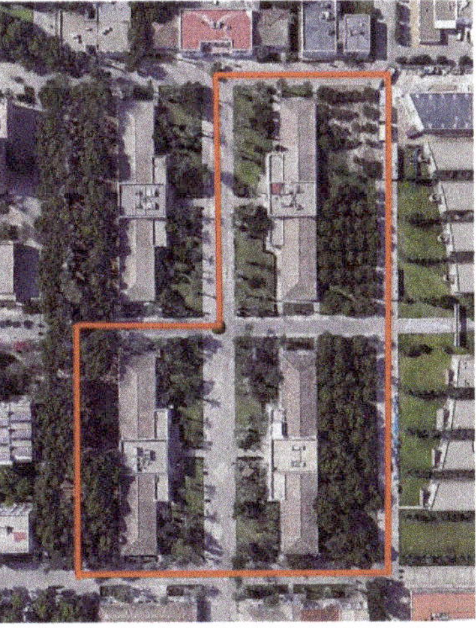

Figure 13. (**left**) path *R3* around the two buildings on the right side of the image. This path leaves the map shown in Figure 11 in the area of the building in the lower right quadrant of the image. (**right**) Path *R4* around the two buildings on the right side of the image, and the building on the lower left side of the image. This is the path that makes the longest journey off the map.

In Figure 14 we can see the trajectories obtained when following the path $R1$, with GNSS standalone (left), and with GNSS RTK floating (right). This is the only journey that we made entirely within the map. In these conditions, the AMCL localization proved to be very robust. This translates into very small variances what makes the KF output to be very similar to the output of the AMCL. As we can see in Figure 14, the red line (KF) is superimposed on the green line (AMCL), as both provide almost the same location. In both cases, we see that the blue line (GNSS) differs from the fusion, although to a lesser extent in the case of RTK.

Figure 15 shows the paths obtained for $R2$. In this case, the behavior in the map area is the same as shown in the previous case. However, in the area where we travel off-map ($x > 75$ m) the fusion adapts to the observations provided by the GNSS-based subsystem. In this area, when the AMCL uncertainty grows exceeding the thresholds, the Kalman filter output corrects the AMCL state estimation so that when the robot re-enters to the mapped area the AMCL subsystem converges again.

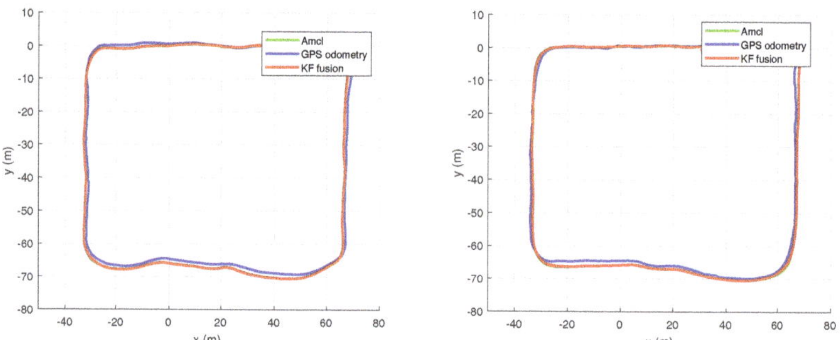

Figure 14. Trajectories through the $R1$ path expressed by localization using the fusion described in Section 5. (**left**) Using GNSS standalone. (**right**) Using GNSS floating point. In both cases, the fusion gives more weight to AMCL, because the robot is inside the map during the whole trajectory.

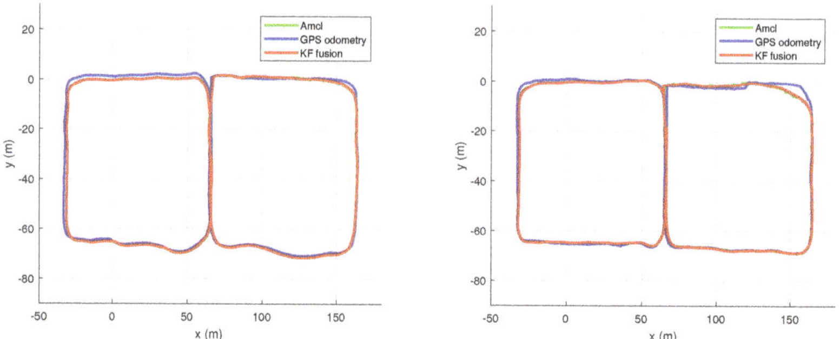

Figure 15. Trajectories through the $R2$ path expressed by localization using the fusion described in Section 5. (**left**) Using GNSS standalone. (**right**) Using GNSS floating point. In both cases, the loop closure of the right quadrant is performed off-map. We see that thanks to the fusion of GNSS, we can maintain the trajectory to re-enter the map area.

Figure 16 illustrates the trajectories made for the path $R3$. In this case, we can observe a behavior similar to that of the path $R2$ where the fusion adapts to the GNSS system in the off-map area ($x > 75$ m). In the case of the path made using the RTK system, we see that the low variances makes the fusion to closely follow the observations. For this reason, we see how the red line (AMCL) is superimposed to the rest of the subsystems in Figure 16 right.

Figure 17 shows the result of performing manual driving through the path $R4$. In this case, the trajectories run off the map in the whole area defined by $x > 75$ m and $y < -90$ m. We see that the behavior is similar to the cases described for the paths $R2$ and $R3$ with the difference that in this case most of the journey is off-map and verifying that the re-entry can be done successfully even after a long journey without mapped references.

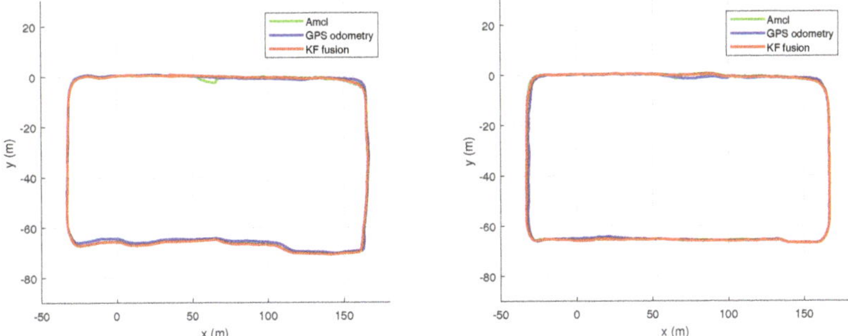

Figure 16. Trajectories through the $R3$ path expressed by localization using the fusion described in Section 5. (**left**) Using GNSS standalone. (**right**) Using GNSS floating point. We see that in the case of the fusion with GNSS real-time kinematic RTK floating point (**right**), in the off-map area the fusion adapts to these observations.

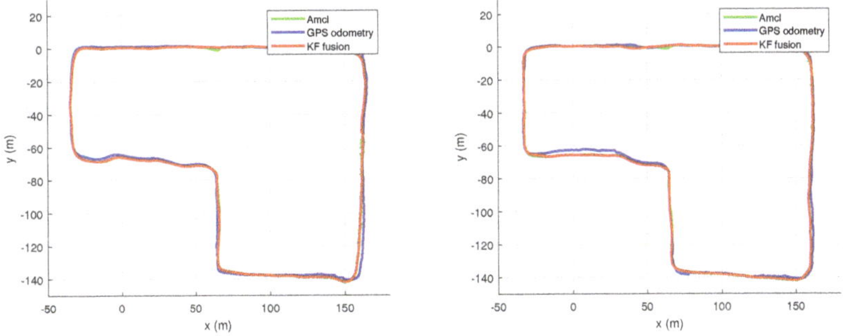

Figure 17. Trajectories through the $R4$ path expressed by localization using the fusion described in Section 5. (**left**) Using GNSS standalone. (**right**) Using GNSS floating point. In this case, the whole area $x > 50$ m is off-map. We see that in both cases we can make the re-entry on the map. However, in the case of the GNSS RTK floating point (**right**) the trajectory seems more consistent except in the curve located in the lower right quadrant, where conditions were very unfavorable due to occlusions and multi-path produced by trees and buildings.

6.4.2. Autonomous Navigation

In the previous section, we described the experiments carried out to test the localization but in that case, the control loop closure was being performed by the users who handle the remote control. Because of this, we are influencing the localization by using our own human localization at the cognitive level. In order to test how location affects control, we integrated into our vehicle the architecture shown in Figure 3 to navigate autonomously in closed circuit. Using the paths recorded during the previous experiments we generated two maps of trajectories through the routes $R3$ and $R4$ (Section 6.4.1), defining two different circuits. For these experiments, the application is configured to run through the specified circuit indefinitely. In this way, we will be able to observe the repetitiveness of the trajectories in each lap. With these conditions, we carried out four different experiments. First, we performed the autonomous navigation along the circuit defined by the path $R3$ using the GNSS in standalone mode. Next, we used the circuit described by the path $R4$ also with the GNSS configured in standalone mode. In order to see how the differential corrections of GNSS can affect navigation, we repeated the two experiments previously described but using the RTK in floating mode.

The system behavior in the paths $R3$ and $R4$ with the standalone GNSS configuration was similar. In both cases, the errors of the standalone GNSS system caused failures in the zones marked with yellow circles in Figure 18. These errors resulted in emergency stops, thanks to the system described in Section 4.2.4. However, it was possible to demonstrate the system recoverability since, after skipping these areas in manual driving we were able to complete the circuit autonomously. During the RTK experiments, the system was able to complete the circuit without interruptions including the off-map areas. The behavior in these experiments was very similar to the described in the previous section: the system shown repeatability and a smooth and stable control without oscillations. However, in these experiments we were not able to circulate indefinitely, since in the second lap, in the areas marked with yellow circles in Figure 18 we lost connection with the base station, disabling the RTK mode and forcing to interrupt the autonomous navigation. As in the previous case, the system proved to be recoverable as it returned to autonomous navigation after passing these critical points.

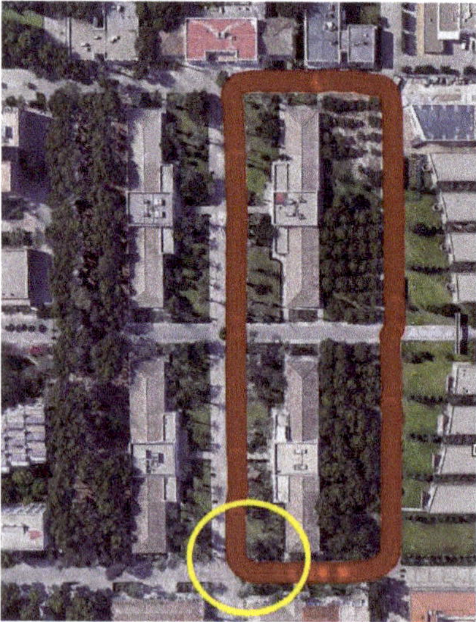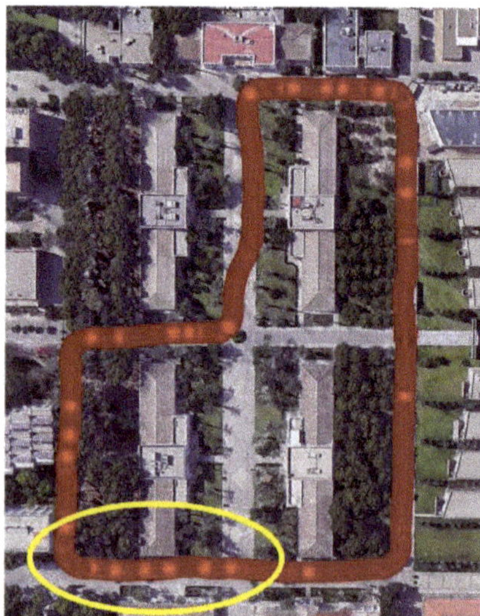

Figure 18. Representation of GNSS fixes in RTK floating point mode, recorded during autonomous navigation. (**right**) Through the path $R3$. (**right**) Through the path $R4$. In both cases, if the link to the base station is available, the closed loop is completed without problems. Areas marked with yellow circles indicate problem zones for the link to the base station.

7. Conclusions and Future Work

In this paper, we presented a framework designed for fast prototyping of autonomous navigation systems. This framework reduces dramatically the amount of work required to implement a whole application, making easier to test and to compare different algorithms in real-world conditions. The proposed architecture permits us to focus the efforts in the desired research topics, while the provided basic set of tools enables the users to generate fully operative autonomous navigation systems to perform experimental tests. To validate the described approach we used the framework and the provided tools to implement an initial basic system that was able to complete successfully several laps around a building autonomously in a challenging outdoor scenario. To demonstrate the easiness of module substitution, we developed a novel algorithm that relies in a Kalman filter to fuse 2D SLAM and GNSS positioning and used it to replace the localization module of the initial basic system. This new module was incorporated just changing a single line of the launch file. Extensive tests for this new localization module were performed navigating autonomously through mixed on-map/off-map trajectories in the University of Alicante campus. This new localization module has shown interesting properties on its own, and thus has become one of the contributions of the present work. The experimental sessions covered more than twenty kilometers in two absolutely different outdoor environments. All the software is publicly available in a GitHub repository (https://github.com/AUROVA-LAB/aurova_framework) with the aim of being a useful tool for research groups interested in any of the fields related to autonomous navigation.

As future work, we want to extend provided set of tools, adding re-planning capabilities, redundant safety modules and terrain analysis algorithms. In the localization part, we plan to integrate a graph SLAM system implementing a tight integration of GNSS raw observables and to use unsupervised learning techniques for landmark detection.

Author Contributions: Conceptualization, M.A.M. and I.d.P.; methodology, M.A.M. and I.d.P.; software, M.A.M. and I.d.P.; validation, M.A.M. and I.d.P.; formal analysis, M.A.M. and I.d.P.; investigation, M.A.M., I.d.P., F.A.C. and F.T.; writing—original draft preparation, M.A.M., I.d.P., F.A.C. and F.T.; writing—review and editing, M.A.M., I.d.P., F.A.C. and F.T.; supervision, F.A.C. and F.T.; project administration, F.T.; funding acquisition, F.A.C. and F.T.

Funding: This work has been supported by InterregV Sudoe and FEDER programs of European Commission through the COMMANDIA project SOE2/P1/F0638, and by the Spanish Government through the FPU grant FPU15/04446 and the research project RTI2018-094279-B-I00.

Conflicts of Interest: The authors declare no conflict of interest.

References

1. Niemueller, T.; Lakemeyer, G.; Ferrein, A. The RoboCup logistics league as a benchmark for planning in robotics. In Proceedings of the International Conference on Automated Planning and Scheduling (ICAPS)—WS on Planning and Robotics (PlanRob), Jerusalem, Israel, 7–11 June 2015; Association for the Advancement of Artificial Intelligence (AAAI): Palo Alto (CA), USA, 2015.
2. King, A. The future of agriculture. *Nature* **2017**, *544*, S21–S23. [CrossRef] [PubMed]
3. Milanés, V.; Bergasa, L.M. Introduction to the Special Issue on "New Trends towards Automatic Vehicle Control and Perception Systems". *Sensors* **2013**, *13*, 5712–5719. [CrossRef] [PubMed]
4. Nilsson, N.J. A Mobile Automaton: An Application of Artificial Intelligence Techniques. In Proceedings of the 1st International Joint Conference on Artificial intelligence (IJCAI'69), Washington, DC, USA, 7–9 May 1969; Morgan Kaufmann Publishers Inc.: San Francisco, CA, USA, 1969; pp. 509–520.
5. Carrio, A.; Sampedro, C.; Rodriguez-Ramos, A.; Campoy, P. A review of deep learning methods and applications for unmanned aerial vehicles. *J. Sens.* **2017**, *2017*, 3296874. [CrossRef]
6. Redmon, J.; Divvala, S.; Girshick, R.; Farhadi, A. You only look once: Unified, real-time object detection. In Proceedings of the IEEE Conference on Computer Vision and Pattern Recognition, Las Vegas, NV, USA, 26 June–1 July 2016; 779 788.
7. Gomes, L. When will Google's self-driving car really be ready? It depends on where you live and what you mean by "ready" [News]. *IEEE Spectr.* **2016**, *53*, 13–14. [CrossRef]
8. Canedo-Rodríguez, A.; Alvarez-Santos, V.; Regueiro, C.V.; Iglesias, R.; Barro, S.; Presedo, J. Particle filter robot localisation through robust fusion of laser, WiFi, compass, and a network of external cameras. *Inf. Fusion* **2016**, *27*, 170–188. [CrossRef]
9. Castro-Toscano, M.J.; Rodríguez-Quiñonez, J.C.; Hernández-Balbuena, D.; Rivas-Lopez, M.; Sergiyenko, O.; Flores-Fuentes, W. Obtención de Trayectorias Empleando el Marco Strapdown INS/KF: Propuesta Metodológica. *Rev. Iberoam. De Automática E Informática Ind.* **2018**, *15*, 391–403. [CrossRef]
10. Alami, R.; Chatila, R.; Fleury, S.; Ghallab, M.; Ingrand, F. An architecture for autonomy. *Int. J. Robot. Res.* **1998**, *17*, 315–337. [CrossRef]
11. Thrun, S.; Montemerlo, M.; Dahlkamp, H.; Stavens, D.; Aron, A.; Diebel, J.; Fong, P.; Gale, J.; Halpenny, M.; Hoffmann, G.; et al. Stanley: The robot that won the DARPA Grand Challenge. *J. Field Robot.* **2006**, *23*, 661–692. [CrossRef]

12. Sanchez-Lopez, J.L.; Pestana, J.; de la Puente, P.; Campoy, P. A reliable open-source system architecture for the fast designing and prototyping of autonomous multi-uav systems: Simulation and experimentation. *J. Intell. Robot. Syst.* **2016**, *84*, 779–797. [CrossRef]

13. Siegwart, R.; Nourbakhsh, I.R.; Scaramuzza, D. *Introduction to Autonomous Mobile Robots*; MIT Press: Cambridge, MA, USA, 2011.

14. The Robotics Data Set Repository (Radish). Available online: http://radish.sourceforge.net (accessed on 5 May 2019).

15. Koenig, N.; Howard, A. Design and use paradigms for gazebo, an open-source multi-robot simulator. In Proceedings of the 2004 IEEE/RSJ International Conference on Intelligent Robots and Systems (IROS 2004), Sendai, Japan, 28 September–2 October 2004; IEEE: Piscataway, NJ, USA, 2004; Volume 3, pp. 2149–2154.

16. Calisi, D.; Iocchi, L.; Nardi, D. A unified benchmark framework for autonomous mobile robots and vehicles motion algorithms (MoVeMA benchmarks). In Proceedings of the Workshop on Experimental Methodology and Benchmarking in Robotics Research (RSS 2008), Zurich, Switzerland, 26–30 June 2008.

17. Kweon, I.S.; Goto, Y.; Matsuzaki, K.; Obatake, T. CMU sidewalk navigation system: A blackboard-based outdoor navigation system using sensor fusion with colored-range images. In Proceedings of the Fall Joint Computer Conference, Dallas, TX, USA, 2–6 November 1986; IEEE: Piscataway, NJ, USA, 1986.

18. Goto, Y.; Stentz, A. Mobile robot navigation: The CMU system. *IEEE Expert* **1987**, *2*, 44–54. [CrossRef]

19. Buehler, M.; Iagnemma, K.; Singh, S. *The 2005 DARPA Grand Challenge: The Great Robot Race*; Springer: Berlin, Germany, 2007; Volume 36.

20. Buehler, M.; Iagnemma, K.; Singh, S. *The DARPA Urban Challenge: Autonomous Vehicles in City Traffic*; Springer: Berlin, Germany, 2009; Volume 56.

21. Montemerlo, M.; Becker, J.; Bhat, S.; Dahlkamp, H.; Dolgov, D.; Ettinger, S.; Haehnel, D.; Hilden, T.; Hoffmann, G.; Huhnke, B.; et al. Junior: The stanford entry in the urban challenge. *J. Field Robot.* **2008**, *25*, 569–597. [CrossRef]

22. Urmson, C.; Anhalt, J.; Bagnell, D.; Baker, C.; Bittner, R.; Clark, M.; Dolan, J.; Duggins, D.; Galatali, T.; Geyer, C.; et al. Autonomous driving in urban environments: Boss and the urban challenge. *J. Field Robot.* **2008**, *25*, 425–466. [CrossRef]

23. Quigley, M.; Conley, K.; Gerkey, B.; Faust, J.; Foote, T.; Leibs, J.; Wheeler, R.; Ng, A.Y. ROS: An open-source Robot Operating System. In Proceedings of the ICRA Workshop on Open Source Software, Kobe, Japan, 17 May 2009; Volume 3, p. 5.

24. Guimarães, R.L.; de Oliveira, A.S.; Fabro, J.A.; Becker, T.; Brenner, V.A. ROS navigation: Concepts and tutorial. In *Robot Operating System (ROS)*; Springer International Publishing: Cham, Switzerland, 2016; Volume 1, pp. 121–160.

25. Rösmann, C.; Hoffmann, F.; Bertram, T. Kinodynamic trajectory optimization and control for car-like robots. In Proceedings of the 2017 IEEE/RSJ International Conference on Intelligent Robots and Systems (IROS), Vancouver, BC, Canada, 24–28 September 2017; IEEE: Piscataway, NJ, USA, 2017; pp. 5681–5686.

26. Conner, D.C.; Willis, J. Flexible navigation: Finite state machine-based integrated navigation and control for ROS enabled robots. In Proceedings of the SoutheastCon 2017, Charlotte, NC, USA, 30 March–2 April 2017; IEEE: Piscataway, NJ, USA, 2017; pp. 1–8.

27. Brahimi, S.; Tiar, R.; Azouaoui, O.; Lakrouf, M.; Loudini, M. Car-like mobile robot navigation in unknown urban areas. In Proceedings of the 2016 IEEE 19th International Conference on Intelligent Transportation Systems (ITSC), Rio de Janeiro, Brazil, 1–4 November 2016; IEEE: Piscataway, NJ, USA, 2016; pp. 1727–1732.

28. Vivacqua, R.; Vassallo, R.; Martins, F. A low cost sensors approach for accurate vehicle localization and autonomous driving application. *Sensors* **2017**, *17*, 2359. [CrossRef] [PubMed]

29. Ferrer, G.; Zulueta, A.G.; Cotarelo, F.H.; Sanfeliu, A. Robot social-aware navigation framework to accompany people walking side-by-side. *Auton. Robot.* **2017**, *41*, 775–793. [CrossRef]

30. Dove, R.; Schindel, B.; Scrapper, C. Agile systems engineering process features collective culture, consciousness, and conscience at SSC Pacific Unmanned Systems Group. *INCOSE Int. Symp.* **2016**, *26*, 982–1001. [CrossRef]

31. Li, X.; Sun, Z.; Cao, D.; Liu, D.; He, H. Development of a new integrated local trajectory planning and tracking control framework for autonomous ground vehicles. *Mech. Syst. Signal Process.* **2017**, *87*, 118–137. [CrossRef]

32. Hernádez Juan, S.; Herrero Cotarelo, F. *Autonomous Navigation Framework for a Car-Like Robot (Technical Report IRI-TR-15-07)*; Institut de Robòtica i Informàtica Industrial (IRI): Barcelona, Spain, 2015.
33. Rodrigues, M.; McGordon, A.; Gest, G.; Marco, J. Developing and testing of control software framework for autonomous ground vehicle. In Proceedings of the 2017 IEEE International Conference on Autonomous Robot Systems and Competitions (ICARSC), Coimbra, Portugal, 26–28 April 2017; IEEE: Piscataway, NJ, USA, 2017; pp. 4–10.
34. Stateczny, A.; Burdziakowski, P. Universal autonomous control and management system for multipurpose unmanned surface vessel. *Pol. Marit. Res.* **2019**, *26*, 30–39. [CrossRef]
35. Huskić, G.; Buck, S.; Zell, A. GeRoNa: Generic Robot Navigation. *J. Intell. Robot. Syst.* **2018**, 1–24. [CrossRef]
36. Hartmann, J.; Klüssendorff, J.H.; Maehle, E. A unified visual graph-based approach to navigation for wheeled mobile robots. In Proceedings of the 2013 IEEE/RSJ International Conference on Intelligent Robots and Systems, Tokyo, Japan, 3–7 November 2013; IEEE: Piscataway, NJ, USA, 2013; pp. 1915–1922.
37. Guzmán, R.; Ariño, J.; Navarro, R.; Lopes, C.; Graça, J.; Reyes, M.; Barriguinha, A.; Braga, R. Autonomous hybrid GPS/reactive navigation of an unmanned ground vehicle for precision viticulture-VINBOT. In Proceedings of the 62nd German Winegrowers Conference, Stuttgart, Germany, 27–30 November 2016.
38. Romay, A.; Kohlbrecher, S.; Stumpf, A.; von Stryk, O.; Maniatopoulos, S.; Kress-Gazit, H.; Schillinger, P.; Conner, D.C. Collaborative Autonomy between High-level Behaviors and Human Operators for Remote Manipulation Tasks using Different Humanoid Robots. *J. Field Robot.* **2017**, *34*, 333–358. [CrossRef]
39. OpenSLAM: GMapping Algorithm. Available online: https://openslam-org.github.io (accessed on 5 May 2019).
40. Thrun, S.; Burgard, W.; Fox, D. *Probabilistic Robotics*; MIT Press: Cambridge, MA, USA, 2005.
41. Moore, T.; Stouch, D. A generalized extended kalman filter implementation for the robot operating system. In Proceedings of the 13th International Conference IAS-13 (Intelligent Autonomous Systems 13), Padova, Italy, 15–18 July 2014; Springer International Publishing: Cham, Switzerland, 2016; pp. 35–348.
42. del Pino, I.; Muñoz-Bañon, M.Á.; Cova-Rocamora, S.; Contreras, M.Á.; Candelas, F.A.; Torres, F. Deeper in BLUE. *J. Intell. Robot. Syst.* **2019**, 1–19. [CrossRef]
43. del Pino, I.; Muñoz Bañón, M.A.; Contreras, M.Á.; Cova, S.; Candelas, F.A.; Torres, F. Speed Estimation for Control of an Unmanned Ground Vehicle Using Extremely Low Resolution Sensors. In Proceedings of the 15th International Conference on Informatics in Control, Automation and Robotics (ICINCO), Portu, Portugal, 29–31 July 2018.
44. Gmapping in ROS. Available online: http://wiki.ros.org/gmapping (accessed on 5 May 2019).
45. Amcl in ROS. Available online: http://wiki.ros.org/amcl (accessed on 5 May 2019).

Article

Smart Obstacle Avoidance Using a Danger Index for a Dynamic Environment

Jiubo Sun [1], Guoliang Liu [1,*], Guohui Tian [1] and Jianhua Zhang [2]

[1]   School of Control Science and Engineering, Shandong University, Jinan 250061, China;
      sunjiubo@mail.sdu.edu.cn (J.S.); g.h.tian@sdu.edu.cn (G.T.)
[2]   School of Mechanical Engineering, Hebei University of Technology, Tianjin 300131, China;
      jhzhang@hebut.edu.cn
*   Correspondence: liuguoliang@sdu.edu.cn

Received: 1 April 2019; Accepted: 12 April 2019; Published: 17 April 2019

Abstract: The artificial potential field approach provides a simple and effective motion planner for robot navigation. However, the traditional artificial potential field approach in practice can have a local minimum problem, i.e., the attractive force from the target position is in the balance with the repulsive force from the obstacle, such that the robot cannot escape from this situation and reach the target. Moreover, the moving object detection and avoidance is still a challenging problem with the current artificial potential field method. In this paper, we present an improved version of the artificial potential field method, which uses a dynamic window approach to solve the local minimum problem and define a danger index in the speed field for moving object avoidance. The new danger index considers not only the relative distance between the robot and the obstacle, but also the relative velocity according to the motion of the moving objects. In this way, the robot can find an optimized path to avoid local minimum and moving obstacles, which is proved by our experimental results.

Keywords: artificial potential field; path planning; obstacle avoidance; dynamic window; danger index

---

1. Introduction

Moving in free space while avoiding collisions within the dynamic environment is known as obstacle avoidance or path planning, which is the backbone of mobile robots. Many well-known path planning algorithms have been proposed in robotics research literature. The reward-based planning algorithms [1–3] propose to give the robot a positive reward when it reaches the target, and give the robot a negative reward when it collides with the obstacle. The path is then obtained by maximizing the cumulative future reward. However, due to the discretization of the state and control space, their trajectories are mostly not smooth [4]. The sampling-base planning algorithms, such as the Rapidly exploring Random Tree (RRT) [5] and the Probabilistic Road Maps (PRM) [6], generally connect a series of randomly sampled points from a barrier-free space, and attempt to establish a path from the initial position to the target position. Although their time cost is less, the random sampling introduces the randomness of the planned path. Therefore, we cannot predict the result of the path planner, which is not optimal in general. The grid-based planning algorithms, such as the A* [7] and the Dijkstra's [8] algorithm, which have been used in the robotic operating systems (ROS), can find the optimal trajectory, but their time cost and memory usage grow exponentially with the dimension of the state space [4]. The novel artificial potential field (APF) developed by Khatib [9–11] is a simple and effective path planning method. The theory behind APF is simple and straightforward: it has an attractive potential field for attracting the robot to the target point, and a repulsive potential field for pushing the robot away from obstacles. The APF approach can be used for global and local path planning.

However, there are some inherent problems in traditional artificial potential field approaches: (1) oscillations in the presence of obstacles due to the large move step of the robot; (2) goal nonreachable with obstacles nearby (GNRON); (3) trapped in a local minimum region; and (4) moving obstacles avoidance. Solving these problems has become an interesting topic in the field of APF based path planning [12,13]. Park et al. [14] deal with the local minimum region by filling it with the virtual obstacle. Zhu et al. [15] introduce simulated annealing to look for the global minimum and escape the local minimum region. Doria et al. [16] use the Deterministic Annealing (DA) approach to avoid local minima in APF by adding a temperature parameter into the cost function of the APF approach. The value of the temperature parameter starts to increase until the robot goes away from the local minimum region. Lee et al. [17] propose to use two modes to control the APF, i.e., one mode is to go directly to the target using traditional APF and the other mode uses a new point APF algorithm to avoid collisions with the static obstacle. Mode 1 can be switched to mode 2 when the robot is trapped in a local minimum region or blocked in front of the obstacle. A new direction is synthesized using the robot motion information and direction between the robot and the goal point, which can guide the robot to escape the local minimum region. Weerakoon et al. [18] propose an improved repulsive force for overcoming a local minima problem, which generates a new repulsive force to the primary force when the robot detects an obstacle within its sensory range. The new repulsive force component turns the robot smoothly away from the obstacles. Kim et al. [19] modify the APF approach with the collision cone approach, which adds an artificial force to the existing attractive and repulsive forces when the relative velocity vector between the robot and obstacle is included in a collision cone. The improved resultant force can guide the robot to reach the goal. These algorithms may help the robot to escape the local minimum region in a simple statistic environment, but require to detect whether the robot is trapped in the local minimum region.

For the dynamic environment, the moving objects can affect the performance of the APF. Ge et al. [20] define a virtual force which is the negative gradient of the potential with respect to both relative position and velocity between the robot and the target or obstacles. The motion of the mobile robot is then determined by the total virtual force through the Newtons Law or steering control depending on the driving type of the robot. Cao et al. [21] propose a simultaneous target tracking and moving obstacle avoidance method using a threat coefficient based force function by taking into account the relative velocities of the target with respect to the robot and the relative velocities of the robot with respect to the obstacles. Montiel et al. [22] introduce a parallel evolutionary artificial potential field for the dynamic environment, which makes possible controllability in complex real-world sceneries with dynamic obstacles if a reachable configuration set exists. These methods make decisions according to the relative position and velocity direction between robot and obstacle, but the magnitude of the obstacle speed is not fully considered.

In this paper, we propose a dynamic APF approach (DAPF). The contributions of our work are the following: first, to predict and avoid the local minimum region, we combine the APF with the dynamic window approach [23], which employs a cost function to evaluate simulated trajectories, such that the local minimum region can be found in the prior step, and the robot can avoid the obstacle in a short path without any oscillations. Second, we propose an evolutionary potential field function based on an improved minimum danger index (DI) [24,25], which can plan an optimal path using not only the relative distance and velocity direction information, but also the magnitude information of the robot and obstacle speeds. The robot can make a smart decision to avoid moving obstacles, e.g., if the obstacle moves fast, the robot goes behind the obstacle. We demonstrate the proposed ideas using a real mobile robot system in static and dynamic environment, respectively, which shows the improved performance using our idea.

2. The Proposed Method

In this section, we analyze the problems of the traditional APF approaches and further introduce the proposed idea in detail.

2.1. Dynamic Window Based Artificial Potential Field (DAPF)

In the artificial potential field, the robot motion is controlled by the attractive force and the repulsive force, i.e., the attractive force is generated by the distance and direction to the target point, whereas the repulsive force is generated by the distance and direction to obstacles. The forces of the robot in the artificial potential field are shown in Figure 1, where F_2 is the repulsive force between the obstacle and the robot F_1 is the attractive force between the target point and the robot, and F is the resultant force for controlling the robot.

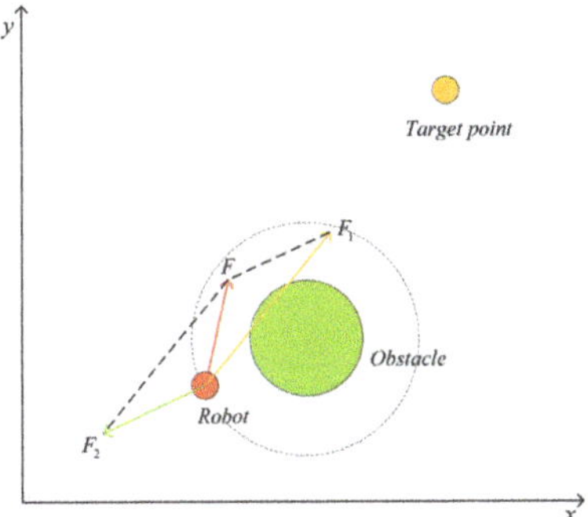

Figure 1. Definition of attractive force and repulsive force in artificial potential field.

The attractive potential field function is defined as:

$$U_{att}(\phi) = \frac{1}{2}k_a(\phi - \phi_g)^2, \tag{1}$$

where $\phi = (x, y)^T$ is the coordinate of the robot, k_a is the coefficient constant of the attractive field, and ϕ_g is the coordinate of the target point. The negative gradient of the attractive potential field function is defined as the attractive force function, which is:

$$F_{att}(\phi) = -\nabla(U_{att}(\phi)) = k_a(\phi_g - \phi), \tag{2}$$

where $F_{att}(\phi)$ is a vector directed toward to ϕ_g, which is a linear correlation with the distance from ϕ to ϕ_g. Obstacles provide repulsive force and repel the robot. When the robot is far enough from obstacles, obstacles will not affect the motion of the robot. As shown in Equation (3), the repulsion potential field function can be defined as:

$$U_{req}(\phi) = \begin{cases} \frac{1}{2}k_r(\frac{1}{d(\phi)} - \frac{1}{d_{max}})^2, & d(\phi) \leq d_{max}, \\ 0, & d(\phi) > d_{max}, \end{cases} \tag{3}$$

where k_r is the coefficient constant of the repulsion field, d_{max} is the maximum impact extent of the single obstacle, and $d(\phi)$ is the current distance between the robot and the obstacle. The repulsive force is the negative gradient of the repulsive potential function, which is

$$F_{req}(\phi) = -\nabla(U_{req}(\phi)), \tag{4}$$

$$F_{req}(\phi) = \begin{cases} k_r(\frac{1}{d(\phi)} - \frac{1}{d_{max}})\frac{1}{d(\phi)^2} \cdot \frac{\partial d(\phi)}{\partial \phi}, & d(\phi) \leq d_{max}, \\ 0, & d(\phi) > d_{max}. \end{cases} \tag{5}$$

The total repulsive potential field is the summation of the repulsive potential fields from all obstacles. The final artificial potential field function is as follows:

$$U(\phi) = U_{att}(\phi) + \sum U_{req}(\phi). \tag{6}$$

The resultant force on the robot is $F(x) = F_{att}(x) + \sum F_{req}(x)$.

In some situations, the goal can be unreachable and the robot can be trapped in a local minimum region using the traditional APF approach as shown in Figure 2. For instance, the robot can not reach the target point when the target point is very close to the obstacle. In addition, the resultant force on the robot can be reduced gradually until the target point is reached. However, the resultant force of the robot is also likely to be zero or minimized, when the attractive force and repulsive force are almost equal and collinear but in the opposite direction. The robot in this situation is trapped in a local minimum region.

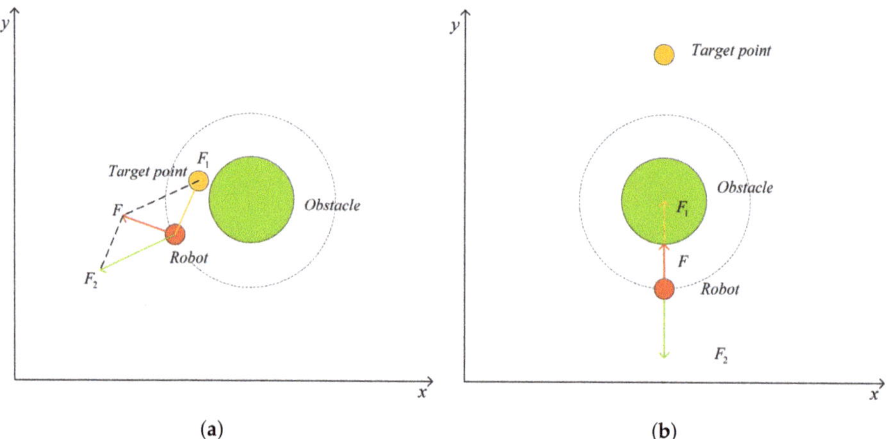

Figure 2. Problems of a traditional artificial potential field (APF): (**a**) unreachable goals with obstacles nearby (GNRON) and (**b**) trapped in a local minimum region.

To solve the GNRON problem, the distance between the current position and the target position can be added to the repulsion field function, such that the repulsive force can be relatively reduced when the robot closes to the target point. The improved repulsive potential field function is defined as

$$U_{req}(\phi) = \begin{cases} \frac{1}{2}k_r(\frac{1}{d(\phi)} - \frac{1}{d_{max}})^2(\phi - \phi_g)^n, & d(\phi) \leq d_{max}, \\ 0, & d(\phi) > d_{max}, \end{cases} \tag{7}$$

where k_r is the repulsion field coefficient constant and n is an arbitrary real number which is greater than zero. The improved repulsive force function is defined as

$$F_{req}(\phi) = \begin{cases} F_{req1}(\phi) + F_{req2}(\phi), & d(\phi) \leq d_{max}, \\ 0, & d(\phi) > d_{max}, \end{cases} \tag{8}$$

where $F_{req1}(\phi)$ and $F_{req2}(\phi)$ are two components of the $F_{req}(\phi)$, which are defined in Equation (9) and (10):

$$F_{req1}(\phi) = k_r(\frac{1}{d(\phi)} - \frac{1}{d_{max}})\frac{1}{d(\phi)^2}(\phi - \phi_g)^n \cdot \frac{\partial d(\phi)}{\partial \phi}, \tag{9}$$

$$F_{req2}(\phi) = \frac{1}{2}k_r n(\frac{1}{d(\phi)} - \frac{1}{d_{max}})^2(\phi - \phi_g)^{n-1} \cdot \frac{\partial(\phi - \phi_g)}{\partial \phi}. \tag{10}$$

The direction of $F_{req1}(\phi)$ is from the obstacle to the mobile robot, and $F_{req2}(\phi)$ is from the robot to the target point.

For solving the local minima problem, we propose a dynamic APF approach (DAPF), which employs the dynamic window approach (DWA) to predict the local minimum region and solve the oscillations. The key idea of the DWA is to sample a number of predicted trajectories according to the current position (x_t, y_t), direction θ_t and velocity v_r as shown in Figure 3. The sampled directions θ_ω are the following:

$$\theta_\omega = \theta_t + (\omega - t)\triangle\theta, \omega = (1, ..., t, ..., n), \tag{11}$$

where θ_t is the current direction of the robot, $\triangle\theta$ is the angular step between samples and n is the number of samples. Using these sampled directions, we simulate a number of positions that the robot might go as

$$\begin{bmatrix} x'_1 & y'_1 \\ ... & ... \\ x'_t & y'_t \\ ... & ... \\ x'_n & y'_n \end{bmatrix} = \begin{bmatrix} x_t & y_t \\ ... & ... \\ x_t & y_t \\ ... & ... \\ x_t & y_t \end{bmatrix} + \begin{bmatrix} v_r\triangle t\cos(\theta_1) & v_r\triangle t\sin(\theta_1) \\ ... & ... \\ v_r\triangle t\cos(\theta_t) & v_r\triangle t\sin(\theta_t) \\ ... & ... \\ v_r\triangle t\cos(\theta_n) & v_r\triangle t\sin(\theta_n) \end{bmatrix}, \tag{12}$$

where $x'_1, ..., x'_n$ and $y'_1, ..., y'_n$ are coordinates of simulated positions, and v_r is the velocity of the robot. The final choice of the position can be defined as the one whose APF cost function is minimum:

$$\begin{bmatrix} x & y \end{bmatrix} = \min(APF(\begin{bmatrix} x'_1 & y'_1 \\ ... & ... \\ x'_t & y'_t \\ ... & ... \\ x'_n & y'_n \end{bmatrix})), \tag{13}$$

where x, y is the coordinate of the position whose resultant force value is minimum.

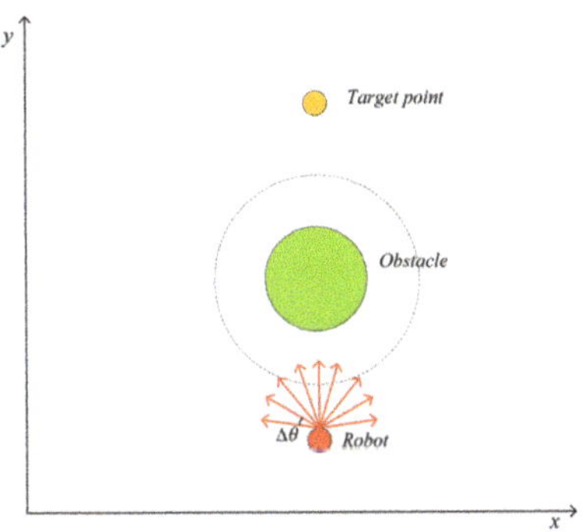

Figure 3. Robot trajectories simulated by a dynamic window approach (DWA).

Based on the improved APF function, the closer the robot is to the target, the smaller the resultant force value is. Before the robot enters the obstacle's influence range, it applies the DWA approach to simulate positions and calculate the resultant force for every simulated position. If the simulated position is in the obstacle range, the resultant force value is less than 0. If the simulated position is outside of the obstacle range, the resultant force of the simulated position that is closer to the target point is less than the positions that are far from the target point. Finally, the robot can obtain a simulated position that is close to the target point but not in the obstacle area, such that the local minimum region can be avoided using the DAPF approach. If the target point is within the obstacle range, the DWA method will be stopped when the distance of the robot from the target point is less than the distance from the obstacle.

2.2. Danger Index Based Artificial Potential Field (DIAPF) for a Dynamic Environment

To avoid the moving obstacles, the APF must consider not only the position information of the obstacles, but also the motion information of the obstacles, e.g., the direction and magnitude of the velocity. In the paper, we define the motion model of the dynamic obstacle as:

$$\delta(t+1) = \delta(t) + v_o(t) \times \triangle t, \tag{14}$$

$$v_o(t+1) = v_o(t) + a_o'(t) \times \triangle t, \tag{15}$$

which can be written in matrix form:

$$p(t+1) = Ap(t) + a_o, \tag{16}$$

where $\delta(t) = (x_o, y_o)^T$ is the coordinate of the dynamic obstacle, $p(t+1) = (\delta(t), v_o(t))^T$ is the state vector, $A = \begin{bmatrix} 1 & \triangle t \\ 0 & 1 \end{bmatrix}$ is the state transition matrix, and $a_o = (0, a_o')^T$ is the acceleration of the dynamic obstacle. In this paper, a uniform motion model is adopted and the a_o is a zero vector.

The APF should not only consider the spatial location of the dynamic obstacle, but also consider the motion state of the dynamic obstacle. Therefore, the velocity repulsive potential field is added to the dynamic window based artificial potential field function (DVAPF) to avoid dynamic obstacles, which is defined as

$$U_v(v_{ro}) = \begin{cases} \frac{1}{2}k_{ro}v_{ro}^2, & 0 \leq \rho_{cm} \leq \rho_{cmax} \cap \alpha \in \left(-\frac{\pi}{2}, \frac{\pi}{2}\right), \\ 0, & otherwise, \end{cases} \tag{17}$$

where k_{ro} is a proportionality constant, $v_{ro} = v_r - v_o$ is the relative velocity between the robot and the dynamic obstacle, α is the angle between the v_{ro} and the vector from the robot to the obstacle, ρ_{cmax} is the maximum range of the local map, and ρ_{cm} is the current distance between the robot and the dynamic obstacle. The velocity repulsive potential field starts to work when the dynamic obstacle enters the local map whose range is defined by the distance parameter ρ_{cmax}, which can be chosen according to the sensor range. The velocity repulsive force is the negative gradient of the velocity repulsive potential function:

$$F_v(v_{ro}) = -\nabla(U_v(v_{ro})), \tag{18}$$

$$F_v(v_{ro}) = \begin{cases} k_{ro}v_{ro}, & 0 \leq \rho_{cm} \leq \rho_{cmax} \cap \alpha \in \left(-\frac{\pi}{2}, \frac{\pi}{2}\right) \\ 0, & otherwise \end{cases}. \tag{19}$$

The DVAPF approach can handle dynamic obstacle avoidance problem when the velocity difference between the robot and the dynamic obstacle is large. The effectiveness of the velocity repulsive force can be decreased if the relative velocity is small. We here propose an improved method using a novel danger index considering relative distance and velocity between the robot and the dynamic obstacle, which includes three factors:

(1) Distance influence factor:

$$f_{cm}(\rho_{cm}) = \begin{cases} \eta\left(\frac{1}{\rho_{cm}} - \frac{1}{\rho_{cmax}}\right), & \rho_{cm} \leq \rho_{cmax}, \\ 0, & \rho_{cm} > \rho_{cmax}, \end{cases} \tag{20}$$

where $\eta = \left(\frac{\rho_{cmin}\rho_{cmax}}{\rho_{cmax} - \rho_{cmin}}\right)$ is the distance scale factor, and ρ_{cmin} is the minimum allowed distance between the robot and the dynamic obstacle.

Distance influence factor is influenced by the distance relationship between the robot and the dynamic obstacle. The closer the distance between the robot and the dynamic obstacle, the larger the distance influence factor.

(2) Speed influence factor:

$$k_o = sgn(\varepsilon \times |v_o| - |v_r|),$$

(21)

where ε is the speed scale factor, $|v_o|$ and $|v_r|$ are the magnitudes of the dynamic obstacle velocity and robot velocity, respectively, and $sgn()$ is the sign function.

Speed influence factor is influenced by the speed relationship between the robot and the dynamic obstacle. If the speed of the dynamic obstacle is ε times the speed of the robot, the speed influence factor is an integer and vice versa.

(3) Danger index:

$$DI = \begin{cases} k_{ro}(v_r - f_{cm}v_o), & k_o > 0 \cap \alpha \in (-\frac{\pi}{2}, \frac{\pi}{2}) \cap f_{cm} > 0, \\ k_{ro}(v_r + f_{cm}v_o), & k_o \leq 0 \cap \beta \in (-\frac{\pi}{2}, \frac{\pi}{2}) \cap f_{cm} > 0, \\ 0, & otherwise, \end{cases}$$

(22)

where k_{ro} is a proportionality constant, f_{cm} is the distance influence factor, α is the angle between the $(v_r - f_{cm}v_o)$ and the vector from the robot to the obstacle. β is the angle between the $(v_r + f_{cm}v_o)$ and the vector from the obstacle to the robot.

The novel danger index includes distance influence factor and speed influence factor. The robot can make smarter avoidance decisions according to the speed influence factor, i.e., the robot can move in front of the moving obstacle if the speed of the robot is not larger than ε times that of the dynamic obstacle, whereas the robot can move behind the moving obstacle for other situations.

If the distance between the robot and the dynamic obstacle is very close, the distance influence factor will be large, such that the robot will move parallel with the dynamic obstacle, which is to ensure the safety of the path. If the distance between the robot and the dynamic obstacle is far, the distance factor will be small, such that the robot will not over-react to the obstacle motion, which can increase the efficiency of the robot motion. Furthermore, if the moving obstacle is outside of the local map, i.e., $\rho_{cm} > \rho_{cmax}$, the danger index is 0, so the robot will not react to such a situation.

The danger index combines the distance influence factor and the speed influence factor to optimize the planned path to avoid dynamic obstacles. The danger index based APF (DIAPF) function is defined as

$$U'(\phi) = U(\phi) + \frac{1}{2}DI^2,$$

(23)

where $U(\phi)$ is the value returned from the DAPF function. To summarize, the whole algorithm of our proposed DIAPF approach in the dynamical environment is shown in Figure 4.

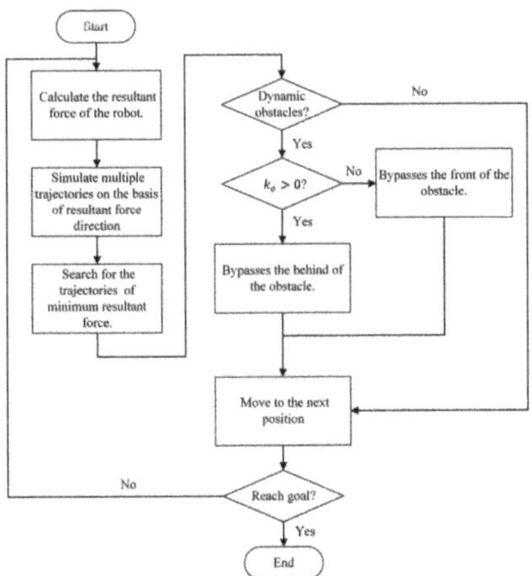

Figure 4. Danger Index Based Artificial Potential Field (DIAPF) based moving obstacle avoidance algorithm.

3. Experiment and Analysis

To evaluate the performance of the proposed idea, we use a Turtlebot robot to demonstrate the algorithm in the real environment where we have built a known map. The nurse robot plays as a moving obstacle in the map. To simplify the problem, the map of the environment and the motion status of the obstacle (nurse robot) are known for the Turtlebot. The Turtlebot robot and experimental environment are shown in Figure 5. We first compare the DAPF to the standard APF approach in the static environment to show how the dynamic window approach can improve the performance of the APF, and then compare the DIAPF with the DAPF, the DVAPF and ROS (robot operating system) path planning algorithm to show how the danger index can help the robot avoid the moving obstacle.

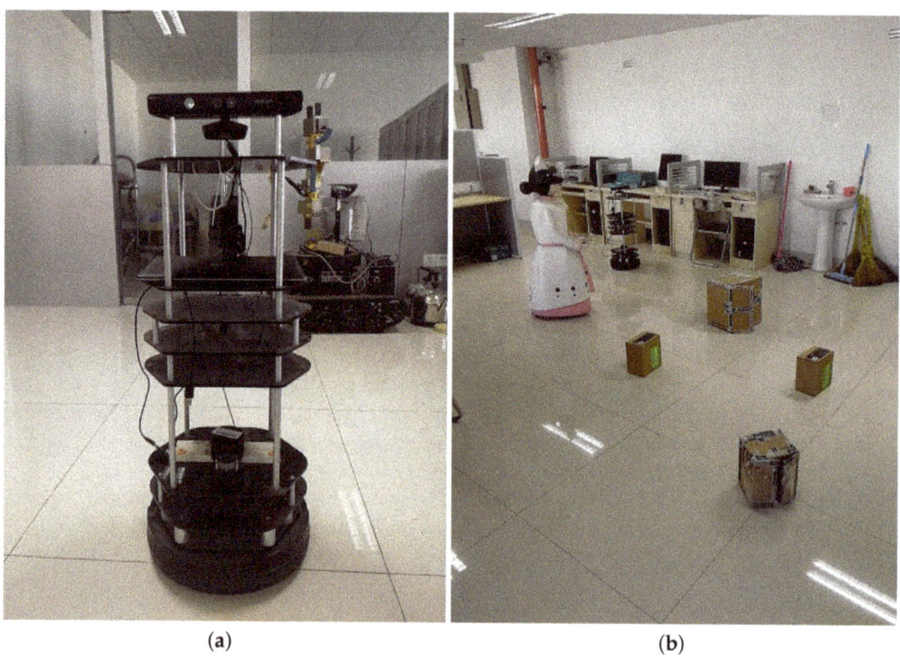

(a)   (b)

Figure 5. (a) Turtlebot robot and (b) experimental environment.

3.1. Static Environment

For a static environment, we place several obstacles randomly in the room. The DAPF and the standard APF algorithms are carried out using the experimental parameters defined in Table 1.

Table 1. The experimental parameters in a static environment.

d_{max}	$\triangle\theta$	k_r	k_a	m	n	Target Point
0.2 m	10°	1	1	1	2	(2.5, 2.5)

The robot has the oscillation problem and can not avoid the local minimum region using the standard APF algorithm as shown in Figure 6a,c, respectively. In contrast to the APF, the proposed DAPF approach has a smooth planned path along the edge of the influence range of obstacles as shown in Figure 6b,d. For the local minimum problem, the DAPF has no need to detect whether the robot has entered a local minimum region, since the DWA can predict and avoid such a situation in advance.

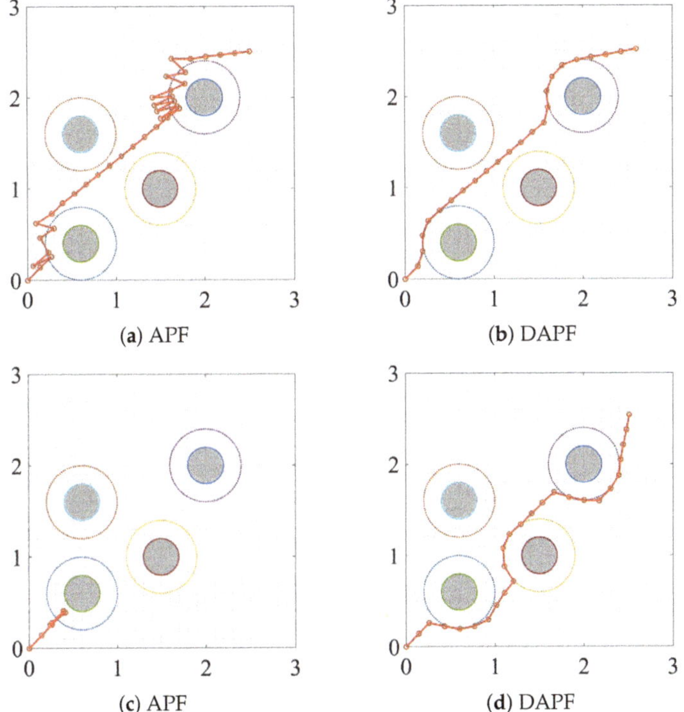

Figure 6. Experimental results using APF (**a,c**) and proposed DAPF (**b,d**) in the static environment where the gray solid circles are obstacles, the circles that around obstacles means the influence range, the small green circle is the robot position, and the red curve is the planned path. The APF has the oscillation problem that occurs in the (**a**) and local minimum problem that occurs in the (**c**), whereas the DAPF solves such problems and achieves better results as shown in (**b,d**), respectively.

3.2. Moving Obstacle

For dynamic environment, we compare three methods to avoid the moving obstacle, i.e., DAPF, DVAPF and DIAPF. We show that these methods have different performance for faster and slow moving obstacles. The experimental parameters are shown in Table 2.

The performance using the DAPF for avoiding the slow and fast obstacles are shown in Figures 7 and 8 respectively. For the slow obstacle, the DAPF has a period of oscillation when the obstacle is in front of the robot. For the fast obstacle, the situation is worse, i.e., the collision between the robot and the obstacle occurs.

Table 2. The experimental parameters for moving obstacle avoidance

k_{ro}	ε	ρ_{max}	ρ_{min}	v_r	v_o (fast)	v_o (slow)
1	2	1.2m	0.3m	0.2m/s	0.3m/s	0.1m/s

The DVAPF is better than the DAPF as shown in Figures 9 and 10, where we can see that the DVAPF prefers a path behind the moving obstacle. Using such a strategy, the robot and the dynamic obstacle will move in opposite directions when the obstacle enters the robot's local map if the dynamic obstacle is faster than the robot, and the robot will continue to move towards the target point when $\alpha \in (-\pi, -\frac{\pi}{2}) \cap (\frac{\pi}{2}, \pi)$ or the obstacle leaves the robot's local map. If the moving obstacle is slower than the robot or their velocity is similar, the planned path can be inefficiency and dangerous as shown in Figure 10, i.e., the robot enters the influence area of the obstacle since the robot's deflection angle is too small to avoid the obstacle.

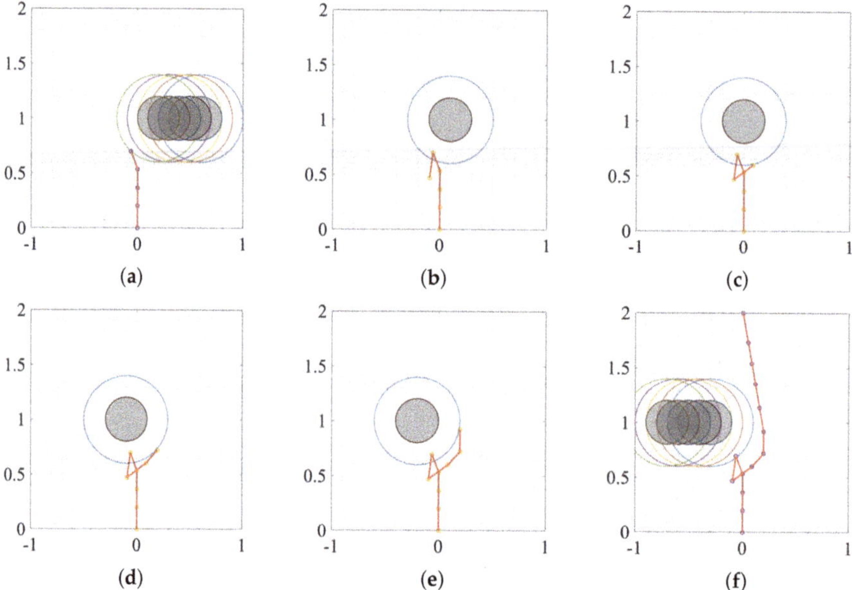

Figure 7. The robot uses the DAPF approach to avoid the slow moving obstacle: (**a**) shows the robot is before obstacle avoidance; (**b–e**) show that the robot is avoiding the obstacle, and (**f**) shows that the robot finishes obstacle avoidance.

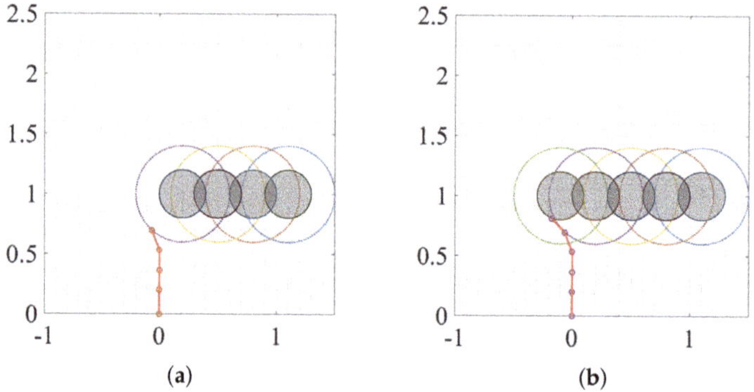

Figure 8. The robot uses the DAPF approach to avoid a fast moving obstacle: (**a**) shows that the robot is before obstacle avoidance, and (**b**) shows that the robot collides with an obstacle.

In contrast to the DVAPF and DAPF, the DIAPF has the best performance according to the experiment results as shown in Figures 11 and 12. From the experimental results, it can be seen that the DIAPF approach can make a prediction according to the state of the moving obstacle: if the obstacle velocity is similar or faster, the robot can plan a path behind the obstacle. If the obstacle velocity is slower, the robot can plan a path in front of the obstacle. This strategy makes the planned path more efficient. In addition, the distance influence factor plays a key role to increase the safety and effectiveness of the planned path, e.g., the distance influence factor can reduce the robot's deviation angle if the moving obstacle is far away, and can increase this angle if the moving obstacle is closer.

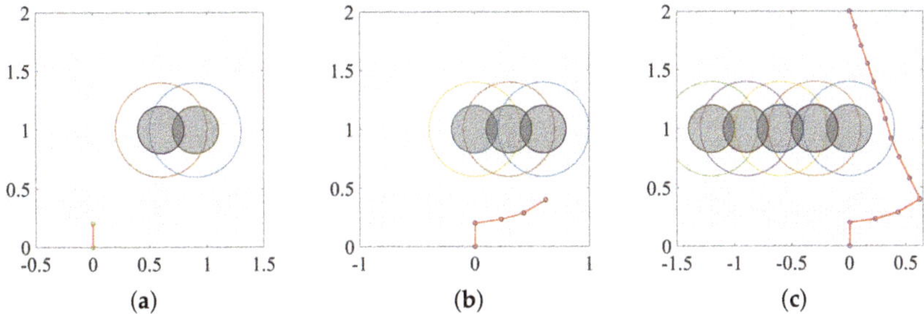

Figure 9. The robot uses the dynamic window based artificial potential field (DVAPF) approach to avoid a fast moving obstacle: (**a**) before avoiding; (**b**) avoiding the obstacle; and (**c**) finish avoiding.

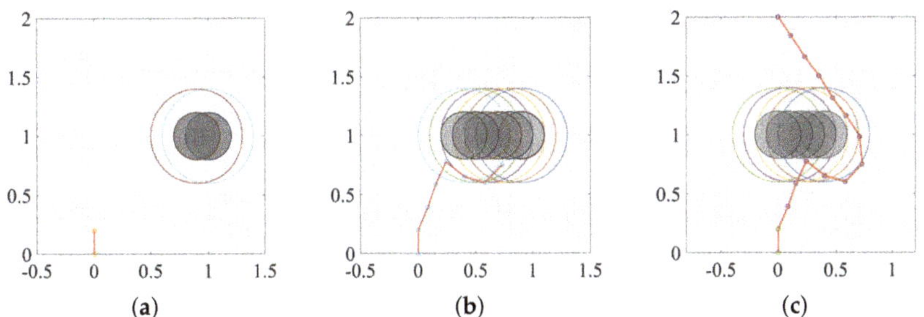

Figure 10. The robot uses the DVAPF approach to avoid a slow moving obstacle: (**a**) before avoiding; (**b**) avoiding the obstacle; (**c**) finish avoiding.

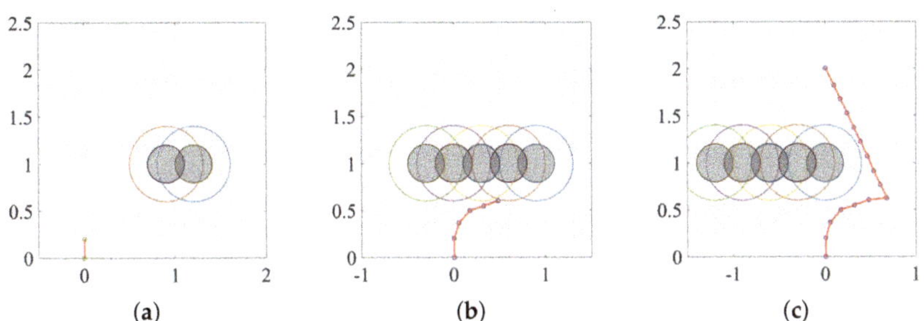

Figure 11. The robot uses the DIAPF approach to avoid a fast moving obstacle: (**a**) before avoiding; (**b**) avoiding the obstacle; and (**c**) finish avoiding.

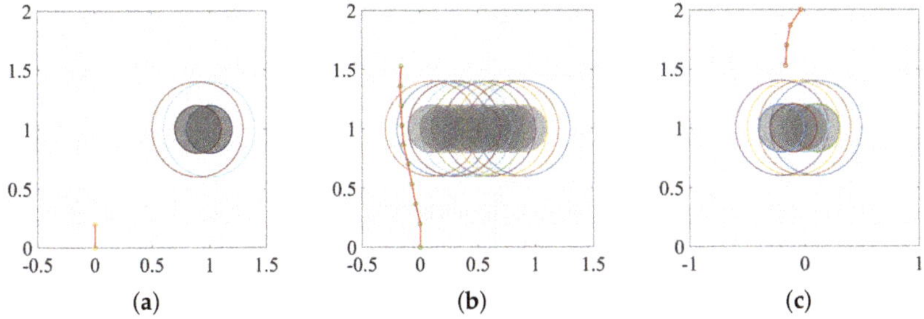

Figure 12. The robot uses the DIAPF approach to avoid a slow moving obstacle: (**a**) before avoiding; (**b**) avoiding the obstacle; and (**c**) finish avoiding.

3.3. Dynamic Environment

In this section, we compare the proposed DIAPF method to the ROS [26] path plan package which employs the DWA and A* algorithm. The experimental environment includes four static obstacles and a moving obstacle. As can be seen from Figure 13, the ROS uses A* algorithm to plan a global path and uses the DWA algorithm to handle nearby obstacles in the local map. If the speed of the moving obstacle is slow, the robot can avoid the obstacle directly as shown in Figure 13a. However, if the obstacle moves fast, the robot needs to re-plan the path according to the position of the obstacle. The global path also needs to replan according to the local path changes as shown in Figure 13b. In contrast to the ROS method, the proposed DIAPF can make decisions in advance according to the relative velocity between the moving obstacle and robot as shown in Figures 14 and 15. The planned path using the DIAPF is more efficient than the one using the ROS module as seen in Figure 16, which shows the changes of the distance between the robot and the target point. The DIAPF shows a linear decrease of the distance while the ROS shows a curve that decreases first and then increases due to a re-planned path.

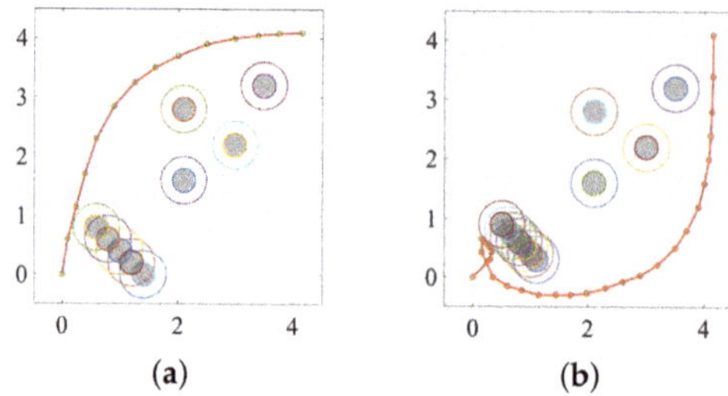

Figure 13. The robot uses path planner in robot operating system (ROS) to avoid a (**a**) slow and (**b**) fast moving obstacle in the dynamic environment.

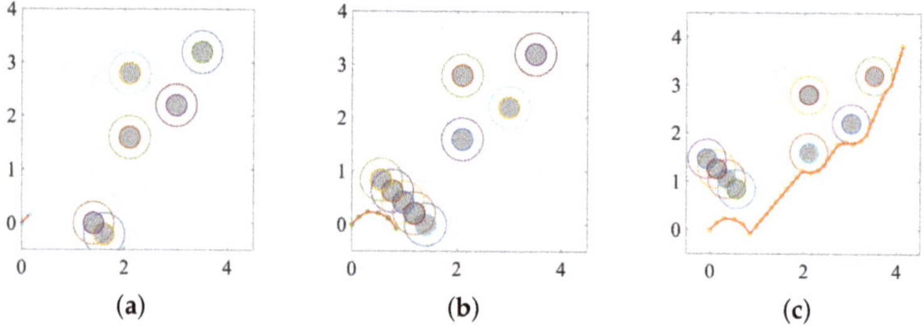

Figure 14. The robot uses the DIAPF approach to avoid a fast moving obstacle in the dynamic environment: (**a**) before avoiding; (**b**) avoiding the dynamic obstacle; and (**c**) finish avoiding.

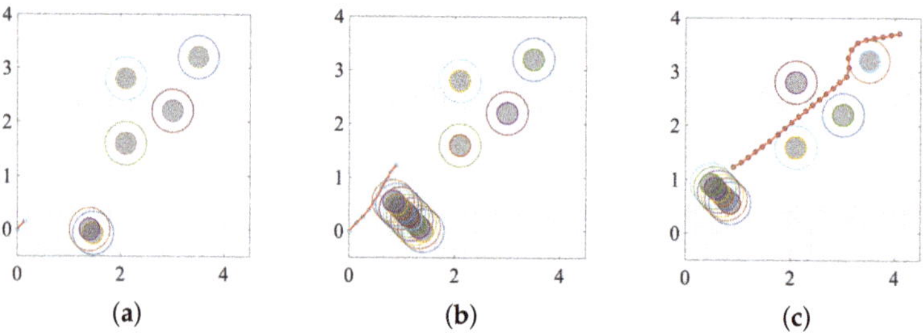

Figure 15. The robot uses the DIAPF approach to avoid a slow moving obstacle in the dynamic environment: (**a**) before avoiding, (**b**) avoiding the dynamic obstacle; and (**c**) finish avoiding.

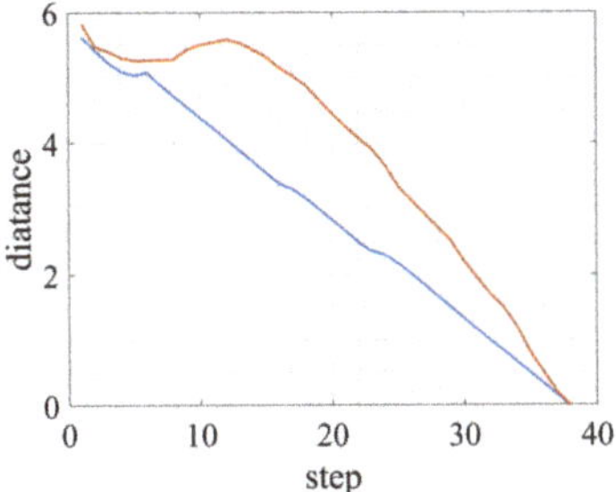

Figure 16. The changes of the distance between the robot and the target point using the proposed DIAPF method (blue) and ROS path plan module (red).

4. Conclusions

In this paper, we introduce a dynamic window approach (DWA) and danger index (DI) based artificial potential field (APF) for static and moving obstacle avoidance. The dynamic window approach based APF (DAPF) can overcome the local minimum problem since the DWA can predict the local minima region and avoid in advance. For moving obstacles, we propose a novel danger index based APF (DIAPF) that considers not only the relative distance, but also the relative velocity between the robot and the moving obstacle. In this way, an optimized strategy can be made according to the motion status of the moving obstacle, which leads to a safer and efficient planned path compared to the state-of-the-art works, e.g., ROS path plan module. Finally, we demonstrate the proposed idea in the real environment using the Turtlebot robot, and the experiment results prove our claims. Future work will be extending current work to the environment with an unknown map and multiple randomly moving obstacles.

Author Contributions: The manuscript was written through the contributions of all authors. All authors discussed the original idea. J.S. and G.L. designed and performed the simulation experiments; J.S. wrote this manuscript with support from G.L., G.T. and J.Z.; all authors provided critical feedback and helped shape the research, analysis, and reviewed the manuscript; G.L. supervised this work.

Funding: This research was supported by the National Key R&D Program of China (2018YFB1306504), the National Natural Science Foundation of China (61603213, 91748115), the Young Scholars Program of Shandong University (2018WLJH71), the Hebei Provincial Natural Science Foundation (F2017202062), and the Taishan Scholars Program of Shandong Province.

Conflicts of Interest: The authors declare no conflict of interest.

References

1. Rafieisakhaei, M.; Tamjidi, A.; Chakravorty, S.; Kumar, P.R. Feedback motion planning under non-Gaussian uncertainty and non-convex state constraints. In Proceedings of the 2016 IEEE International Conference on Robotics and Automation (ICRA), New York, NY, USA, 15–19 August 2016; pp. 4238–4244.
2. Al-Sabban, W.H.; Gonzalez, L.F.; Smith, R.N. Wind-energy based path planning for unmanned aerial vehicles using markov decision processes. In Proceedings of the 2013 IEEE International Conference on Robotics and Automation (ICRA), Karlsruhe, Germany, 6–10 May 2013; pp. 784–789.
3. Sun, W.; van den Berg, J.; Alterovitz, R. Stochastic extended LQR for optimization-based motion planning under uncertainty. *IEEE Trans. Autom. Sci. Eng.* **2018**, *13*, 437–447. [CrossRef] [PubMed]
4. Chi, W.; Wang, C.; Wang, J.; Meng, M.Q. Risk-DTRRT-Based optimal motion planning algorithm for mobile robots. *IEEE Trans. Autom. Sci. Eng.* **2008**, 142–149. [CrossRef]
5. LaValle, S.M. *Rapidly-Exploring Random Trees: A New Tool for Path Planning*; TR 98-11, Tech. Rep.; Iowa State University: Ames, IA, USA, 1998.
6. Kayraki, L.; Svestka, P.; Latombe, J.C.; Overmars, M. Probabilistic roadmaps for path planning in high-dimensional configurations space. *IEEE Trans. Robot. Autom.* **1996**, *12*, 566–580. [CrossRef]

7. Hart, P.E.; Nilsson, N.J.; Raphael, B.A. Formal basis for the heuristic determination of minimum cost paths. *IEEE Trans. Syst. Sci. Cybern.* **1968**, *4*, 100–107.
8. Dijkstra, E.W. A note on two problems in connexion with graphs. *Numer. Math.* **1959**, *1*, 269–271. [CrossRef]
9. Janabi-Sharifi, F.; Vinke, D. Integration of the artificial potential field approach with simulated annealing for robot path planning. In Proceedings of the 8th IEEE International Symposium on Intelligent Control, Chicago, IL, USA, 25–27 August 1993; pp. 539–544.
10. Lee, S.; Park, J. Cellular robotic collision-free path planning. In Proceedings of the Fifth International Conference on Advanced Robotics' Robots in Unstructured Environments, Pisa, Italy, 19–22 June 1991; pp. 536–541.
11. Khatib, O. Real-Time Obstacle Avoidance System for Manipulators and Mobile Robots. In Proceedings of the 1985 IEEE International Conference on Robotics and Automation, St. Louis, MO, USA, 25–28 March 1985; pp. 500–505.
12. Volpe, R.; Khosla, P. Manipulator control with superquadric artificial potential functions: theory and experiments. *IEEE Trans. Syst. Man Cybern.* **1990**, *20*, 1423–1436. [CrossRef]
13. Mohanty, P.K.;Parhi, D.R. A new hybrid intelligent path planner for mobile robot navigation based on adaptive neuro-fuzzy inference system. *Aust. J. Mech. Eng.* **2015**, *13*, 195-207. [CrossRef]
14. Min, G.P.; Jeon, J.H.; Min, C.L. Obstacle avoidance for mobile robots using artificial potential field approach with simulated annealing. In Proceedings of the 2001 IEEE International Symposium on Industrial Electronics Proceedings, Pusan, Korea, 12–16 June 2001; pp. 1530–1535.
15. Zhu, Q. and Yan, Y. and Xing, Z. Robot path planning based on artificial potential field approach with simulated annealing. In Proceedings of the Sixth International Conference on Intelligent Systems Design and Applications, Jinan, China, 16–18 October 2006; pp. 622–627.
16. Doria, N.S.F.; Freire, E.O.; Basilio, J.C. An algorithm inspired by the deterministic annealing approach to avoid local minima in artificial potential fields. In Proceedings of the 2013 16th International Conference on Advanced Robotics (ICAR), Montevideo, Uruguay, 25–29 November 2013; pp. 1–6.
17. Lee, D.; Jeong, J.; Kim, Y.H.; Park, J.B. An improved artificial potential field method with a new point of attractive force for a mobile robot. In Proceedings of the 2017 2nd International Conference on Robotics and Automation Engineering (ICRAE), Shanghai, China, 29–31 December 2017; pp. 63–67.
18. Weerakoon, T.; Ishii, K.; Nassiraei, A.A.F. An artificial potential field based mobile robot navigation method to prevent from deadlock. *J. Artif. Intell. Soft Comput. Res.* **2015**, *5*, 189–203. [CrossRef]
19. Kim, Y.H.; Son, W.-S.; Park, J.B.; Yoon, T.S. Smooth path planning by fusion of artificial potential field method and collision cone approach. *MATEC Web Conf.* **2016**, *75*, 05004. [CrossRef]
20. Ge, S.S.; Cui, Y.J. Dynamic motion planning for mobile robots using potential field method. *Auton. Robots* **2002**, *13*, 207–222. [CrossRef]
21. Cao, Q.; Huang, Y.; Zhou, J. An evolutionary artificial potential field algorithm for dynamic path planning of mobile robot. In Proceedings of the 2006 IEEE/RSJ International Conference on Intelligent Robots and Systems, Beijing, China, 9–15 October 2006; pp. 3331–3336.
22. Montiel, O.; Sepúlveda, R. and Orozco-Rosas, U. Optimal path planning generation for mobile robots using parallel evolutionary artificial potential field. *J. Intell. Robot. Syst.* **2015**, *79*, 237–257. [CrossRef]
23. Fox, D.; Burgard ,W.; Thrun, S. The dynamic window approach to collision avoidance. *IEEE Robot. Autom. Mag.* **1997**, *4*, 23–33. [CrossRef]
24. Kulić, D.; Croft, E.A. Safe planning for human-robot interaction. *J. Robot. Syst.* **2005**, *22*, 383–396. [CrossRef]
25. Kulić, D.; Croft, E.A. Affective state estimation for human–robot interaction. *IEEE Trans. Robot.* **2007**, *23*, 991–1000. [CrossRef]
26. Marder-Eppstein, E.; Berger, E. ; Foote, T. ; Gerkey, B. ; Konolige, K. The Office Marathon: Robust Navigation in an Indoor Office Environment. In Proceedings of the 2010 IEEE International Conference on Robotics and Automation (ICRA), Anchorage, AK, USA, 3–7 May 2010; pp. 300–307.

Article

Integrating a Path Planner and an Adaptive Motion Controller for Navigation in Dynamic Environments

Junjie Zeng, Long Qin *, Yue Hu, Quanjun Yin and Cong Hu

College of Systems Engineering, National University of Defense Technology, Changsha 410073, China;
zengjunjie13@nudt.edu.cn (J.Z.); huyue.cse@gmail.com (Y.H.); yin_quanjun@163.com (Q.Y.);
hccz95@163.com (C.H.)
* Correspondence: qldbx2007@sina.com

Received: 8 March 2019; Accepted: 29 March 2019; Published: 2 April 2019

Abstract: Since an individual approach can hardly navigate robots through complex environments, we present a novel two-level hierarchical framework called JPS-IA3C (Jump Point Search improved Asynchronous Advantage Actor-Critic) in this paper for robot navigation in dynamic environments through continuous controlling signals. Its global planner JPS+ (P) is a variant of JPS (Jump Point Search), which efficiently computes an abstract path of neighboring jump points. These nodes, which are seen as subgoals, completely rid Deep Reinforcement Learning (DRL)-based controllers of notorious local minima. To satisfy the kinetic constraints and be adaptive to changing environments, we propose an improved A3C (IA3C) algorithm to learn the control policies of the robots' local motion. Moreover, the combination of modified curriculum learning and reward shaping helps IA3C build a novel reward function framework to avoid learning inefficiency because of sparse reward. We additionally strengthen the robots' temporal reasoning of the environments by a memory-based network. These improvements make the IA3C controller converge faster and become more adaptive to incomplete, noisy information caused by partial observability. Simulated experiments show that compared with existing methods, this JPS-IA3C hierarchy successfully outputs continuous commands to accomplish large-range navigation tasks at shorter paths and less time through reasonable subgoal selection and rational motions.

Keywords: autonomous navigation; dynamic environments; Deep Reinforcement Learning; geometrical path planner

---

1. Introduction

Navigation in dynamic environments plays an important role in computer games and robotics [1], such as generating realistic behaviors of Non-Player Characters (NPCs) and meeting practical applications of mobile robots in the real world. In this paper, we focus on the navigation problems of nonholonomic mobile robots [2] with continuous control in dynamic environments, as this is an important capability for widely used mobile robots.

Conventionally, sampling-based and velocity-based methods are used to support the navigation of mobile robots [1]. Sampling-based methods, such as Rapidly Exploring Trees (RRTs) [3] and Probabilistic Roadmap (PRM) [4], deal with environmental changes by reconstructing pre-built abstract representations at high time costs. Velocity-based methods [5] compute avoidance maneuvers by searching over a tree of avoidance maneuvers, which require high time consumption in complex environments.

Alternatively, many researchers focus on learning-based methods—mainly those including Deep Learning (DL) and Deep Reinforcement Learning (DRL). Although DL achieves great performance in robotics [6], it is hard to apply DL in this navigation problem, since collecting considerable amounts of labeled data requires much time and energy for researchers. By contrast, DRL does not need supervised samples, and has been widely used in video games [7] and robot manipulation [8]. There are two categories of DRL (i.e., value-based DRL and policy-based DRL). Compared with valued-based DRL methods, policy-based methods are more suitable for us to handle continuous action spaces. Asynchronous Advantage Actor-Critic (A3C), which is a policy-based method, is widely used in video games and robotics [9–11], since it can greatly decrease training time and meanwhile increase performance [12]. However, several challenges need to be tackled if A3C is used to deal with robot navigation in dynamic environments.

First, robots trained by A3C are vulnerable to local minima such as box canyons and long barriers, which most DRL algorithms cannot effectively resolve. Second, rewards are sparsely distributed in the environment, for there may be only one goal location [9]. Third, robots with limited visibility can only gather incomplete and noisy information from current environmental states due to partial observability [13], and in this paper, we are concerned with robots that just know the relative positions of obstacles within a small range.

Given that an individual DRL approach can hardly drive robots out of regions of local minima and navigate them in changing surroundings, this paper proposes a hierarchical navigation algorithm based on a two-level architecture, whose high-level path planner efficiently searches for a sequence of subgoals placing at the exits of those regions, while the low-level motion controller learns how to tackle moving obstacles in local environments. Since JPS+ (P), which is a variant of JPS+, on average answers a path query faster by up to two orders of magnitude over traditional A* [14], the path planner uses it to find subgoals between which the circumstances are relatively simple and easy for A3C to train and converge to a motion policy.

To tackle two other challenges regarding learning, we propose an improved A3C (IA3C) for the motion controller by making some improvements to A3C. IA3C combines a modified curriculum learning with reward shaping to build a novel reward function framework in order to solve the problem of sparse rewards. In the framework, the modified curriculum learning adds prior experiences to set different difficulty levels of navigation tasks and adjusts the frequencies of different tasks by considering features of navigation. Reward shaping plays an auxiliary role in each task to speed up training efficiency. Moreover, IA3C builds a network architecture based on long short-term memory (LSTM), which integrates current observation and hidden state from historical observations to estimate current state. Briefly, the proposed method integrates JPS+ (P) and IA3C, named JPS-IA3C; this model can realize long-range navigation in complex and dynamic environments by addressing mentioned challenges.

The rest of this paper is organized as follows: Section 2 discusses some related work on model-based methods, DL, and DRL. Section 3 presents our hierarchical navigation approach. In Section 4, its performance is evaluated in simulation experiments. Finally, this paper is concluded in Section 5.

2. Related Work

This section will introduce related works regarding sampling-based and velocity-based methods, DL and DRL. Sampling-based and velocity-based methods are classical navigation methods that are not suitable for dealing with environmental changes. Learning-based methods such as DL and DRL can improve the learning ability of mobile robots to quickly adapt to new environments. DL requires considerable amounts of labeled data, which is difficult to be collected, while DRL does not need prior knowledge.

2.1. Sampling-Based and Velocity-Based Methods

Autonomous navigation in dynamic environments is challenging, since the planner is forced to frequently adjust its planned results for handling dynamic objects such as moving obstacles. Velocity obstacle (VO) [5] can simply consider dynamic constraints to compute trajectories of robots via the concept of velocity obstacles that denote the robot's velocities causing a collision with obstacles soon. However, computing maneuvers by velocity obstacles has low efficiency for real-time applications. Sampling-based methods such as PRM and RRT are popular in robotics, benefitting from advantages regarding efficiency and robustness [15]. This kind of method can solve high-dimensional planning problems by approximating C-space. Nevertheless, in dynamic environments, PRM needs to recheck edge connection [16], and RRT is required to modify the pre-built exploring tree [15], which further decrease time efficiency when environments become more dynamic. Consequently, the above-mentioned model-based methods still suffer from essential disadvantages that cannot realize time-saving navigation in dynamic environments, so as to hinder the robot's navigation capabilities.

2.2. Deep Learning

In recent years, as deep neural networks show great potential for solving complex estimation problems, a rapidly growing trend of utilizing DL techniques has appeared in robotics tasks. For navigation based on visual information, Chen et al. [17] used deep neural networks to learn how to recognize objects from the image, and then determined discrete actions such as turning left and turning right according to the identified information. Gao et al. [18] proposed a novel deep neural network architecture, Intention-Net, to track the planned path by mapping monocular images to four discrete actions such as going forward. Since the action is discrete, the navigation behaviors in above-mentioned researches are rough and likely to be unfeasible. For navigation based on range information, Pfeiffer et al. [19] used demonstrative data to train a DL model that can control the robot through inputting the data from laser sensors and the target position. However, since a large number of labeled data are required for training the above DL models, it may be unpractical for real-world applications.

2.3. Deep Reinforcement Learning

Autonomous navigation demands two essential building blocks: perception and control. Similar to the above-mentioned studies about DL, many research studies [20] about applying DL in navigation are pure perception, which means that agents passively receive observations and infer desired information. Compared with pure perception, control goes one step further, seeking to actively interact with the environment by executing actions [21]. Then, navigation becomes a problem of sequential decisions. DRL is very suitable for solving it, which is proved by reference [7] in game and reference [8] in robotics. Since DRL methods can deal with better full observable states than partial states, lots of studies [9,10,22] about applying DRL in navigation focus on static environments. Regarding dynamic environments, Chen et al. [23] proposed a time-efficient approach based on DRL for socially aware navigation. However, to calculate the reward based on social norms, the motion information of pedestrians are known in advance, which are not reliable in real-world applications, since it is not precise to estimate information by sensor readings.

Given that it is difficult for an individual DRL method to solve the navigation problem, hierarchical approaches are widely researched in the literature [24–26]. Lei et al. [26] combined A* and least-squares policy iteration for mobile robot navigation in complex environments. Aleksandra et al. [25] integrated sampling-based path planning with reinforcement learning (RL) agents for indoor navigation and aerial cargo delivery. The aforementioned methods can only handle navigation in static environments. Kato et al. [24] combined value-based DRL with A* on topological maps to solve navigation in environments with pedestrians. However, the experiment results are not well applied in dynamic environments, since the learned policy is reactive and the way of taking observations as states is inaccurate, which may lead to irrational behaviors and even collisions with obstacles.

Compared with the above hierarchical methods, our hierarchical navigation algorithm can generate better waypoints by JPS+ (P) and navigate in dynamic environments via IA3C.

3. The Methodology

In this section, we at first briefly introduce the problem and model the virtual tracked robot. Next, we present the architecture of the proposed navigation method. Finally, we describe the global path planner based on JPS+ (P) and the local motion controller based on IA3C in detail.

3.1. Problem Description

In this paper, the problem of navigation can be stated as follows: on an initially known map, the mobile tracked robot that is only equipped with local laser sensors autonomously navigate without colliding with moving obstacles via continuous control. Since environments with moving obstacles are dynamic and uncertain, this navigation task is difficult to be dealt with, and is an non-deterministic polynomial-hard problem [27].

In this paper, we adopt the Cartesian coordinate system. The structure of the tracked robot is shown in Figure 1. The position state of the tracked robot is defined as $s = (x_r, y_r, \theta_r)$ by the global coordinate of the map, where x_r and y_r are the tracked robot's horizontal and vertical coordinates, respectively, and θ_r is the angle between the forward orientation of the tracked robot and the abscissa axis. Twelve laser sensors equipped at the front of the tracked robot can sense the nearby surroundings 360 degrees. The detection angle of each laser sensor is 30 degrees. The max detection distance is D_{max}. The robot is controlled by the left and right tracks. The kinematic equations were as follows:

$$
\begin{pmatrix} \dot{x}_r \\ \dot{y}_r \\ \dot{\theta}_r \end{pmatrix} = R \begin{bmatrix} \frac{cos\theta_r}{2} & \frac{cos\theta_r}{2} \\ \frac{sin\theta_r}{2} & \frac{sin\theta_r}{2} \\ \frac{1}{l_B} & \frac{-1}{l_B} \end{bmatrix} \begin{pmatrix} \omega_l \\ \omega_r \end{pmatrix},
\tag{1}
$$

$$
\begin{pmatrix} \dot{\omega}_l \\ \dot{\omega}_r \end{pmatrix} = \begin{pmatrix} \frac{\partial \omega_l}{\partial t} \\ \frac{\partial \omega_r}{\partial t} \end{pmatrix},
\tag{2}
$$

where R is the radius of the driving wheels in the tracks; ω_l and ω_r denote the angular velocity (rad·s^{-1}) of left tracks and right tracks; $\dot{\omega}_l$ and $\dot{\omega}_r$ respectively denote the angular acceleration of left tracks and right tracks (rad·s^{-2}), and l_B denotes the distance between the left and right tracks.

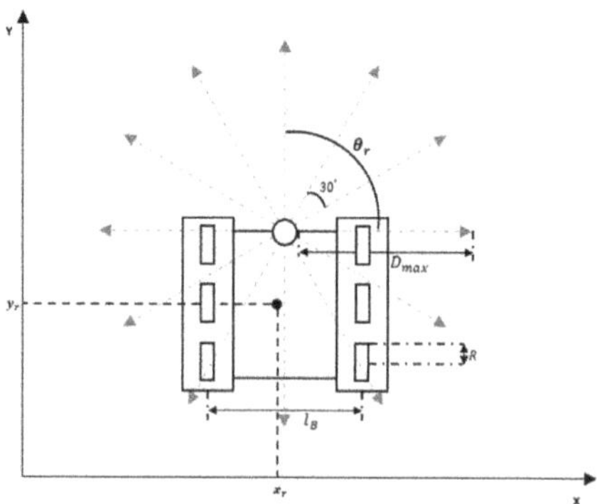

Figure 1. Schematic diagram of the tracked robot. Gray dash-dotted lines denote the senor lines of the robot.

3.2. The Architecture of the Proposed Algorithm

As mentioned above, JPS-IA3C resolves the robot navigation in a two-level manner: the first for the efficient global path planner, and the second for the robust local motion controller. Figure 2 shows the architecture of the proposed algorithm.

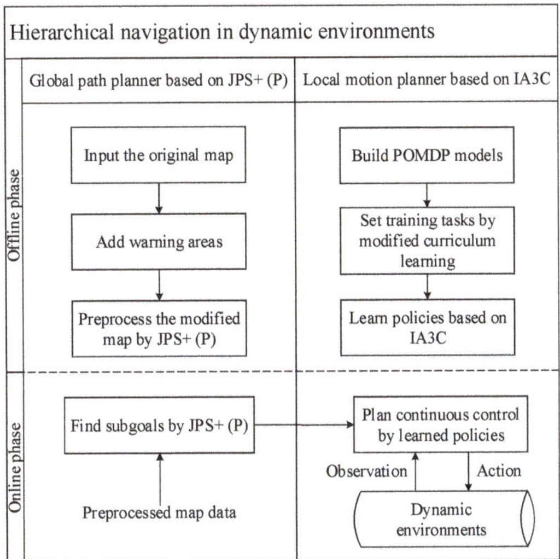

Figure 2. Flowchart of the Jump Point Search improved Asynchronous Advantage Actor-Critic (JPS-IA3C) method.

As shown in Figure 2, at the offline phase of the path planner whose goal is to efficiently find subgoals, we first define a warning circled area for every obstacle by taking the self-defined safety distance and sizes of robots into consideration. It is perilous for robots to touch these areas, since they are close to obstacles. Then, based on the modified map, JPS+ (P) calculates and stores the topology information of a map by preprocessing. In the online phase, JPS+ (P) uses canonical ordering known as diagonal-first to efficiently find subgoals based on preprocessed map data.

At the offline step of the motion controller whose task is planning feasible paths between adjacent subgoals, we firstly extract key information about robots and environments to build a partially observable Markov decision process (POMDP) model, which is a kind of formal description about partially observable environments for IA3C. Then, we quantify the difficulty of navigation and set training plans via a modified curriculum learning. Next, IA3C is used to learn navigation policies with high training efficiencies. At the online step, which is guided by subgoals, the robot receives sensor data about local environmental observations and then plans continuous control by learned policies. Next, the robot executes actions in dynamic environments and simultaneously updates its trajectories by kinematic equations. The online phase of the local motion planner denotes the interaction between robots and environments. Note that IA3C works on the original map, because warning areas cannot be detected by sensors.

There are two advantages of our method for robot navigation. The first advantage is that its high-level path planner can plans subgoals by preprocessing data and canonical ordering averagely within dozens of microseconds, so as to quickly respond to online tasks and avoid the so-called first-move lags [28]. Besides, it decomposes a task of long-range navigation into a number of local controlling phases between every two consecutive subgoals. Actually, traversing each of these segments, where there are no twisty terrains, rids our RL-driven controller of local minima. Therefore, benefitting from the path planner, the proposed algorithm can adapt to large-scale and complex environments.

The second advantage is that IA3C can learn near-optimal policies to navigate agents through adjacent subgoals in dynamic environments with kinematic and task constraints satisfied. There are two strengths of IA3C regarding learning to navigate. (1) To estimate complete environmental states, IA3C takes the time dimension into consideration by constructing an LSTM-based network architecture that can memorize information at different timescales. (2) Curriculum learning is useful for agents to gradually master complex skills, but conventional curriculum learning has been proven to be ineffective at training LSTM layers [29]. Therefore, we improve curriculum learning by adjusting distributions of different difficult samples over time and combine it with reward shaping [30] to build a novel reward function framework that can address the challenge of spare rewards. Moreover, due to the generalization of the local motion controller, the proposed algorithm can navigate robots to unseen environments without retraining.

3.3. Global Path Planner Based on JPS+ (P)

Figure 3a–c shows the offline work of the path planner, which includes constructing warning areas and identifying subgoals by preprocessing. Without the warning areas computed based on distance maps, the subgoals would be too close to the obstacles for moving robots, which increase the probabilities of colliding with walls.

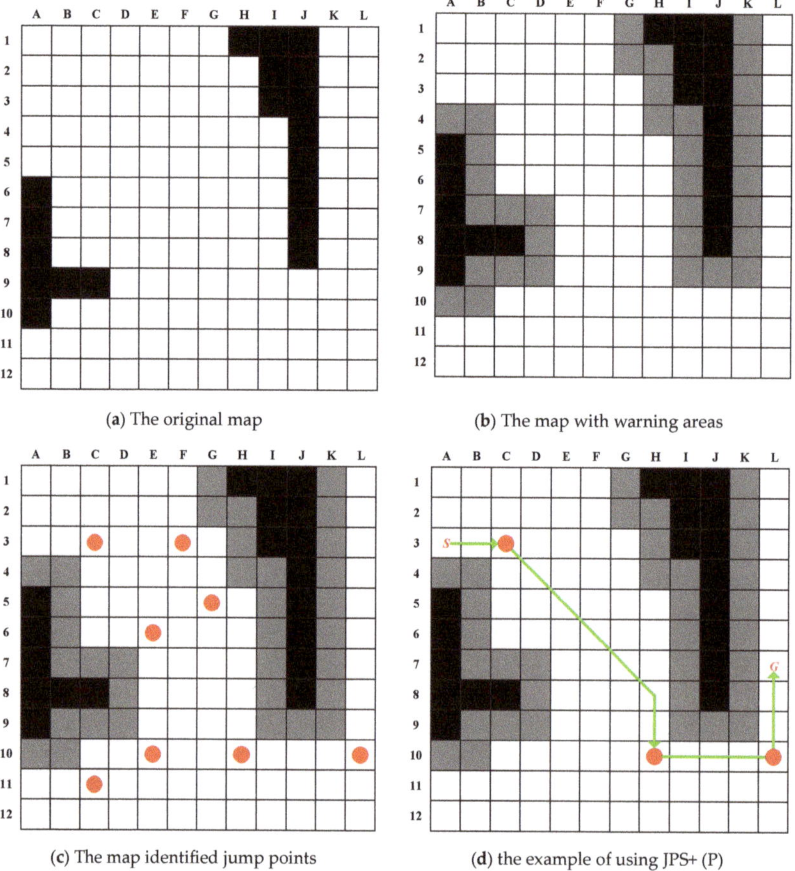

Figure 3. A graphical representation of how Jump Point Search+ (Prune) (JPS+ (P)) preprocesses an initially known grid-based map and what the paths that it plans look like. (**a**) An original map in which the white and black cells are unblocked and blocked locations, respectively. (**b**) Gray cells denote warning areas. (**c**) Jump points are marked as red. (**d**) Green lines denote an optimal path planned by JPS+ (P).

During the process of constructing warning areas, the dynamic brushfire algorithm [31] is used to build the Euclidean distance map (EDM) based on the original map. The minimum distance between robots and obstacles is limited by the physical radius of robots; thus, a user-defined safety distance is required to increase insurance for robots that carry out important tasks. Therefore, the size of warning areas is determined by the above two factors (Figure 3b).

In the JPS family, jump points are the intermediate points that are necessary to travel through for at least one optimal path [14]. To put it simply, JPS and JPS+ call locations where a cardinal search can branch itself to wider areas due to obstacle disappearance in the incoming direction straight jump points. For example, in Figure 3c, an eastern search from A3 wraps around obstacles at C3, and then broadens itself to the southeastern areas. Besides straight nodes, they identify diagonal jump points linking two straight ones where the incoming diagonal search from the first can turn cardinally to reach the second. JPS+ improves JPS through exploiting diagonal-first ordering in the offline phase, and storing for each traversable cell the distance to the nearest jump point or obstacle that can be reached in every cardinal or diagonal direction. More details can be found in [14,32].

This paper applies JPS+ (P), which is an enhanced version of JPS+ with an intermediate pruning trick; thereby, the optimal path in Figure 3d will not generate and expand the diagonal jump point H8. JPS+ (P) requires the same preprocessing overheads to JPS+, but has stronger online efficiency.

To sum up, JPS+ (P)—which we borrow from [32]—has obvious advantages over other high-level planners of hierarchical methods in the literature [24–26], in terms of its low precomputation costs and outstanding online performance. Moreover, it serves as an ideal for the location distribution of jump points, providing a DRL-based controller with meaningful subgoals that can completely throw the problem of local minima out of consideration.

3.4. Local Motion Controller Based on IA3C

3.4.1. Construction of Navigation POMDP

In [26], the problem of local navigation is decomposed into two subproblems (i.e., approaching targets and avoiding obstacles), which easily leads to local optimal policies due to simplifying the problem by adding prior experiences. In JPS-IA3C, to acquire optimal navigation policies, the motion controller directly builds models for the entire navigation problem.

According to [24], current observations are regarded as states, whose representation is non-Markovian in the proposed navigation problem, since current observations including sensor readings do not contain all of the states of dynamic environments, such as the velocities of obstacles. To tackle the above problems, a POMDP is built to describe the process of navigating robots toward targets without colliding obstacles. It can capture dynamics in environments by explicitly acknowledging that sensations received by agents are only partial glimpses of the underlying system state [13].

A POMDP is formally defined as a tuple of (S, A, P, R, O, Ω), where S, A, and O, are the state, action, and observation spaces, respectively. The state-transition function, $P(s, a, s')$, denotes the probability of transferring the current state s to next state s' by executing action a. The reward function, $R(s, a, s')$, denotes the immediate reward obtained by agents, when the current state s is transferred to the next state s' by taking action a. The observation function $\Omega(o, s, a)$ denotes the probability of receiving observation o after taking action a in state s [33].

In dynamic and partially observable environments, the robot cannot directly obtain exact environmental states. However, it can construct its own state representation about environments, which are called belief states Bs. There are three ways of constructing belief states, including the complete history, beliefs of the environment state, and recurrent neural networks. Owing to the powerful nonlinear function approximator, we adopt a typical recurrent neural network: LSTM. The equation of constructing Bs via LSTM is described as follows:

$$Bs_t^a = \sigma\left(w_s * Bs_{t-1}^a + w_o * o_t\right) \tag{3}$$

where $\sigma()$ denotes the activation function, w_s and w_o denote the weight parameters of neural networks, and o_t denotes the observation at time t.

The definitions of this POMDP are described in Table 1.

Table 1. Definitions of the Partially Observable Markov Decision Process (POMDP).

Observations	$(S_1, S_2, \ldots, S_{12}, d_g, a_g, \omega_l, \omega_r, \dot{\omega}_l, \dot{\omega}_r)$	
Actions	$\dot{\omega}_l \epsilon [-0.5, 0.5]$ $\dot{\omega}_r \epsilon [-0.5, 0.5]$	
Reward	+10	if $d_g < d_k$
	−5	if $S_{min} < d_u$
	0	else

In this POMDP, observation S_i ($i \in (1, 2, 3 \ldots, 12)$) denotes the reading of sensor i. The range of S_i is [0, 5]. Observation d_g denotes the normalization value of the distance between the robot and the target. The absolute value of observation a_g denotes the angle between the direction of the robot toward the goal and the robot's orientation. The range of a_g is $[-\pi, \pi]$. When the goal is on the left side of the robot's direction, the value of a_g is positive; otherwise, it is negative. To support the navigation ability of our approach, the observation space of the robot is expanded to include the last angular accelerations and angular velocities. Observations $\omega_l, \omega_r, \dot{\omega}_l$, and $\dot{\omega}_r$ are described in Section 3.1. The ranges of ω_l and ω_r are both $[-0.5, 0.5]$.

The action space is continuous, including the angular accelerations of the left and right tracks, so that the robot has great potential to perform flexible and complex actions. Besides, taking continuous angular accelerations as actions, trajectories updated by kinematic equations can preserve G2 continuity [34].

In the reward function, d_k denotes the goal tolerance. $S_{min} = min(S_1, S_2, \ldots, S_{12})$. d_u denotes the brake distance for the robot. When the distance between the robot and the subgoal is less than d_k, the robot gets a +10 scalar reward. When the robot collides with obstacles, it gets a −5 scalar reward. Otherwise, the reward is 0. However, the reward function is not suitable for this POMDP, since fewer samples with positive rewards are collected during the initial learning phase, which incurs difficult training and even non-convergence in complex environments. Some improvements in reward functions will be discussed in Section 3.4.3.

3.4.2. A3C Method

In DRL, there are some key definitions [30]. The value function $V(s)$ denotes the expected return for the agent in a given state s. The policy function $\Pi(s, a)$ denotes the probability of an action executed by the agent in a given state s.

Compared with valued-based DRL methods such as Deep Q Network (DQN) and its variations, A3C as a representative policy-based method can deal with continuous state and action spaces, and be more time-efficient. A3C is an on-policy and online algorithm that can use samples generated by current policies to efficiently update network parameters and increase robustness [12].

As shown in Figure 4, A3C asynchronously parallel executes multiple agents on multiple instances of the environment. All of the agents have the same local network architecture as the global network that is used to approximate policy functions and value functions. After the agent collects certain samples to calculate the gradient of its network, it uploads this gradient to the global network. Then, the agent downloads the global network to substitute its current network, after the global network updates its parameters by the uploaded gradient. Note that agents cannot upload their gradient simultaneously.

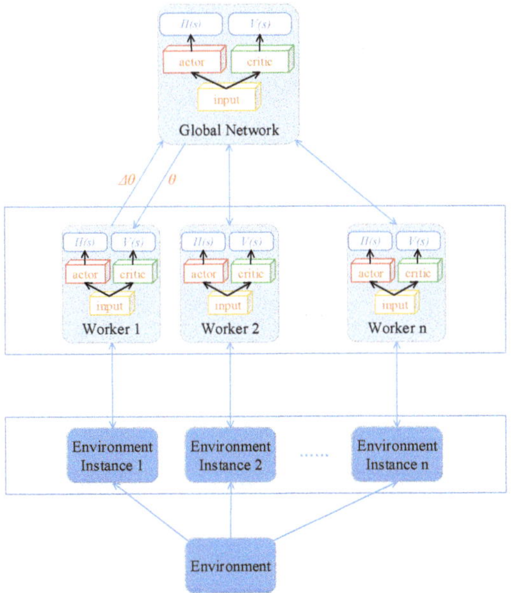

Figure 4. The flowchart of Asynchronous Advantage Actor-Critic (A3C).

It is difficult for the predecessor of A3C, Actor-Critic (AC), to converge, since consecutive sequences of samples used to update networks have strong temporal relationships. To solve this problem, A3C adopts an asynchronous framework to construct independent and identically distributed training data. Compared with Deep Deterministic Policy Gradient (DDPG) [8], which uses experience replay to reduce the correlation between training samples, A3C can better decorrelate the agents' data by using asynchronous actor-learners interacting with environments in parallel.

A3C includes two neural networks, actor-network and critic-network. Actor-network is used to approximate the policy function (i.e., $\Pi(a_t|s_t;\theta_a)$); it outputs the expectation and variance of the action distribution in continuous action spaces, and then samples an action from the built distribution. Critic-network used to approximate the value function (i.e., $V(s_t;\theta_v)$); it outputs the estimation of the value function of the current state to optimize probabilities of choosing actions. As seen in Algorithm 1 [12], each actor-learner can share experiences through communicating with the global network, which can greatly increase the training efficiency. The condition of updating is every t_{max} step or reaching the end of episode. The gradient of the actor-network is:

$$\nabla_{\theta_a} log\Pi(a_t|s_t;\theta_a)A(s_t,a_t;\theta_v) + \beta\nabla_{\theta_a}H(\Pi(s_t;\theta_a)) \tag{4}$$

where H is the entropy of the policy Π, and the hyper-parameter β controls the strength of the entropy regularization term. $A(s_t,a_t;\theta_v)$ is an estimate of the advantage function given by:

$$\sum_{i=0}^{k-1}\gamma^i r_{t+i} + \gamma^k V(s_{t+k};\theta_v) - V(s_t;\theta_v) \tag{5}$$

where k can vary from state to state, and is upper-bounded by t_{max} [12].

Algorithm 1. Pseudo-code of each actor-learner in Asynchronous Advantage Actor-Critic (A3C).

1: // θ_a, θ_v: global shared parameters of actor-network and critic-network
2: // θ_a', θ_v': thread-specific parameters of actor-network and critic-network
3: // t: thread step counter
4: // t_{max}: max value for t
5: // T: global shared counter
6: // T_{max}: max value for T
7: // env: an instance of the environment
8: $T \leftarrow 0$
9: Repeat
10: $\theta_a' \leftarrow \theta_a$
11: $\theta_v' \leftarrow \theta_v$
12: $t \leftarrow 0$
13: $s_t = $ env.reset()
14: While True
15: $a_t = \Pi(a_t|s_t; \theta_a')$
16: $r_t, s_{t+1} = $ env.step(a_t)
17: $t \leftarrow t + 1$
18: $T \leftarrow T + 1$
19: If terminal s_t or $t \% t_{max} = 0$
20: upload the gradients of θ_a', θ_v' to θ_a, θ_v
21: update θ_a, θ_v
22: download the parameters of θ_a, θ_v to θ_a', θ_v'
23: If terminal s_t
24: break
25: until $(T > T_{max})$

3.4.3. IA3C Method

Although A3C can be successfully applied in Atari Game and robot manipulation [12], it cannot be directly used in this navigation scenario due to the mentioned challenges. In Section 3.3, we discussed the challenge of local minima solved by the high-level path planner in JPS-IA3C. In this section, IA3C is proposed to address other two challenges, which are described as follows.

Firstly, an LSTM-based network structure can solve the problem of incomplete and noisy state information caused by partial observability. The original A3C is designed to solve MDPs that cannot precisely describe partially observable environments, but POMDPs can. In the proposed POMDP, the robot can only get range data about nearby environments by laser sensors, while velocities of obstacles as part of state information are unknown for the robot. The LSTM-based network architecture is constructed to solve the built POMDP, which takes current observation and abstract history information as input to estimate complete states. Since LSTM can store abstract information at different time scales, the proposed network can add the time dimension to infer hidden states. As shown in Figure 5, the observation o_t is taken as the input of the network. The first hidden layer is a fully-connected (FC) layer with 215 nodes, and the second is an FC layer with 32 nodes. The last hidden layer is an LSTM layer with 32 nodes. The output layer outputs the expectation and variance of the Gaussian distributions of action spaces. Instead of directly outputting actions, the way of building distribution can introduce more randomness to increase the diversities of samples.

In the network architecture, FC layers can abstract key features from current observations. Then, an LSTM layer with recurrence mechanism takes current abstracted features and the last output of it as input, and then extracts high-level features based on complete state information. The last output layer is an FC layer that is used to determine the action distribution. In conclusion, the learning procedure of this network structure is first to abstract observation features, secondly to estimate complete state features, and finally to make decisions, which is similar to the process of memory and inference in human beings.

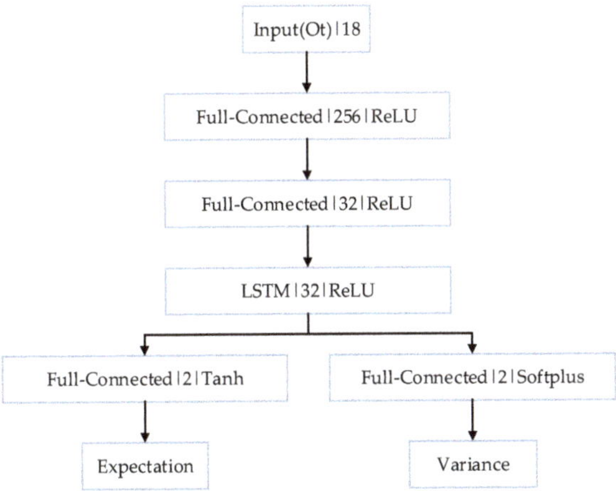

Figure 5. The network structure of IA3C.

Secondly, to solve the problem of sparse reward, we make another improvement on A3C by building a novel reward function framework that integrates a modified curriculum learning and reward shaping. In the task of navigation, the robot only gets a positive reward when it reaches the goal, which means that positive rewards are sparsely distributed in large-scale environments. Curriculum learning can be used to speed up the process of training networks by gradually increasing the samples' complexity in supervised learning [35]. In DRL, it can be used to tackle the challenge of sparse reward. Since original curriculum learning is not appropriate to train an LSTM-based network, we improve it by dividing whole learning tasks into two parts. The first part of learning tasks' complexity is gradually increased, and the second part's complexity is randomly chosen and not less than that of the first part. The navigation task is parameterized by distance (i.e., the initial distance between the goal and the robot) and the number of moving obstacles. Four kinds of learning tasks are designed through considering suitable difficult gradients, and each learning task has its own learning episodes (Table 2). Reward shaping is also a kind of useful way to reduce the side effects of the sparse reward problem. As shown in Table 3, the reward shaping means giving the robot early rewards before the end of the episode, but the early rewards must be much smaller than the reward received at the end of the episode [30].

Table 2. Learning plan of modified curriculum learning.

Number of Obstacles	Distance	Training Episodes
1	3	5000
2	5	7000
3	9	8000
3	13	9000

Table 3. Reward shaping.

	$+10$	if $d_g < d_k$
Reward	-5	if $S_{min} < d_u$
	$p_r\left(d_{g_{t-1}} - d_{g_t}\right)$	else

Where p_r denotes a hyper-parameter used to control the effect of early rewards, and d_{g_t} denotes d_g at time t.

In summary, the above-mentioned reward function framework can increase the density of reward distribution in dynamic environments and improve the training efficiency to achieve better performance.

4. Experiments and Results

JPS-IA3C combines a geometrical path planner and a learning-based motion planner to navigate robots in dynamic environments. To verify the performances of the path planner, we evaluate JPS+ (P)'s ability to find subgoals on complex maps, and then compare it with the ability of A*. Then, to evaluate the effectiveness of the motion planner, we validate the capabilities of the LSTM-based network architecture and the novel reward function framework in the training process. Finally, the performance of JPS-IA3C will be evaluated in large-scale and dynamic environments. To acquire the map data of tested environments, we firstly constructed grayscale maps via real data from laser sensors, and then transform them to grid maps (if a cell is occupied by an obstacle, this cell is set to be untraversable; otherwise, it is traversable). The distance measurement accuracy can be improved by optimization methods [36]. The dynamics of environments are caused by moving obstacles that imitate the simple behaviors of people (i.e., linear motion with constant speed).

For simulation settings, we adopt numerical methods to compute the robot's position via kinematic equations. Although smaller steps can achieve more precise results, they also lead to more computational consumptions and slower computation speed. To acquire high efficiencies, we select the step of simulation time as 0.1 s.

The discounted factor γ is 0.99. The learning rates of the actor-network and critic-network are 0.00006 and 0.0004, respectively. The unrolling step is 7. The subgoal tolerance is 1.5, and the final goal tolerance is 1.0. Since subgoals roughly guide the robot, the subgoal tolerance is relatively large. In the training, we set the time interval of action choice to 2 s, while the time interval decreases to 0.5 s in the evaluation.

We run experiments on a 3.4-GHz Intel Core i7-6700 CPU with 16 GB of RAM.

4.1. Training Environment Settings

The training environment is a simple indoor environment whose size is 20 by 20 (Figure 6). A start point, a goal point, and the motion states of moving obstacles are randomly initialized in every episode so as to cover the entire state space as soon as possible. Moving obstacles denote people, robots, and other moving objects, which adopt the uniform rectilinear motion. The range of moving obstacles' velocities is [0.08, 0.1]. The initial position is at the edge of the maximum detection distance of the robot, and the initial direction is roughly toward the robot. When the robot arrives at the goal or collides with obstacles, the current episode ends, and the next episode starts.

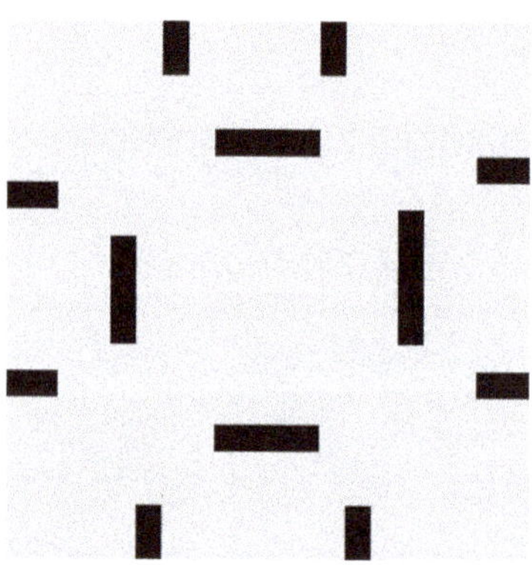

Figure 6. Training environment—20 by 20.

4.2. Evaluation of Global Path Planning

In Figure 7a, subgoals planned by A* are not placed at the exit of local minima due to ignoring the map topology, such as for example the local minimum caused by the concave canyon between subgoal 1 and subgoal 2. In Figure 7b, the number of subgoals of A* is over two times that of JPS-IA3C. Although the subgoals generated by A* cover all of the exits of local minima, they also include many unnecessary subgoals, which will hinder the ability to avoid moving obstacles. In conclusion, the subgoals planned by JPS-IA3C are essential waypoints that can help the motion controller handle the problem of local minima and simultaneously give full play to capabilities of the motion controller.

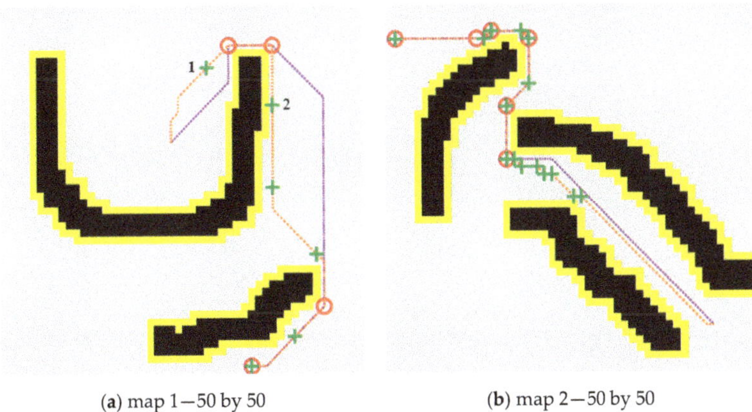

(a) map 1—50 by 50   (b) map 2—50 by 50

Figure 7. (a–b) Orange lines denote paths planned by A*, and purple lines denote paths generated by the path planner in JPS-IA3C. Small red circles denote subgoals planned by JPS-IA3C. Yellow regions denote warning areas. (a) A* find subgoals (i.e., green crosses) by sampling optimal paths at fixed intervals. (b) A* plans subgoals (i.e., green crosses) through taking the inflection points of optimal paths.

To avoid randomness, A* and JPS-IA3C are tested 100 times on large-scale maps (Figure 8), and the inflection points of optimal paths are regarded as subgoals of A*. First-move lags of A* include the time of planning optimal paths and finding subgoals by checking inflection points, while JPS-IA3C can directly finding subgoals without generating optimal paths. In Table 4, the first-move lags of A* are 271, 1309, and 860 times those of JPS-IA3C on these tested environments. Obviously, compared with A*, JPS-IA3C can more efficiently generate subgoals by the cached information about map topology and the canonical ordering way of search in JPS+ (P). Besides, the first-move lag of JPS-IA3C is below 1 millisecond on large-scale environments, which proves that the proposed approach can plan subgoals at low time costs to provide users with excellent experiences.

Table 4. First-Move Lag of A* and JPS-IA3C.

Method	Environment 1 (ms)	Environment 2 (ms)	Environment 3 (ms)
JPS-IA3C	0.097	0.134	0.230
A*	26.287	175.36	197.71

The first-move lag denotes the amount of planning time the robot takes before deciding on the first move (i.e., time of finding subgoals).

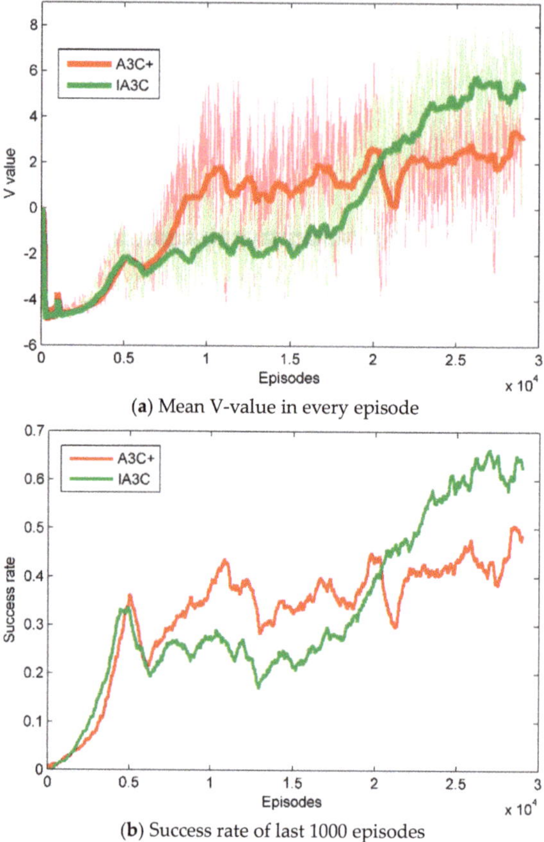

(**a**) Mean V-value in every episode

(**b**) Success rate of last 1000 episodes

Figure 8. The performances of IA3C and A3C+ in the training phase.

4.3. Evaluation of Local Motion Controlling

To evaluate the LSTM-based network, we compare IA3C with A3C+, which uses an FC layer to substitute an LSTM layer of IA3C's network in the training process. State values denote the outputs of the critic-network, which can reflect the performances of learned policies [7]. Success rates are probabilities of accomplishing training tasks in the last 1000 episodes.

In Figure 8a, when the episode is less than 20,000, A3C+ performs better than IA3C. It demonstrates that A3C+ can learn reactive policies to deal with simple learning tasks that only require a robot to navigate for short distances while avoiding a few moving obstacles, and the learning efficiency of LSTM-based networks is slower than that of FC-based networks. When the complexity of learning tasks increases further (i.e., the episode increases to 20,000), the performance of IA3C is better than that of A3C+. It proves that the reactive policies generated by A3C+ only consider current observations, while IA3C can integrate current observations and abstract history information to learn more rational policies. Besides, modified curriculum learning can transfer experiences from learned tasks to new complex tasks, which is more suitable to accelerate the training efficiencies of LSTM-based networks. In Figure 8b, the data of success rates presents the same trend as that of the mean V-value, which further demonstrates the former conclusion.

To evaluate the effectiveness of the novel reward function framework, we remove individual reward shaping or single modified curriculum learning from the reward framework, but there is no way to converge. It shows that the proposed reward function framework can solve the problem of sparse reward and be beneficial for learning useful policies.

4.4. Performance on Large-Scale Navigation Tasks

Finally, we evaluate the performance of JPS-IA3C on large-scale navigation tasks. JPS-A3C+ replaces IA3C with A3C+ in the proposed method. As shown in Figure 9, JPS-IA3C can navigate the nonholonomic robot with continuous control by dealing with both complex static obstacles and moving obstacles. Furthermore, navigation policies learned by JPS-IA3C have the notable generalization ability and can handle unseen tested environments that are almost three orders of magnitude larger than the training environment (Figure 6). We do not show trajectories generated by JPS-A3C+ for clear representation.

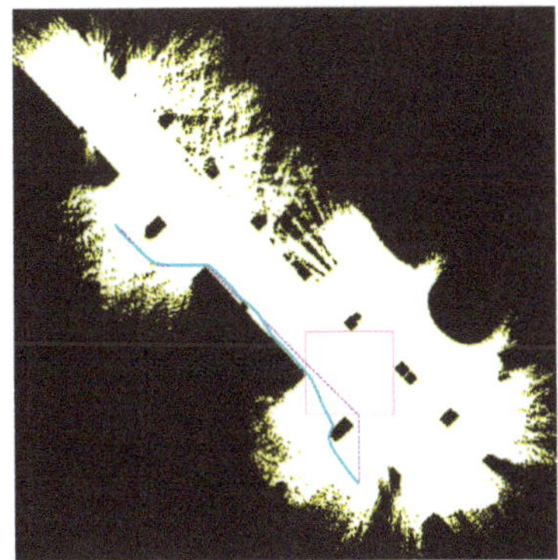

(a) Environment 1—722 by 722

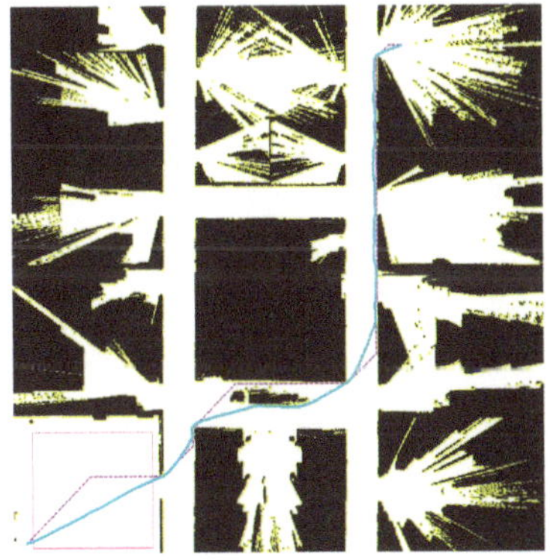

(b) Environment 2—624 by 624

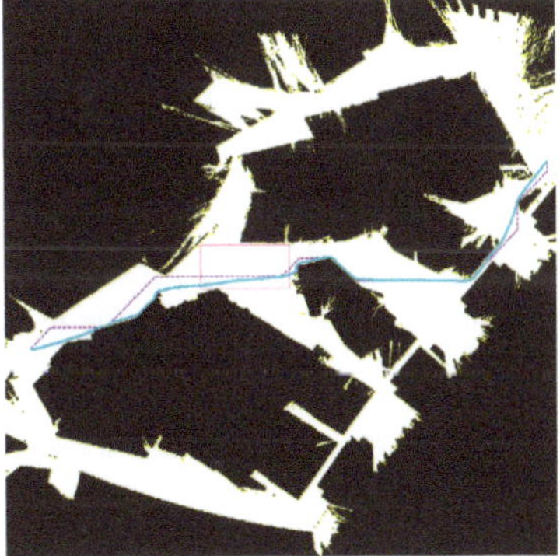

(c) Environment 3—1084 by 1084

Figure 9. (a–c) Given the start position and the goal position, the purple line is the path planned by the Jump Point Search+ (Prune) (JPS+ (P)), and the blue line is the trajectory generated by JPS-IA3C. Black regions are unpassable, while light white regions are traversable for robots. Yellow regions close to obstacles denote warning areas. Moving obstacles appear in pink square areas, where there are 30 random moving obstacles.

In Figure 10a, JPS-IA3C can achieve success rates of more than 94% in all of the tested environments and even 99.8% in environment 3, while the success rates of JPS-A3C+ are below 42% in all of the evaluations and even 0% in environment 2. Guided by the same subgoals, JPS-A3C+ only considers current observations and ignores hidden states (i.e., moving obstacles' velocities and changes of map topology) from history observations, leading to collisions with complex static obstacles or moving obstacles. However, JPS-IA3C can use the LSTM-based network to construct accurate belief states by storing historical information over a certain length, which can handle environmental changes. In Figure 10b,c, compared with JPS-A3C+, JPS-IA3C can navigate robots in large-scale environments with shorter path lengths and less execution time, which concludes that constructing the LSTM-based network architecture can greatly improve the performance of the proposed algorithm by endowing the robot with certain memory.

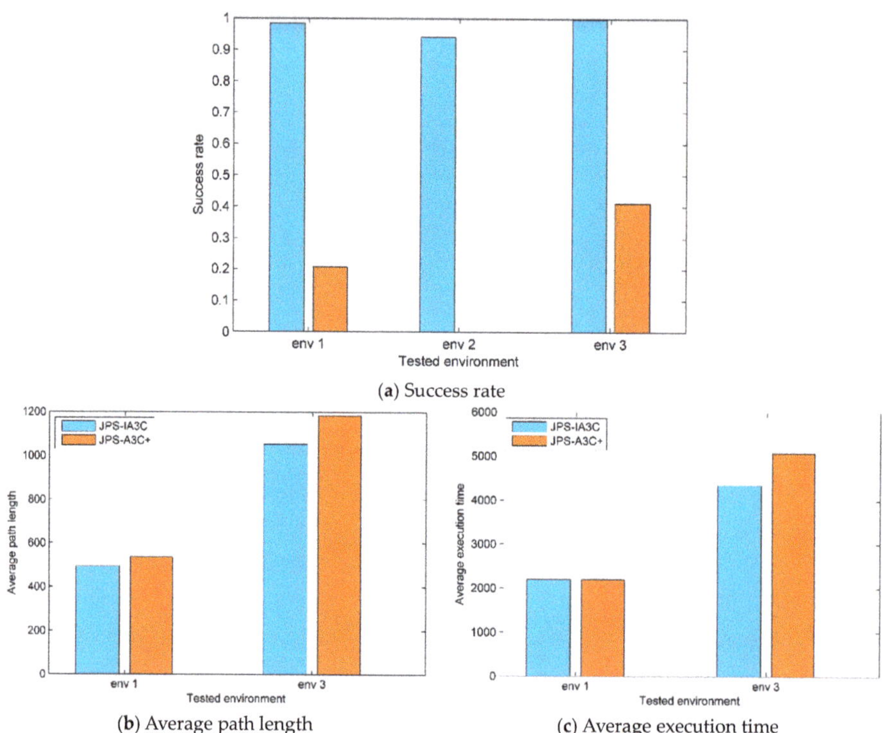

(**a**) Success rate

(**b**) Average path length

(**c**) Average execution time

Figure 10. We generate 500 navigation tasks in each tested environment (in sum 1500) whose starting and goal points are randomly chosen. The distance between the initial starting and goal position is greater than half of the width of the environment. (**a**) Success rate denotes the probability of successfully finishing tasks. (**b**) Path length denotes the traveling length between the starting and goal position. (**c**) Execution time denotes the total execution time for the motion controller.

5. Conclusions

In this paper, we proposed a novel hierarchical navigation approach to solving the problem of navigation in dynamic environments. It integrates a high-level path planner based on JPS+ (P) and a low-level motion controller based on an improved A3C (IA3C). JPS+ (P) can efficiently plan subgoals for the motion controller, which can eliminate first-move lag and address the challenge of the local minima trap. To meet the kinetic constraints and deal with moving obstacles, IA3C can learn near-optimal control policies to plan feasible trajectories between adjacent subgoals via continuous control. IA3C combines a modified curriculum learning and reward shaping to build a novel reward function framework, which can avoid learning inefficiency because of sparse reward. We additionally strengthen robots' temporal reasoning about the environments by a memory-based network. These improvements make an IA3C controller converge faster and more adaptive to incomplete, noisy information caused by partial observability.

In simulation experiments, compared with A*, JPS-IA3C can plan more essential subgoals placed at the exit of local minima. Besides, JPS-IA3C's first-move lag is between 271–1309 times shorter than A*'s on large-scale maps. Compared with A3C+ that remove IA3C's memory-based network, IA3C can integrate current observation and abstract history information to achieve higher success rates and mean V-values in the training process. Finally, comparing JPS-A3C+, JPS-IA3C can navigate robots in unseen and large-scale environments with shorter path lengths and less execution time, with more than a 94% success rate.

In future work, we consider the more complex behaviors of moving obstacles in the real world. For example, in large shopping malls, people with different intentions can generate more complicated trajectories that cannot be predicted easily. The above problems can be solved by adding an external memory network [37] and feedback control mechanisms [38] into the network architectures of DRL agents, which estimates state-action values more accurately.

Author Contributions: J.Z. and L.Q. conceived and designed the paper structure and the experiments; J.Z. performed the experiments; Y.H., Q.Y., and C.H. contributed materials and analysis tools.

Funding: This work was sponsored by the National Science Foundation of Hunan Province (NSFHP) project 2017JJ3371, China.

Acknowledgments: The work described in this paper was sponsored by the National Natural Science Foundation of Hunan Province under Grant No. 2017JJ3371. We appreciate the fruitful discussion with the Sim812 group: Qi Zhang and Yabing Zha.

References

1. Mohanan, M.G.; Salgoankar, A. A survey of robotic motion planning in dynamic environments. *Robot. Auton. Syst.* **2018**, *100*, 171–185. [CrossRef]
2. Mercorelli, P. Using Fuzzy PD Controllers for Soft Motions in a Car-like Robot, Advances in Science. *Technol. Eng. Syst. J.* **2018**, *3*, 380–390.
3. Lavalle, S.M. Rapidly-exploring random trees: Progress and prospects. In Proceedings of the 4th International Workshop on Algorithmic Foundations of Robotics, Hanover, Germany, 16–18 March 2000.
4. Kavraki, L.E.; Švestka, P.; Latombe, J.C.; Overmars, M.H. Probabilistic roadmaps for path planning in high-dimensional configuration spaces. *IEEE Trans. Robot. Autom.* **1994**, *12*, 566–580. [CrossRef]
5. Van Den Berg, J.; Guy, S.J.; Lin, M.; Manocha, D. Reciprocal nbody collision avoidance. In *Robotics Research*; Springer: Berlin/Heidelberg, Germany, 2011; pp. 3–19.
6. Lecun, Y.; Bengio, Y.; Hinton, G. Deep learning. *Nature* **2015**, *521*, 436. [CrossRef] [PubMed]
7. Mnih, V.; Kavukcuoglu, K.; Silver, D.; Rusu, A.; Veness, J.; Bellemare, M.G.; Graves, A.; Riedmiller, M.; Fidjeland, A.K.; Ostrovski, G.; et al. Human-level control through deep reinforcement learning. *Nature* **2015**, *518*, 529. [CrossRef] [PubMed]
8. Lillicrap, T.P.; Hunt, J.; Pritzel, A.; Heess, N.; Erez, T.; Tassa, Y.; Silver, D.; Wierstra, D. Continuous control with deep reinforcement learning. *arXiv* **2016**, arXiv:1509.02971.
9. Mirowski, P.; Pascanu, R.; Viola, F.; Soyer, H.; Ballard, A.J.; Banino, A.; Denil, M.; Goroshin, R.; Sifre, L.; Kavukcuoglu, K.; et al. Learning to Navigate in Complex Environments. *arXiv* **2016**, arXiv:1611.03673. 2016.
10. Mirowski, P.; Grimes, M.; Malinowski, M.; Hermann, K.M.; Anderson, K.; Teplyashin, D.; Simonyan, K.; Zisserman, A.; Hadsell, R. Learning to Navigate in Cities Without a Map. *arXiv* **2018**, arXiv:1804.00168.
11. Zhu, Y.; Mottaghi, R.; Kolve, E.; Lim, J.J.; Gupta, A.; Fei-Fei, L.; Farhadi, A. Target-driven visual navigation in indoor scenes using deep reinforcement learning. In Proceedings of the International Conference on Robotics and Automation, Singapore, 29 May–3 June 2017; pp. 3357–3364.

12. Mnih, V.; Badia, A.P.; Mirza, M.; Graves, A.; Lillicrap, T.; Harley, T.; Silver, D.; Kavukcuoglu, K. Asynchronous methods for deep reinforcement learning. In Proceedings of the International Conference on Machine Learning, New York, NY, USA, 19–24 June 2016; pp. 1928–1937.

13. Hausknecht, M.J.; Stone, P. Deep Recurrent Q-Learning for Partially Observable MDPs. *arXiv* **2015**, arXiv:1507.06527.

14. Rabin, S.; Silva, F. JPS+ An Extreme A* Speed Optimization for Static Uniform Cost Grids. In *Game AI Pro*, 2nd ed.; Rabin, S., Ed.; CRC Press: Boca Raton, FL, USA, 2015; Volume 3, pp. 131–143, ISBN 9781482254792.

15. Otte, M.; Frazzoli, E. RRT-X: Real-time motion planning/replanning for environments with unpredictable obstacles. In Proceedings of the International Workshop on Algorithmic Foundations of Robotics (WAFR), Istanbul, Turkey, 3–5 August 2014.

16. Kallman, M.; Mataric, M.J. Motion planning using dynamic roadmaps. In Proceedings of the IEEE International Conference on Robotics and Automation (ICRA), New Orleans, LA, USA, 26 April–1 May 2004; pp. 4399–4404.

17. Chen, C.; Seff, A.; Kornhauser, A.; Xiao, J. Deepdriving: Learning affordance for direct perception in autonomous driving. In Proceedings of the IEEE International Conference on Computer Vision, Santiago, Chile, 7–13 December 2015; pp. 2722–2730.

18. Gao, W.; Hsu, D.; Lee, W.S.; Shen, S.; Subramanian, K. Intention-Net: Integrating Planning and Deep Learning for Goal-Directed Autonomous Navigation. *arXiv* **2017**, arXiv:1710.05627.

19. Pfeiffer, M.; Schaeuble, M.; Nieto, J.; Siegwart, R.; Cadena, C. From perception to decision: A data-driven approach to end-to end motion planning for autonomous ground robots. *arXiv* **2016**, arXiv:1609.07910.

20. Guo, Y.; Liu, Y.; Oerlemans, A.; Lao, S.; Wu, S.; Lew, M.S. Deep learning for visual understanding: A review. *Neurocomputing* **2016**, *187*, 27–48. [CrossRef]

21. Tai, L.; Zhang, J.; Liu, M.; Boedecker, J.; Burgard, W. A Survey of Deep Network Solutions for Learning Control in Robotics: From Reinforcement to Imitation. *arXiv* **2016**, arXiv:1612.07139.

22. Tai, L.; Paolo, G.; Liu, M. Virtual-to-real deep reinforcement learning: Continuous control of mobile robots for mapless navigation. In Proceedings of the 2017 IEEE/RSJ International Conference on Intelligent Robots and Systems (IROS), Vancouver, BC, Canada, 24–28 September 2017.

23. Chen, Y.F.; Everett, M.; Liu, M.; How, J.P. Socially aware motion planning with deep reinforcement learning. *arXiv* **2017**, arXiv:1703.08862.

24. Kato, Y.; Kamiyama, K.; Morioka, K. Autonomous robot navigation system with learning based on deep Q-network and topological maps. In Proceedings of the IEEE/SICE International Symposium on System Integration, Taipei, Taiwan, 11–14 December 2017; pp. 1040–1046.

25. Faust, A.; Oslund, K.; Ramirez, O.; Francis, A.; Tapia, L.; Fiser, M.; Davidson, J. PRM-RL: Long-range Robotic Navigation Tasks by Combining Reinforcement Learning and Sampling-Based Planning. In Proceedings of the International Conference on Robotics and Automation, Brisbane, QLD, Australia, 21–25 May 2018; pp. 5113–5120.

26. Zuo, L.; Guo, Q.; Xu, X.; Fu, H. A hierarchical path planning approach based on A* and least-squares policy iteration for mobile robots. *Neurocomputing* **2015**, *170*, 257–266. [CrossRef]

27. Canny, J. *The Complexity of Robot Motion Planning*; MIT Press: Cambridge, MA, USA, 1988.

28. Bulitko, V.; Lee, G. Learning in real-time search: A unifying framework. *J. Artif. Intell. Res.* **2006**, *25*, 119–157. [CrossRef]

29. Zaremba, W.; Sutskever, I. Learning to Execute. *arXiv* **2014**, arXiv:1410.4615.

30. Sutton, R.S.; Barto, A.G. *Reinforcement Learning: An Introduction*, 2nd ed.; MIT Press: Boston, MA, USA, 2018.

31. Lau, B.; Sprunk, C.; Burgard, W. Efficient grid-based spatial representations for robot navigation in dynamic environments. *Robot. Autom. Syst.* **2013**, *61*, 1116–1130. [CrossRef]

32. Harabor, D.; Grastien, A. Improving jump point search. In Proceedings of the International Conference on Automated Planning and Scheduling, Portsmouth, NH, USA, 21–26 June 2014; pp. 128–135.

33. Karkus, P.; Hsu, D.; Lee, W.S. QMDP-Net: Deep Learning for Planning under Partial Observability. *arXiv* **2017**, arXiv:1703.06692.

34. Ravankar, A.; Ravankar, A.A.; Kobayashi, Y.; Takanori, E. SHP: Smooth Hypocycloidal Paths with Collision-Free and Decoupled Multi-Robot Path Planning. *Int. J. Adv. Robot. Syst.* **2016**, *13*, 133. [CrossRef]

35. Bengio, Y.; Louradour, J.; Collobert, R.; Weston, J. Curriculum learning. In Proceedings of the 26th Annual International Conference on Machine Learning, ICML 2009, Montreal, QC, Canada, 14–18 June 2009.

36. Garcia-Cruz, X.M.; Sergiyenko, O.Y.; Tyrsa, V.; Rivas-Lopez, M.; Hernandez-Balbuena, D.; Rodriguez-Quiñonez, J.C.; Basaca-Preciado, L.C.; Mercorelli, P. Optimization of 3D laser scanning speed by use of combined variable step. *Opt. Lasers Eng.* **2014**, *54*, 141–151. [CrossRef]
37. Parisotto, E.; Salakhutdinov, R. Neural Map: Structured Memory for Deep Reinforcement Learning. *arXiv* **2017**, arXiv:1702.08360.
38. Oh, J.; Chockalingam, V.; Singh, S.; Lee, H. Control of memory, active perception, and action in minecraft. *arXiv* **2016**, arXiv:1605.09128.

Article

Multi-Robot Path Planning Method Using Reinforcement Learning

Hyansu Bae [1], Gidong Kim [2], Jonguk Kim [3], Dianwei Qian [4] and Sukgyu Lee [1,*]

[1]   Department of Electrical Engineering Yeungnam University, Gyeongsan 38541, Korea
[2]   TycoAMP, Gyeongsan 38541, Korea
[3]   Korea Polytechnic VII, Changwon 51519, Korea
[4]   School of Control and Computer Engineering, North China Electric Power University, Beijing 102206, China
[*]   Correspondence: sglee@ynu.ac.kr; Tel.: +82-053-810-3923

Received: 29 June 2019; Accepted: 26 July 2019; Published: 29 July 2019

Abstract: This paper proposes a noble multi-robot path planning algorithm using Deep q learning combined with CNN (Convolution Neural Network) algorithm. In conventional path planning algorithms, robots need to search a comparatively wide area for navigation and move in a predesigned formation under a given environment. Each robot in the multi-robot system is inherently required to navigate independently with collaborating with other robots for efficient performance. In addition, the robot collaboration scheme is highly depends on the conditions of each robot, such as its position and velocity. However, the conventional method does not actively cope with variable situations since each robot has difficulty to recognize the moving robot around it as an obstacle or a cooperative robot. To compensate for these shortcomings, we apply Deep q learning to strengthen the learning algorithm combined with CNN algorithm, which is needed to analyze the situation efficiently. CNN analyzes the exact situation using image information on its environment and the robot navigates based on the situation analyzed through Deep q learning. The simulation results using the proposed algorithm shows the flexible and efficient movement of the robots comparing with conventional methods under various environments.

Keywords: reinforcement learning; multi-robots; cooperation; Deep q learning; Convolution Neural Network

---

1. Introduction

Machine learning, which is one of the fields of artificial intelligence and is a technology to implement functions such as human learning ability by computer, has been studied in various concessions mainly used in the field of signal processing and image processing [1–3], such as speech recognition [4,5], natural language processing [6–8], and medical treatment [9,10]. Since the performance using machine learning shows good results, it will play an important role in the 4th industrial revolution as is well known in Alpha machine learning [11]. Reinforced learning is the training of machine learning models to make a sequence of decisions so that a robot learns to achieve a goal in an uncertain, potentially complex environment by selecting the action to be performed according to the environment without an accurate system model. When learning data is not provided, some actions are taken to compensate the system for learning. Reinforcement learning, which includes actor critic [12–14] structure and q learning [15–21], has many applications such as scheduling, chess, and robot control based on image processing, path planning [22–29] and etc. Most of the existing studies using reinforcement learning exclusively the performance in simulation or games. In multi-robot control, reinforcement learning and genetic algorithms have some drawbacks that have to be compensated for. In contrast to the control of multiple motors in a single robot arm, reinforcement learning of the multiple robots for solving one task or multiple tasks is relatively inactive.

This paper deals with information and strategy around reinforcement learning for multi-robot navigation algorithm [30–33] where each robot can be considered as a dynamic obstacle [34,35] or cooperative robot depending on the situation. That is, each robot in the system can perform independent actions and simultaneously collaborate with each other depending on the given mission. After the selected action, the relationship with the target is evaluated, and rewards or penalty is given to each robot to learn.

The robot learns the next action based on the learned data when it selects the next action, and after several learning, it moves to the closest target. By interacting with the environment, robots exhibit new and complex behaviors rather than existing behaviors. The existing analytical methods suffer from adaptation to complex and dynamic systems and environments. By using Deep q learning [36–38] and CNN [39–41], reinforcement learning is performed on the basis of image, and the same data as the actual multi-robot is used to compare it with the existing algorithms.

In the proposed algorithm, the global image information in the multi-robot provides the robots with higher autonomy comparing with conventional robots. In this paper, we propose a noble method for a robot to move quickly to the target point by using reinforcement learning for path planning of a multi-robot system. In this case, reinforcement learning is a Deep q learning that can be used in a real mobile robot environment by sharing q parameters for each robot. In various 2D environments such as static and dynamic environment, the proposed algorithm spends less searching time than other path planning algorithms.

2. Reinforcement Learning

Reinforcement learning is the training of machine learning models to make a sequence of decisions through trial and error in a dynamic environment. The robots learn to achieve a goal in an uncertain, potentially complex environment through programming the object by reward or penalty.

Figure 1 shows a general reinforcement learning, a robot which selects the action a as output recognizes the environment and receives the state, s, of the environment. The state of the behavior is changed, and the state transition is delivered to the individual as a reinforcement signal called a reinforcement signal. The behavior of the individual robot is selected in such a way as to increase the sum of the enhancement signal values over a long period of time.

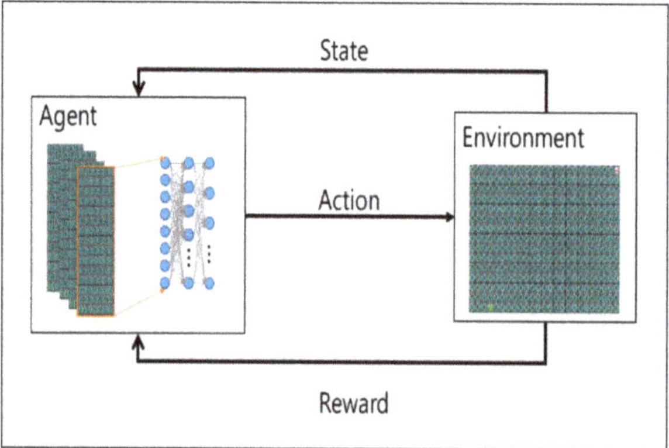

Figure 1. Q Learning concept.

One of the most distinguishable differences of reinforcement learning from supervised learning is that it does not use the representation of the link between input and output. After choosing an action, it is rewarded and knows the following situation. Another difference is that the online features of reinforcement learning are important since the computation of the system takes place at the same time as learning. One of the reinforcement learning features, Deep q learning, is characterized by the temporal difference method, which allows learning directly by experience and does not need to wait until the final result to calculate the value of each state like dynamic programming.

When a robot is moving in a discrete, restrictive environment, it chooses one of a set of definite behaviors at every time interval and assumes that it is in Markov state; the state changes to the probability s of different.

$$\Pr[s_{t+1}] = s'[s_t, a_t] = \Pr[a_t] \tag{1}$$

At every time interval t, a robot can get status s from the environment and then take action a_t. It receives a stochastic prize r, which depends on the state and behavior of the expected prize R_{st} to find the optimal policy that an entity wants to achieve.

$$R_{st}(a_i) = E\left\{ \sum_{i=0}^{\infty} Y^j r_{t+j} \right\} \tag{2}$$

The discount factor means that the rewards received at t time intervals are less affected than the rewards currently received. The operational value function Va is calculated using the policy function π and the policy value function Vp as shown in Equation (3). The state value function for the expected prize when starting from state s and following the policy is expressed by the following equation.

$$V_a(s_t) \equiv R_s(\pi(s_t)) + Y \sum_u Pxy[\pi(s_t)]V_p(s_t) \tag{3}$$

It is proved that there is at least one optimal policy as follows. The goal of Q-learning is to set an optimal policy without initial conditions. For the policy, define the Q value as follows.

$$Q_p(s_t, a_t) = R_s(a_t) + Y \sum_u Pxy[\pi(s_t)]V_p(y) \tag{4}$$

$Q(s_t, a_t)$ is the newly calculated $Q(s_{t-1}, a_{t-1})$, and $Q(s_{t-1}, a_{t-1})$ corresponding to the next state by the current $Q(s_{t-1}, a_{t-1})$ value and the current $Q(s_{t-1}, a_{t-1})$.

3. Proposed Algorithm

Figure 2 is the form of the proposed. The proposed algorithm uses empirical representation technique. The learning experience that occurs at each time step through multiple episodes to be stored in the dataset is called memory regeneration. The learning data samples are used for updating with a certain probability in the reconstructed memory each time. Data efficiency can be improved by reusing experience data and reducing correlations between samples.

Without treating each pixel independently, we use the CNN algorithm to understand the information in the images. The transformation layer transmits the feature information of the image to the neural network by considering the area of the image and maintaining the relationship between the objects on the screen. CNN extracts only feature information from image information. The reconstructed memory to store the experience basically stores the agent's experience and uses it randomly when learning neural networks. Through this process, which prevents learning about immediate behavior in the environment, the experience is sustained and updated. In addition, the goal value is used to calculate the loss of all actions during learning.

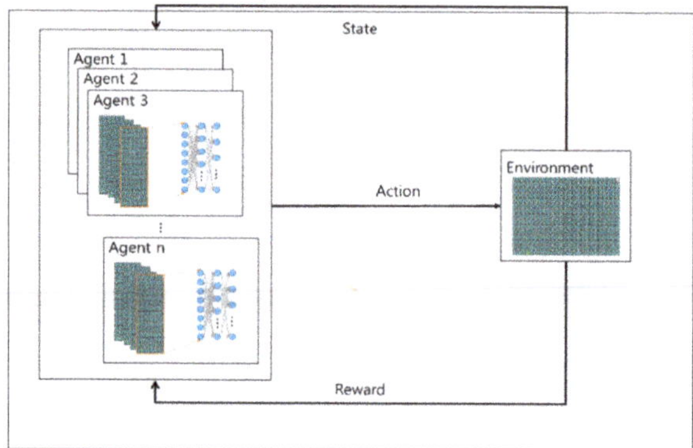

Figure 2. The concept of the proposed algorithm.

The proposed algorithm uses empirical data according to the roles assigned individually. In learning how to set goal value by dividing several situations, different expectation value for each role before starting to learn is set. For example, assuming that the path planning should take less time than the A* algorithm in any case and that it is a success factor for each episode, the proposed algorithm designates an arbitrary position and number of obstacles in the map of a given size and use the A*. This time is used as a compensation function of Deep q algorithm.

The agent learns so that the compensation value always increases. If the searching time increases more than the A* algorithm, the compensation value decreases and learning is performed so that the search time does not increase. Figure 3 shows the learning part of the overall system, which consists of preprocessing part for finding outlier in video using CNN and post-processing part for learning data using singular point.

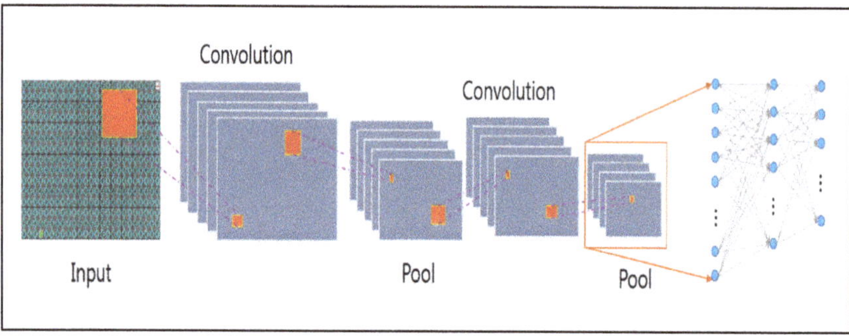

Figure 3. Simulation environment.

In the preprocessing part, the features of the image are searched using input images, and these features are collected and learned. In this case, the Q value is learned for each robot assigned to each role, but the CNN value has the same input with a different expected value. Therefore, the Q value is shared when learning, and used for the learning machine. In order to optimize the updating of Q value, it is necessary to define an objective function as defined as error of target value and prediction value of Q value. Equation (5) describes an objective function.

$$L = \frac{1}{2}[r + max_{a'}Q(s', a') - Q(s,a)]^2 \tag{5}$$

The basic information for obtaining a loss function is a transition <s, a, r, s'>. Therefore, first, the Q-network forward pass is performed using the state as an input to obtain an action-value value for all actions. After obtaining the environment return value <r, s'> for the action a, the action-value value for all action a' is obtained again using the state s'. Then, we get all the information to get the loss function, which updates the weight parameter so that the Q-value updates for the selected action converges; that is, as close as possible to the target value and the prediction value. Algorithm 1 is a function of compensation, which greatly increases compensation if the distance to the current target point is reduced before, and decreases the compensation if the distance becomes longer and closer.

Algorithm 1 Reward Function

---

if $distance_{t-1}$ > $distance_t$
then reward = 0.2
else
then reward = −0.2
if action == 0
then reward = reward + 0.1
else if action == 1 or action == 2
then reward = reward − 0.1
else if action == 3 or action == 4
then reward = reward − 0.2
if collision
then reward = −10
else if goal
then reward = 100
return reward

The two Q networks, such as target Q network and Q network, which is used mandatorily, are employed in the algorithm. Both networks are exactly the same with different weight parameters. In order to smooth convergence in DQN, the target network is updated not continuously, but periodically.

The adaptive learning rate method, RMSProp method, was used as the optimizer and the learning rate was adjusted according to the parameter gradient. This means that, in a situation where the training set is continuously changing, unlike the case of a certain training set, it is necessary to continuously change certain parameters.

The algorithm is simply stated as follows.

When entering the state s_t, randomly selects a behavior a that maximizes Q $(s_t, a; \theta_t)$.

Obtain state s_{n+1} and a complement r_t by action a.

Enter state s_{t+1} and find Max Q $(s_{t+1}, a; \theta)$.

Update the variable θ_t with the correct answer.

$$Yi = r_t + Y \, Max \, Q(s_{t+1}, a; \theta) \tag{6}$$

$$L(\theta_t) = 1/2 \, X \, (y_i - Q(s_t, a; \theta))^2 \tag{7}$$

$$Q_{i+1} = Qi - \alpha \nabla_{\theta i} L(\theta_i) \tag{8}$$

4. Experiment

4.1. Experiment Environment

Experiments were conducted on Intel Core i5-6500 3.20GHz, GPU GTX 1060 6GB, RAM 8GB, OS Ubuntu 16.04 and Window 7 Python and C++.

On a regular PC, we built a simulator using C++ and Python in a Linux environment. The robot has four random positions and its arrival position is always in the upper right corner. In a static and dynamic obstacle environment in the simulator, the number and location of obstacles are chosen randomly. Under each environment, the proposed algorithm is compared with A* and algorithm D* algorithm, respectively.

The simulation was conducted under the environment without considering the physical engine of the robot, assuming that the robot made ideal movements.

4.2. Learning Environment

The experiment result compares the proposed algorithm with the existing A* algorithm. The proposed algorithm learns based on the search time of the A* algorithm and assumes learning success when it reaches the target position more quickly than the A* algorithm. Figure 4 shows the simulation configuration diagram of the proposed algorithm. The environmental information image used by CNN is 2560 × 2000 and is designed with 3 × 3 filters with a total of 19 layers (16conv. and 3 FC layers) in the form of VGGNet [42].

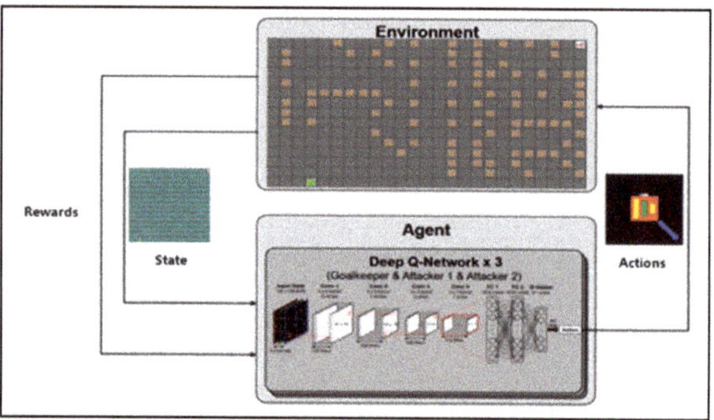

Figure 4. Simulation configuration diagram of proposed algorithm.

Figure 5 shows the score of the Q parameter when learning by episode. In the graph, the red line depicts the score of the proposed algorithm and the blue line is the score when learning Q parameters per mobile robot. The experiment result confirms that the score of the proposed algorithm is slightly slower at the beginning of learning but reaches the final goal score.

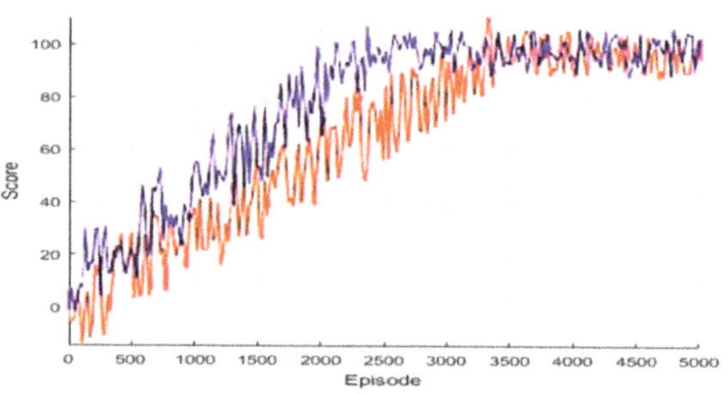

Figure 5. Score depending on algorithms according to episode.

The proposed algorithm, as in Figure 6, which shows the target arrival speed of the episode-based algorithm, has slower learning progression speed than the Q parameter of each model with a similar target position.

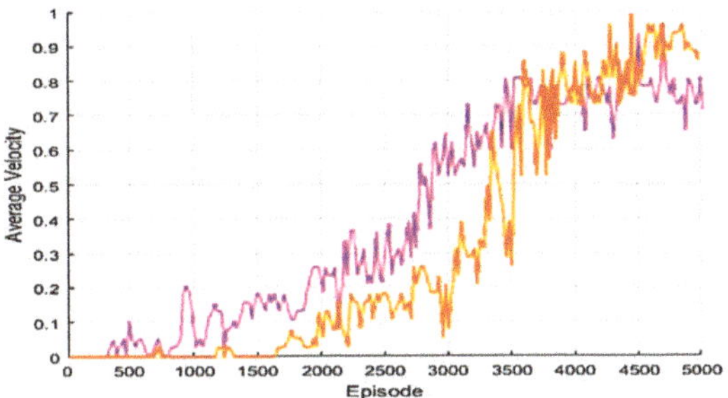

Figure 6. The rate at which episode goals are reached.

Figure 7 depicts that the average score of each generation increases gradually as learning progresses. As the learning progresses, it gradually gets better results.

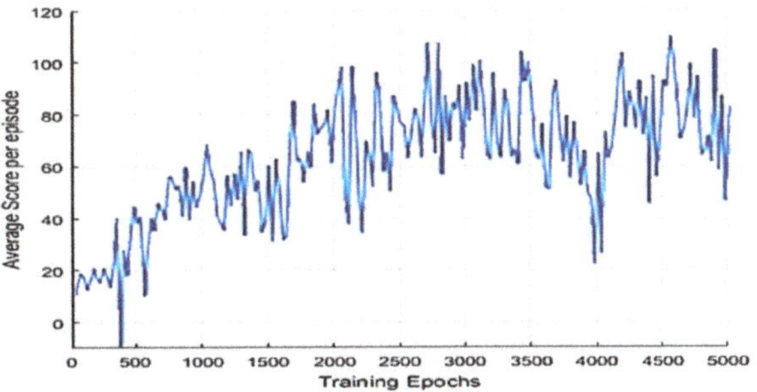

Figure 7. The rate of reaching the goal by Epochs.

4.3. Experimental Results in a Static Environment

In Section 4.3, the proposed algorithm is compared with the A* algorithm. Because the A* algorithm always finds the shortest path, it can see how quickly the proposed algorithm can find the shortest path. The proposed algorithm and the A* algorithm were tested by four robots to find a path with an arbitrary numbers of obstacles in random initial positions. The test was repeated 100 times on the generated map, which is changed 5000 times to obtain the route visited by the robot and the shortest path for each robot to reach the target point. In the first case, the search area and the time to reach the target point of the proposed algorithm is smaller than the A* algorithm, but the actual travel distance of the robot is increased, as shown in Figures 8 and 9. Figures 8 and 9 depict the results of the A* algorithm and the proposed algorithm, respectively. The average search range of A* algorithm is 222 node visitation. The shortest paths of robots 1 to 4 were 35, 38, 43, and 28 steps, respectively, and the proposed algorithm, where each robot path are 41, 39, 44, and 31 steps, had 137 visited nodes.

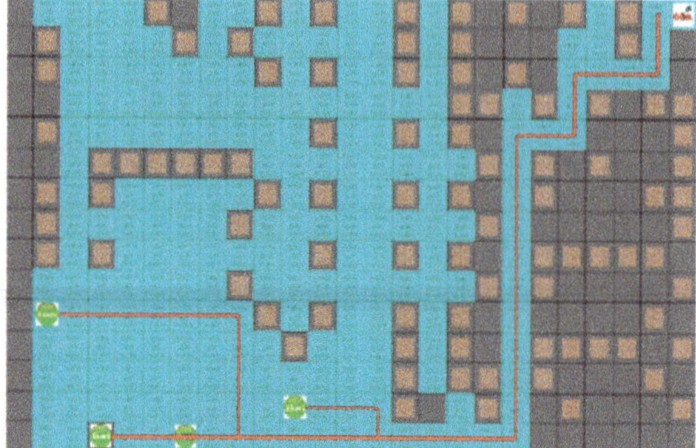

Figure 8. The simulation result of A* algorithm for the first situation.

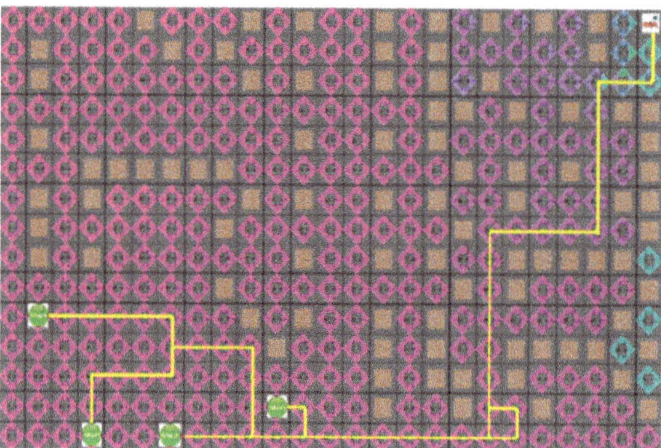

Figure 9. The simulation result of proposed algorithm for the first situation.

As shown in Figures 10 and 11, which are the result of the A* algorithm and the proposed algorithm, respectively for the second case, the search range of the proposed algorithm is similar to that of the A* algorithm. The average search range of A* algorithm is 80 node visitation. The shortest paths of robots 1 to 4 were 33, 29, 28, and 21 steps, and the proposed algorithm had 65 visited nodes, where each robot pathare 34, 36, 32, and 25 steps, respectively.

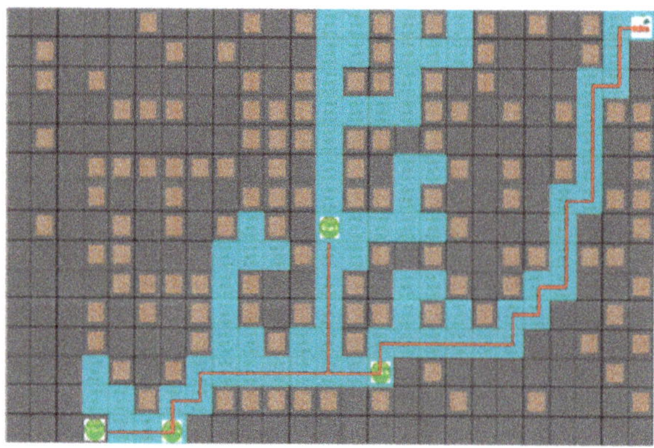

Figure 10. The simulation result of A* algorithm for the second situation.

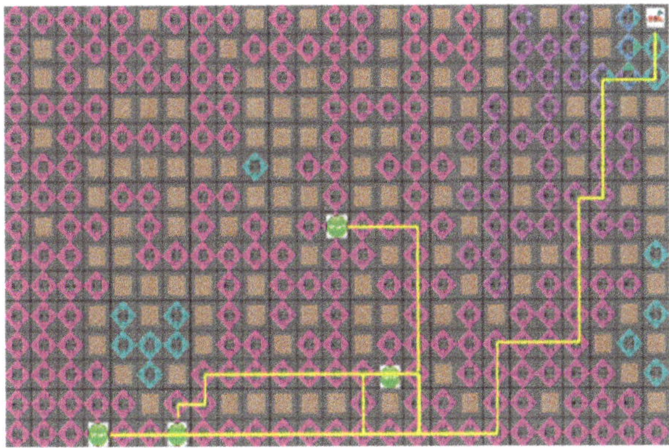

Figure 11. The simulation result of proposed algorithm for the second situation.

In the third case without obstacles in environment, as shown in Figures 12 and 13, the search range of two algorithms is the same, and the movement path of the robot is similar. The average search range of the A* algorithm was 66 node visitation. The shortest paths of robots 1 to 4 were 33, 29, 28, 21, and 66 steps; the proposed algorithm had 66 visited nodes, and each robot path was 34, 31, 28, and 22 steps, respectively.

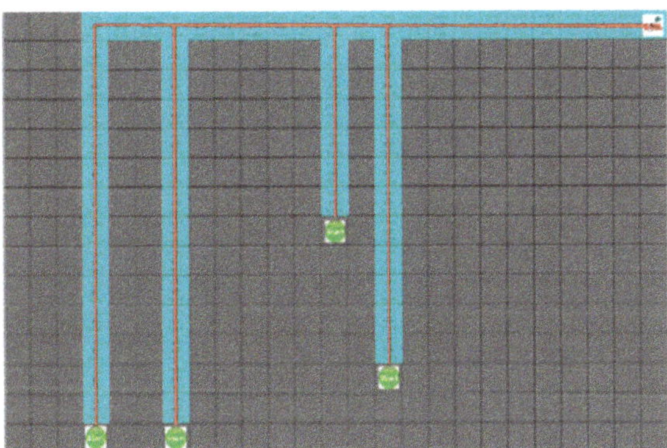

Figure 12. The simulation result of A* algorithm result for the third situation.

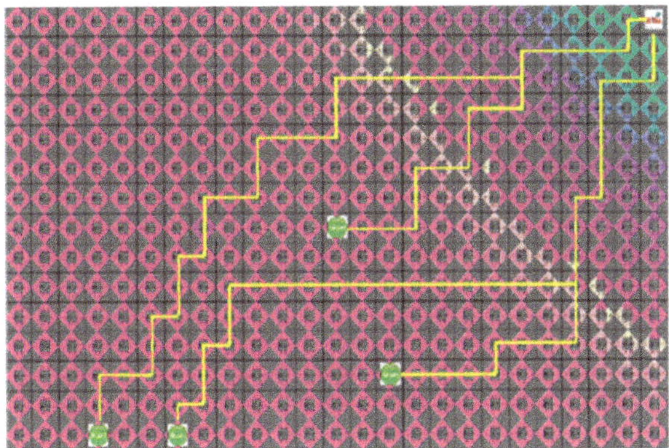

Figure 13. The simulation result of proposed algorithm for the third situation.

In the fourth case, the search range of both algorithms is similar, and the movement path of the robot is the same. Figures 14 and 15 show the results of the A* algorithm and the proposed algorithm, respectively. In the average search range of A* algorithm, robot visited 121 nodes, and the shortest paths of robots 1 to 4 were 40, 38, 36, and 23 steps respectively. And the proposed algorithm had 100 visited nodes, and the shortest path of each robot is 40, 39, 36, and 23 steps, respectively.

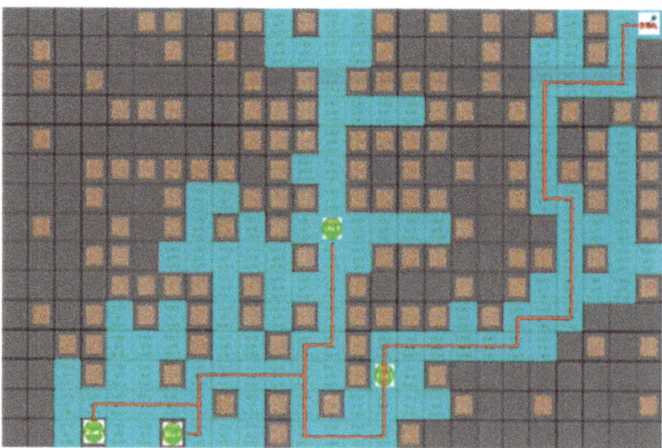

Figure 14. The simulation result of A* algorithm for the fourth situation.

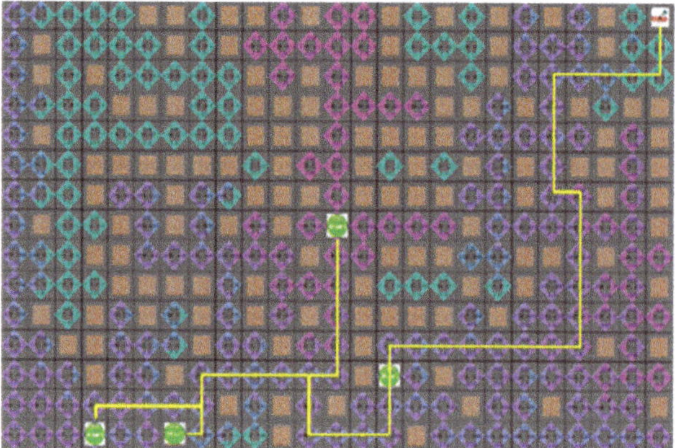

Figure 15. The simulation result of proposed algorithm for the fourth situation.

Table 1 shows the search range of the proposed algorithm and A* algorithm for each situation. The search range is one of the most important factors on effective robot navigation to the target point. As shown in Table 1, the proposed algorithm has a smaller search range than the A* algorithm.

Table 1. The average search range of A* and the proposed algorithm.

Situation	A* Algorithm	Proposed Algorithm
Situation 1	258	155
Situation 2	83	67
Situation 3	69	69
Situation 4	120	101

Table 2 shows the generated path for each robot according to the situation for both proposed and A* algorithm. The proposed algorithm results in no less number of steps than A* in an average generation. In general, A* algorithm is not always searching for the optimal path because it always searches the optimal path in a static environment. However, the proposed algorithm generates the path faster than A*.

Table 2. The robot average generation path of A* and the proposed algorithm.

Situation	A* Algorithm	Proposed Algorithm
Situation 1	35	36
Situation 2	28	30
Situation 3	30	31
Situation 4	32	32

Table 3 shows the frequency of occurrence of each situation when a total of 5000 experiments were conducted. Although the search area is small, the situation 1 where the robot generation path is long shows about 58% of occurrence, and the situation 2 where the search area and travel distance are similar is about 34%. According to the simulation result, the search range of the proposed method is always smaller and the robot generation path is the same or similar to the robot path generated by the A* algorithm with a probability of about 34.76%.

Table 3. Frequency of occurrence by situation.

Situation	Frequency of Occurrence
Situation 1	3262
Situation 2	1556
Situation 3	102
Situation 4	80

4.4. Experimental Results in a Dynamic Environment

Experiments based on the proposed algorithm in a dynamic environment is performed to compare the learning results of the D* algorithm. D* [43] and its variants have been widely used for mobile robot navigation because of its adaptation on dynamic environment. During the navigation of the robot, new information is used for a robot to replan a new shortest path from its current coordinates repeatedly. Under the 5000 different dynamic environments consisting of the smallized static environment as done in Section 4.3, and dynamic obstacles with random initial positions, the simulation was performed around 100 times per environment.

For the D* algorithm, if an obstacle is locates on the robot path, the path will be checked again after the path of movement. In the same situation, different paths may occur depending on the circumstances of the moving obstacle. Figure 16 shows that the yellow star is a dynamic obstacle that creates the shortest path, as shown in Figure 16, or the path to the target point rather than the shortest path, as shown in Figure 17. This shows that the situation of the moving obstacle affects the robot's path.

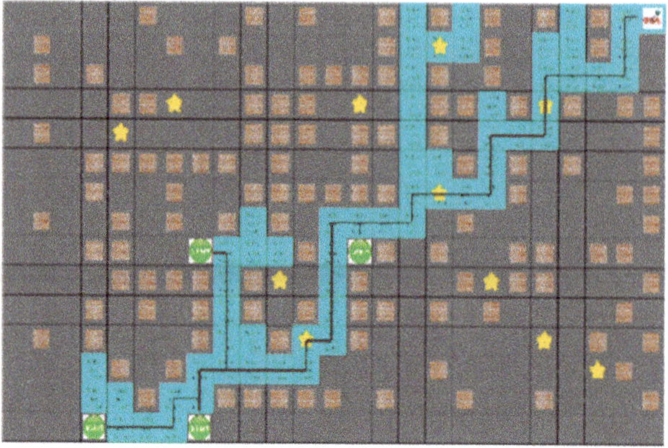

Figure 16. The shortest path generated without collision with dynamic obstacles using D*.

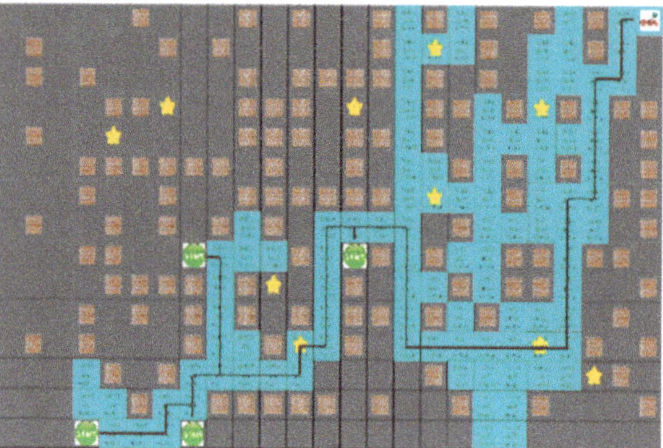

Figure 17. Path using D* in a mobile obstacle environment the route returned by an obstacle.

In the proposed algorithm under a dynamic environment where moving obstacles are located on the traveling path of the robot, they do not block the robot, as shown in Figure 18. It guarantees a short traveling distance of each robot.

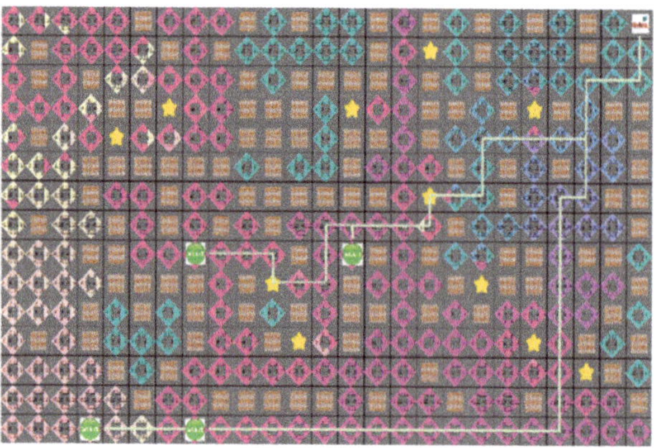

Figure 18. The generated path using proposed algorithm in dynamic environment.

As shown in Figure 19, the robot moves downward to avoid collision with obstacles on the path of the robot. In Figure 19, the extent to which the path created by D* and the proposed algorithm was not moved to the existing path by the mobile yellow star and explored to generate the bypass path.

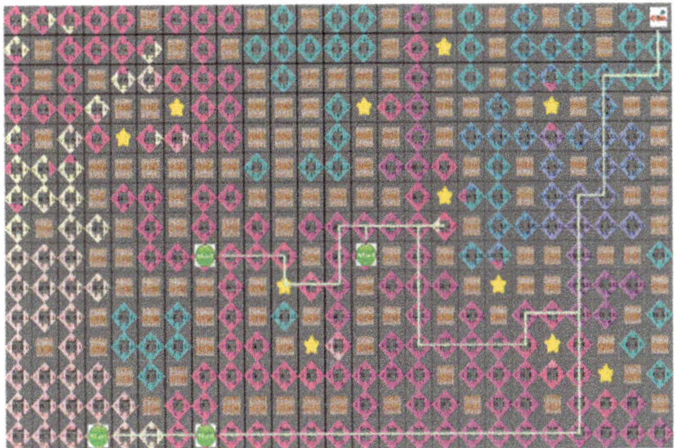

Figure 19. Robot bypass path generation using the proposed algorithm in dynamic environment.

Since the robot moves actively to the next position depending on the results learned step by step, it is possible to create a path to the target point in a dynamic environment.

Table 4 shows under a dynamic environment, the first bypass path of D* and the proposed algorithm have 84 and 65 search ranges to move to the target point, respectively. In addition, for the second bypass, each method has 126 and 83 search range, respectively.

Table 4. Total search scope under dynamic environment for D* and proposed algorithm.

Situation	D* Algorithm	Proposed Algorithm
First bypass	84	65
Second bypass	126	83

Table 5 shows the minimum search range, the maximum search range, and the average search range of the simulation for the D* algorithm and the proposed algorithm for the overall simulation results. All of the three indicators show that the search range of the proposed algorithm is no bigger than D*. Since the search range of the proposed method is small when searching for the bypass path, efficient path planning is possible to the target point. The proposed algorithm used the A* algorithm as a comparison algorithm, which is not available in a dynamic environment. The proposed algorithm was learned by comparing it with the A* algorithm, but unlike the A*, path generation is possible even in a dynamic environment.

Table 5. Minimum, maximum, and average exploration range during the full simulation of D* and proposed algorithms.

Situation	D* Algorithm	Proposed Algorithm
Minimum Exploration Range	66	66
Maximum Exploration Range	294	237
Average Exploration Range	188	141

5. Conclusions

The problem of multi-robot path planning is motivated by many practical tasks because of its efficiency for performing given missions. However, since each robot in the group operates individually or cooperatively depending on the situation, the search area of each robot is increased. Reinforcement learning in the robot's path planning algorithm is mainly focused on moving in a fixed space where each part is interactive.

The proposed algorithm combines the reinforcement learning algorithm with the path planning algorithm of the mobile robot to compensate for the demerits of conventional methods by learning the situation where each robot has mutual influence. In most cases, existing path planning algorithms are highly depends on the environment. Although the proposed algorithm used an A* algorithm that could not be used in a dynamic environment as a comparison algorithm, it also showed that path generation is possible even in a dynamic environment. The proposed algorithm is available for use in both a static environment and in a dynamic environment. Since robots share the memory used for learning, they learn the situation by using one system. Because learning is slow and the result includes errors of each robot at the beginning of learning, each robot has the same result as using each parameter after learning progress to a certain extent. The proposed algorithm including A* for learning can be applied in real robots by minimizing the memory to store the operations and values.

A* algorithm always finds the shortest distance rather than the search time. Under the proposed algorithm, which shows diverse results, the search area of each robot is similar to A*-based learning. However, in the environment where the generated path is simple or without obstacles, an unnecessary movement occurs. To enhance the proposed algorithm, research on the potential field algorithm is undergoing. In addition, the proposed algorithm did not take into account the dynamics of robots and obstacles [44–46] and performed simulations in situations in which robots and obstacles always made ideal movements without taking into account the dynamics of them. Based on the simulation result, the research considering the actual environment and physical engine is on the way.

Author Contributions: Conceptualization, H.B.; software, H.B. and G.K; validation, J.K., D.Q., and G.K.; data curation, J.K.; writing—original draft preparation, H.B.; writing—review and editing, D.Q.; visualization, G.K.; supervision, S.L.; project administration, S.L.

Funding: This work was supported by the Basic Science Research Program, through the National Research Foundation of Korea, Ministry of Science, under Grant 2017R1D1A3B04031864.

Conflicts of Interest: The authors declare no conflict of interest.

References

1. Nasser, M. Pattern Recognition and Machine Learning. *J. Electron. Imaging* **2007**, *16*, 4.
2. Yu, D.; Li, D. Deep learning and its applications to signal and information processing [exploratory dsp]. *IEEE Signal Process. Mag.* **2010**, *28*, 145–154. [CrossRef]
3. Hinton, G.; Deng, L.; Yu, D.; Dahl, G.E.; Mohamed, A.R.; Jaitly, N.; Kingsbury, B. Deep neural Networks for acoustic modeling in speech recognition. *IEEE Signal Process. Mag.* **2012**, *29*, 82–97. [CrossRef]
4. Graves, A.; Abdel-rahman, M.; Geoffrey, H. Speech recognition with deep recurrent neural networks. In Proceedings of the 2013 IEEE International Conference on Acoustics, Speech and Signal Processing, Vancouver, BC, Canada, 26–31 May 2013; pp. 6645–6649.
5. Kumar, A. Ask me anything: Dynamic memory networks for natural language processing. In Proceedings of the International Conference on Machine Learning, New York, NY, USA, 19–24 June 2016; pp. 1378–1387.
6. Machine Learning and Natural Language Processing. Available online: http://l2r.cs.uiuc.edu/~{}danr/Teaching/CS546-13/Papers/marquez-LNLP00.pdf (accessed on 11 July 2010).
7. Manning, C. The Stanford CoreNLP natural language processing toolkit. In Proceedings of the 52nd Annual Meeting of the Association for Computational Linguistics: System Demonstrations, Baltimore, MD, USA, 22–27 June 2014; pp. 55–60.

8. Collobert, R.; Jason, W. A unified architecture for natural language processing: Deep neural networks with multitask learning. In Proceedings of the 25th International Conference on Machine Learning, Helsinki, Finland, 5–9 July 2008; pp. 160–167.

9. Kononenko, I. Machine learning for medical diagnosis: History, state of the art and perspective. *Artif. Intell. Med.* **2001**, *23*, 89–109. [CrossRef]

10. Shvets, A.A.; Rakhlin, A.; Kalinin, A.A.; Iglovikov, V.I. Automatic instrument segmentation in robot-assisted surgery using deep learning. In Proceedings of the 2018 17th IEEE International Conference on Machine Learning and Applications (ICMLA), Orlando, FL, USA, 17–20 December 2018; pp. 624–628.

11. Bottou, L. Large-scale machine learning with stochastic gradient descent. In Proceedings of the COMPSTAT'2010, Paris, France, 22–27 August 2010; pp. 177–186.

12. Peters, J.; Stefan, S. Natural actor-critic. *Neurocomputing* **2008**, *71*, 1180–1190. [CrossRef]

13. Bhasin, S.; Kamalapurkar, R.; Johnson, M.; Vamvoudakis, K.G.; Lewis, F.L.; Dixon, W.E. A novel actor-identifier architecture for approximate optimal control of uncertain nonlinear systems. *Automatica* **2013**, *49*, 82–92. [CrossRef]

14. Vamvoudakis Kyriakos, G.; Frank, L.L. Online actor-critic algorithm to solve the continuous-time infinite horizon optimal control problem. *Automatica* **2010**, *46*, 878–888. [CrossRef]

15. Florensa, C.; Degrave, J.; Heess, N.; Springenberg, J.T.; Riedmiller, M. Self-supervised learning of image embedding for continuous control. *arXiv* **2019**, arXiv:1901.00943.

16. Lillicrap Timothy, P. Continuous control with deep reinforcement learning. *arXiv* **2015**, arXiv:1509.02971.

17. Watkins Christopher, J.; Peter, D. Q-learning. *Mach. Learn.* **1992**, *8*, 279–292. [CrossRef]

18. Littman, M.L. Markov games as a framework for multi-agent-reinforcement learning. In Proceedings of the Eleventh International Conference on Machine Learning, New Brunswick, NJ, USA, 10–13 July 1994; Volume 157, pp. 157–163.

19. Foster, D.; Peter, D. Structure in the space of value functions. *Mach. Learn.* **2002**, *49*, 325–346. [CrossRef]

20. Kofinas, P.; Dounis, A.I.; Vouros, G.A. Fuzzy Q-Learning for multi-agent decentralized energy management in microgrids. *Appl. Energy* **2018**, *219*, 53–67. [CrossRef]

21. Keselman, A.; Ten, S.; Ghazali, A.; Jubeh, M. Reinforcement Learning with A* and a Deep Heuristic. *arXiv* **2018**, arXiv:1811.07745.

22. Stentz, A. Optimal and efficient path planning for partially known environments. In *Intelligent Unmanned Ground Vehicles*; Springer: Boston, MA, USA, 1997; pp. 203–220.

23. Ge, S.S.; Yan, J.C. New potential functions for mobile robot path planning. *IEEE Trans. Robot. Autom.* **2000**, *16*, 615–620. [CrossRef]

24. Zhang, Y.; Gong, D.-W.; Zhang, J.-H. Robot path planning in uncertain environment using multi-objective particle swarm optimization. *Neurocomputing* **2013**, *103*, 172–185. [CrossRef]

25. Tharwat, A.; Elhoseny, M.; Hassanien, A.E.; Gabel, T.; Kumar, A. Intelligent Bézier curve-based path planning model using Chaotic Particle Swarm Optimization algorithm. *Clust. Comput.* **2018**, 1–22. [CrossRef]

26. Elhoseny, M.; Tharwat, A.; Hassanien, A.E. Bezier curve based path planning in a dynamic field using modified genetic algorithm. *J. Comput. Sci.* **2018**, *25*, 339–350. [CrossRef]

27. Hu, X.; Chen, L.; Tang, B.; Cao, D.; He, H. Dynamic path planning for autonomous driving on various roads with avoidance of static and moving obstacles. *Mech. Syst. Signal Process.* **2018**, *100*, 482–500. [CrossRef]

28. Alomari, A.; Comeau, F.; Phillips, W.; Aslam, N. New path planning model for mobile anchor-assisted localization in wireless sensor networks. *Wirel. Netw.* **2018**, *24*, 2589–2607. [CrossRef]

29. Li, G.; Chou, W. Path planning for mobile robot using self-adaptive learning particle swarm optimization. *Sci. China Inf. Sci.* **2018**, *61*, 052204. [CrossRef]

30. Thrun, S.; Wolfram, B.; Dieter, F. A real-time algorithm for mobile robot mapping with applications to multi-robot and 3D mapping. In Proceedings of the ICRA, San Francisco, CA, USA, 24–28 April 2000; Volume 1, pp. 321–328.

31. Bruce, J.; Manuela, V. Real-time randomized path planning for robot navigation. In Proceedings of the IEEE/RSJ International Conference on Intelligent Robots and Systems, Takamatsu, Japan, 31 October–5 November 2002; Volume 3, pp. 2383–2388.

32. Indelman, V. Cooperative multi-robot belief space planning for autonomous navigation in unknown environments. *Auton. Robot.* **2018**, *42*, 353–373. [CrossRef]

33. Fan, T.; Long, P.; Liu, W.; Pan, J. Fully distributed multi-robot collision avoidance via deep reinforcement learning for safe and efficient navigation in complex scenarios. *arXiv* **2018**, arXiv:1808.03841.

34. Van Den Berg, J.; Dave, F.; James, K. Anytime path planning and replanning in dynamic environments. In Proceedings of the 2006 IEEE International Conference on Robotics and Automation, Orlando, FL, USA, 15–19 May 2006; pp. 2366–2371.

35. Raja, P.; Sivagurunathan, P. Optimal path planning of mobile robots: A review. *Int. J. Phys. Sci.* **2012**, *7*, 1314–1320. [CrossRef]

36. Van, H.; Hado, A.G.; David, S. Deep reinforcement learning with double q-learning. In Proceedings of the Thirtieth AAAI Conference on Artificial Intelligence, Phoenix, AZ, USA, 12–17 February 2016.

37. Mnih, V. Playing atari with deep reinforcement learning. *arXiv* **2013**, arXiv:1312.5602.

38. Mnih, V. Asynchronous methods for deep reinforcement learning. In Proceedings of the International Conference on Machine Learning, New York, NY, USA, 19–24 June 2016; pp. 1928–1937.

39. Ren, S. Faster r-cnn: Towards real-time object detection with region proposal networks. In Proceedings of the Twenty-ninth Conference on Neural Information Processing Systems, Montréal, QC, Canada, 7–12 December 2015; pp. 91–99.

40. Kalchbrenner, N.; Edward, G.; Phil, B. A convolutional neural network for modelling sentences. *arXiv* **2014**, arXiv:1404.2188.

41. Abdel-Hamid, O. Convolutional neural networks for speech recognition. *IEEE/ACM Trans. Audio Speech Lang. Process.* **2014**, *22*, 1533–1545. [CrossRef]

42. Simonyan, K.; Zisserman, A. Very deep convolutional networks for large-scale image recognition. *arXiv* **2014**, arXiv:1409.1556.

43. Hart, P.E.; Nilsson, N.J.; Raphael, B. A formal basis for the heuristic determination of minimum cost paths. *IEEE Trans. Syst. Sci. Cybern.* **1968**, *4*, 100–107. [CrossRef]

44. Borboni, A.; Lancini, M. Commanded motion optimization to reduce residual vibration. *J. Vib. Acoust.* **2015**, *137*, 031016. [CrossRef]

45. Alonso-Mora, J.; Montijano, E.; Nägeli, T.; Hilliges, O.; Schwager, M.; Rus, D. Distributed multi-robot formation control in dynamic environments. *Auton. Robots* **2019**, *43*, 1079–1100. [CrossRef]

46. Yu, J.; Ji, J.; Miao, Z.; Zhou, J. Neural network-based region reaching formation control for multi-robot systems in obstacle environment. *Neurocomputing* **2019**, *333*, 11–21. [CrossRef]

Article

A Method of Path Planning on Safe Depth for Unmanned Surface Vehicles Based on Hydrodynamic Analysis

Shuai Liu [1], Chenxu Wang [1,*] and Anmin Zhang [1,2,*]

[1] School of Marine Science and Technology, Tianjin University, Tianjin 300072, China
[2] Tianjin Port Environmental Monitoring Engineering Center, Tianjin 300072, China
* Correspondence: chenxu.wang@tju.edu.cn (C.W.); zhanganmin@sina.com (A.Z.)

Received: 30 June 2019; Accepted: 31 July 2019; Published: 7 August 2019

Abstract: The depth of water is of great significance to the safe navigation of unmanned surface vehicles (USV)in shallow waters, such as islands and reefs. How to consider the influence of depth on the safety of USV navigation and path planning is relatively rare. Under the condition of ocean disturbance, the hydrodynamic characteristics of unmanned surface vehicles will affect its draft and depth safety. In this paper, the hydrodynamic model of unmanned surface vehicles is analyzed, and a water depth risk level A* algorithm (WDRLA*) is proposed. According to the depth point of the electronic navigation chart (ENC), the gridding depth can be obtained by spline function interpolation. The WDRLA* algorithm is applied to plan the path, which takes hydrodynamic characteristics and navigation errors into account. It is compared with the traditional A* shortest path and safest path. The simulation results show that the WDRLA* algorithm can reduce the depth hazard of the shortest path and ensure the safety of navigation.

Keywords: unmanned surface vehicles; path planning; hydrodynamics; electronic navigation chart; numerical simulation; water depth risk

---

1. Introduction

As a sort of autonomous ocean vehicles, USV (unmanned surface vehicles) can autonomously perform tasks in various environments without human intervention [1]. USV with non-linear dynamic characteristics have better endurance ability and payload ability than other unmanned vehicles, which makes them to be essential tools in the area of marine scientific research, marine resources development, marine environment monitoring and national marine security [2–4]. Autonomy and intellectualization are the crucial manifestations of the development of USV. Research on intelligent planning and controlling can improve the working performance of USV [5].

In the area of intelligent planning and control, path planning of single objectives, such as shortest distance, time optimality, energy consumption optimality, smoothness optimality and minimal risk, has gradually been studied furtherly [6]. Multi-objective motion planning and intelligent decision-making [7,8] considering dynamic constraints of mobile robots is the difficulty of current research, where it is necessary to obtain more comprehensive and accurate environmental information by using environmental comprehensive perception methods, such as maps to optimize the path planning results. In this context, it is of great significance to study the path planning algorithm based on ENC (electronic navigation chart) considering the hydrodynamic characteristics of USVs for improving the USV's autonomous environmental awareness, optimizing the path planning results and improving the autonomy and intelligence level of USVs.

A star algorithm, (A* algorithm) is a deterministic algorithm with outstanding consistency and completeness which is widely applied in the research of path planning. It was firstly proposed by

Hart P. et al. [9] and usually combined with grid method [10]. The minimum cost function is used to search the minimum cost path from the starting node to the end node. The grid-based A* path search methods are mainly based on the center point of the grid [11–16] and grid vertexes [7,10,17–19]. Grid method can be divided into a uniform grid and non-uniform grid with diverse shapes and sizes of the grid. Aiming at the deficiencies of the grid-based A* algorithm, such as huge memory expenditure, large path turning angle and unsmooth path, studies have been done to improve it.

Larson et al. [20] found the optimal path with the least nodes in limited time by using A*. Svec et al. [21] combined the A* and locally bounded plan, Naeem et al. [22] proved A* by direction priority sequential selection (DPSS), Improving path quality and computational capacity. Svec et al. [23] introduced a model prediction algorithm and combined it with A* algorithm. The above work concentrates on improving the computational performance and path quality of traditional A* algorithm.

Daniel, K [16] and Soulignac, M [24] carried out the Theta* which released the limited that the center points of the grid can only moves along the center points within A*. Kim et al. [13] solved the problem of unlimited direction with the theory of limit-cycle circle set. Nash, A et al. [15] carried out Lazy Theta*, it was found that vertexes visibility check decreases by using this method. Improved dynamic search D* was put forward by Yang, J.M. et al. [17] based on A*, it was demonstrated that this algorithm fits the dynamic map very well.

Wu, P.P et al. [7] carried out a multi-objective A*, it improves the real-time performance of A*. Naus et al. [25] applied A* to search the shortest and safest trajectory in electronic chart. Kim et al. [26] carried out a curvature path planning algorithm which considering the turning ability of USV through perfecting A*. Liu, C.G. et al. [18] established the surface obstacles risk model considering current, traffic separation and maneuverability and improved the practicability of A* algorithm, but ignored the effect of water depth on navigation safety. In the above work, the A* algorithm is applied to the area of path planning for USVs.

Previous researches perfected the calculating quality of A* and expanded the range where A* could be involved in, and they focused on the estimation of the criticality of surface obstacles in the research concerning safest path planning of USVs, rarely discussed the hazard, due to the shallow depth. Depth is a crucial factor affecting navigation safety and performance for unmanned surface vehicles operation in shallow waters, which is more vulnerable to ocean disturbances, such as wind, waves and currents, [27]. On the other hand, the hydrodynamic characteristics can well reveal the navigation stability of small USVs in shallow waters [28,29], and then evaluate the safe depth of navigation.

In this paper, a method of safe path planning is carried out to improve the depth safety of the shortest path of USV, which guarantees navigation safety in different conditions. In this paper, the navigation stability of USVs in various situations is studied, and the minimum safe water depth is calculated according to the results of hydrodynamics simulation. On this basis, the safer path of USVs is calculated by considering the risk of water depth. The depth safety of USV, which followed the shortest path, could be improved by using the information of depth in ENC, the water depth risk level (WDRL) of shortest path could be reduced as well.

2. Modeling

2.1. Hydrodynamic Modeling

Small unmanned surface vehicles are insufficient to resisting wind and waves. When sailing in constrained waters, due to the blockage effect of water boundary, the reflected wave and the generated wave of the boundary overlap or decrease each other, which results in pressure distribution difference on the hull surface, generating heave and trim motion and deducing sticking on bottom, shore touching and capsizing of the vessel and endangering the safety of navigation. Our group has developed an unmanned surface vessel with an adjustable scale shown as Figure 1 and Table 1 for marine surveying and mapping. The hull is made of two inflatable protons. The width of the equipment bracket can be adjusted and quickly disassembled/assembled in 30 minutes. The navigation status of this unmanned surface vehicle is studied with three kinds of velocity in irregular waves as an example, and then the criticality of the draft variation is estimated. The parameters of the USV are as follows:

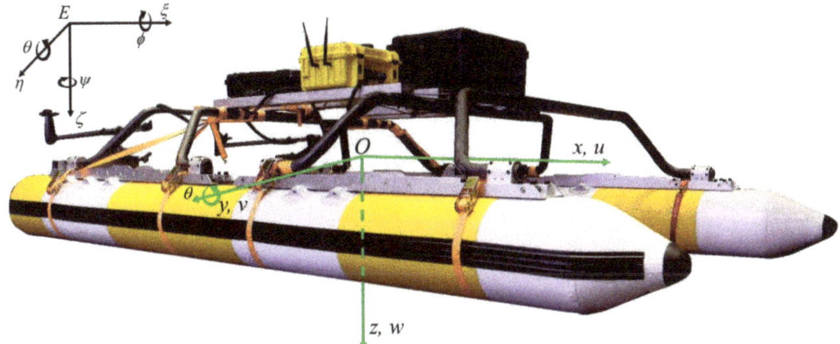

Figure 1. Multi-functional Marine mapping unmanned surface vehicles.

Table 1. Table of the unmanned surface vehicle parameters.

Parameters	Value
Length	3.2 m
Molded breadth	2.2 m
Displacement	92 kg (full load 140 kg)
Location of gravity center	(1.285, 0.000, 0.600)
Moment of inertia of Y axis	400 kg/m^2
Draft	0.5 m

The ship's navigation status mainly contains three states: Heave, pitch and speed. Nine cases were set up with three speeds (5 kn, 8 kn, 11 kn) and three wave heights (0.5 m, 1 m, 1.5 m) when wave period is 2 s (shown in Table 2).

Table 2. Cases list.

Case Number	Speed (kn)	Wave Height (m)	F_{rV} [1]
1	5	0.5	0.42
2	5	1.0	0.42
3	5	1.5	0.42
4	8	0.5	0.63
5	8	1.0	0.63
6	8	1.5	0.63
7	11	0.5	0.84
8	11	1.0	0.84
9	11	1.5	0.84

[1] F_{rV}: Volume Froude number, a non-dimensional parameter which presents the shape and the displacement of the ship.

The governing equations were solved by RANS (standard k-ε) with STAR-CCM+. Assuming that the fluid flow is viscous and incompressible, the governing equations are established as below:

$$\mathrm{div}\boldsymbol{u} = 0, \tag{1}$$

$$\frac{\partial u}{\partial t} + \mathrm{div}(\boldsymbol{uu}) = -\frac{1}{\rho}\frac{\partial p}{\partial x} + v\,\mathrm{div}(\mathrm{grad}\boldsymbol{u}), \tag{2}$$

$$\frac{\partial v}{\partial t} + \mathrm{div}(\boldsymbol{vu}) = -\frac{1}{\rho}\frac{\partial p}{\partial y} + v\,\mathrm{div}(\mathrm{grad}\boldsymbol{v}), \tag{3}$$

$$\frac{\partial w}{\partial t} + \mathrm{div}(\boldsymbol{wu}) = -\frac{1}{\rho}\frac{\partial p}{\partial z} + v\,\mathrm{div}(\mathrm{grad}\boldsymbol{w}), \tag{4}$$

where u, v, w are respectively the components of velocity on the directions of x, y, z (see Figure 1).

The waves were generated by Pierson-Moskowitz spectra, which defines the motion of irregular waves—it is written as:

$$S_{PM}(\omega) = \frac{5}{16}\left(H_S^2 \omega_P^4\right)\omega^{-5}\exp\left(-\frac{5}{4}\left(\frac{\omega}{\omega_P}\right)^{-4}\right),$$

(5)

where ω is the angular frequency of waves. $\omega_P = (2\pi)/(T_P)$ represents the peak frequency of the spectrum. T_P is the peak wave period. H_S is significant wave height (m).

The velocity distribution of the flexible wave generator plate is employed to generate a simulated incident, waves at the inlet boundary, the wave surface equation of irregular waves can be expressed as follows:

$$\eta = H_s\cos(mx - \omega t),$$

(6)

in which η is energy, and m is the number of waves. The velocity field of irregular wave is as below:

$$\begin{cases} U = \omega H e^{kz}\cos(mx - \omega t) \\ \quad V = 0 \\ W = \omega H e^{kz}\sin(mx - \omega t) \end{cases},$$

(7)

where U and W are velocity profiles on x and z direction. H is the wave amplitude, k is wave-number $(2\pi/m)$. x-axis is the direction of wave propagation, and the z-axis is the direction of wave fluctuation. The wave propagation direction is parallel to the heading direction of USV.

The numerical simulation method is shown in Table 3.

Table 3. Numerical simulation methods.

Discretization Method	Finite Volume Method
Scheme	SIMPLE
Solver	Implicit Unsteady
Gradient	Green-Guass Cell Based
Pressure	Second Order
Momentum	Second Order Upwind
Transient Formulation	Second Order Implicit
Mesh motion	Overset motion

The computational domain is shown in Figure 2:

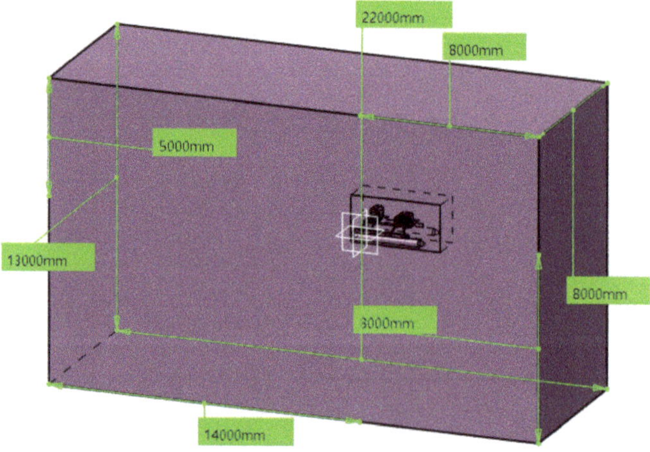

Figure 2. Computational Domain.

The boundary conditions are as follows: The initial pressure of the whole domain basin is zero, the inlet boundary is Velocity-Inlet, and the velocity direction is perpendicular to the flow. Pressure-Outlet is used at the outlet boundary. The initial pressure value is zero. The flow is fully developed. The velocity gradient is zero in the normal direction of the boundary. Velocity-Inlet is used at the upper and lower boundaries of the basin, and symmetrical boundary conditions are used at the side walls.

Three grid numbers were tested on case 1. The numbers are 674921,1373495,1925630. The mean coefficient of drag force $\overline{C}_d$ at the monitoring point of the hull surface is taken to calculate the error E, which is written as:

$$\overline{C}_d = \frac{\int_0^T C_d(t)dt}{T},$$ (8)

where T is monitoring period.

As shown in Table 4, the difference between Mesh 2 and Mesh 3 is extremely small, considering with saving the calculating time and source, Mesh 2 was selected. Its three-dimensional mesh generation and grid distribution around the wall are shown in Figure 3a,b

Table 4. Mesh independency study.

Mesh	Mesh 1	Mesh 2	Mesh 3
Number of meshs	774,921	1,373,495	1,925,630
$E = \left\|1 - \overline{C}_{di}/\overline{C}_{d3}\right\|$	1.13×10^{-3}	4.35×10^{-3}	–

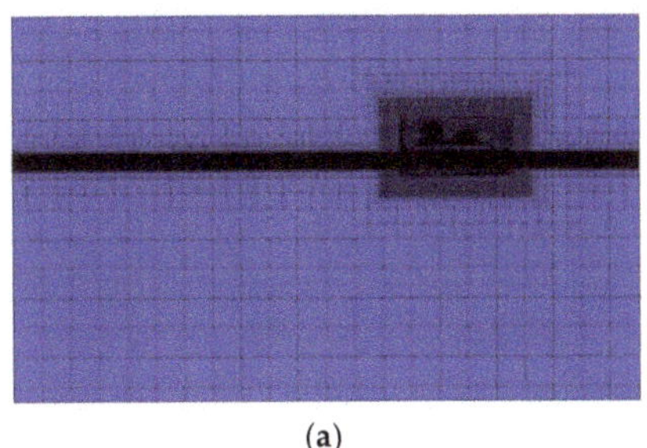

(a)   (b)

Figure 3. Mesh Generation and Distribution. (a) Three-dimensional Mesh Generation; (b) Grid Distribution Around the Wall.

2.2. Path Planning Environment Modeling

Establishing a real ocean environment model is essential so that unmanned surface vehicles can perceive the external environment before path planning. Binarization processing is applied to build environment model via binarize remote sensing satellite is the majority of path planning [7,17,19,26,30]. Nevertheless, the depth of water is not considered, thus navigation along the planned path may be stranding. Vector electronic navigation chart, shown as Figure 4a, supports a variety of intelligent functions and provides water depth information with advantages of minute storage, swift display speed and high accuracy. It is suitable for marine environment modeling and processing for further development [15,20]. Raster data has a simpler data structure than other environmental modeling methods and can contain terrain cost information, such as elevation [16]. In this paper, Vector electronic navigation Chart is applied to describe the real marine environment information. Uniform square gird is selected to process the electronic chart, and global static obstacle and depth of water extracted.

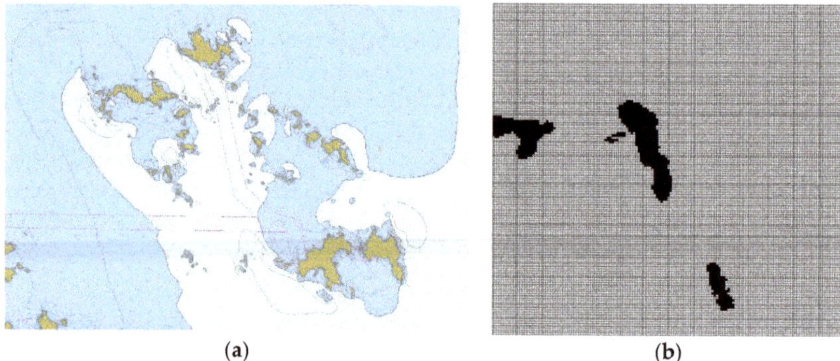

Figure 4. S57 Environmental Modeling Map (**a**) Study Area of S57 Electronic Chart. (**b**) Regional Grid Marine Environment Map with a resolution of 25 m × 25 m. In the picture, the white meshes represent the feasible region, while the black meshes are a barrier.

In vector electronic navigation charts, spatial objects, such as land area, island reef, and navigation aids, are stored and expressed in the form of point, line and area features. In this paper, feasible space [25] is obtained by selecting objects which could risk on navigation.

Obstacle region $O_{bs} \subset R^2$ is calculated by Equation (9):

$$O_{bs} = O_{Aera} \cup O_{Line} \cup O_{Point},\qquad(9)$$

in which, $O_{Aera}, O_{Line}, O_{Point}$ represent surface, line and point obstacles respectively. Feasible space is calculated by Equation (10),

$$O_f = \Omega_D - O_{bs},\qquad(10)$$

where, $\Omega_D \subset R^2$ is the two-dimensional configuration space represented by "DEPARE" coverage in S57 ENC file stands for the area covered with water shown as Figure 5c, where free space Ω_f is obtained by Formula (10) in the form of a polygon. According to the ship maneuverability standard, the minimum turning radius of large ships is equal to five times of the ship length, and that of small ships is three times of the ship length [10]. Considering the size of the vessel (3.2 m), localization error (5 m), buffer zone distance (5 m) and error of electronic navigation chart (5 m), the grid size is set to 25 m × 25 m . The mesh size partition considers the kinematics constrictions of unmanned surface vehicles fully, so that unmanned surface vehicles can make starboard and port side turns up to 180 degrees within a grid cell. Non-feasible space and feasible space can be obtained by (9) and (10) as shown in Figure 5c. The hazard cost of water depth of blocked grids is set to 1.

In electronic navigation charts, water depth is stored and expressed in the form of depth point, depth contour and depth area. The depth contour interval extracted from a chart of various plotting scale is different, as shown in Table 5.

Table 5. Interval table of electronic navigation chart (ENC) with different scales.

Plotting Scale	Depth Contour Interval
≥1:150,000	0/2/5/10/20/30/50/100/200/500
≥1:1,000,000 and ≤1:150,000	0/5/10/20/30/50/100/200/500/1000
≤1:1,000,000	0/10/50/200/500/1000/2000/4000/6000

In order to obtain the water depth information of the planned sea area, we extract water depth point, contour and depth area coverage shown as Figure 5d from S57 vector electronic navigation chart shown as Figure 5a. As shown in Figure 5b, because the depth information is only in the discrete depth area, shown in Table 5, the depth risk of each grid cannot be evaluated more accurately. Therefore, accurate discrete depth points are chosen to be processed to obtain a raster prediction depth, and then evaluates the depth risk of the planned path.

Spline function method with obstacles is utilized to interpolate from discrete water depth points. In this method, the two-dimensional minimum curvature spline method is used to interpolate points into rater surface, and the generated rater surface passes through all of the input depth points from electronic navigation charts. As shown in Figure 5c,d, the prediction depth generated by the spline function with obstacles can be well-matched with the contour line. Through this method, the water depth of each grid in feasible space can be obtained, and then the water depth risk of each grid can be calculated, as detailed in Section 3.2.

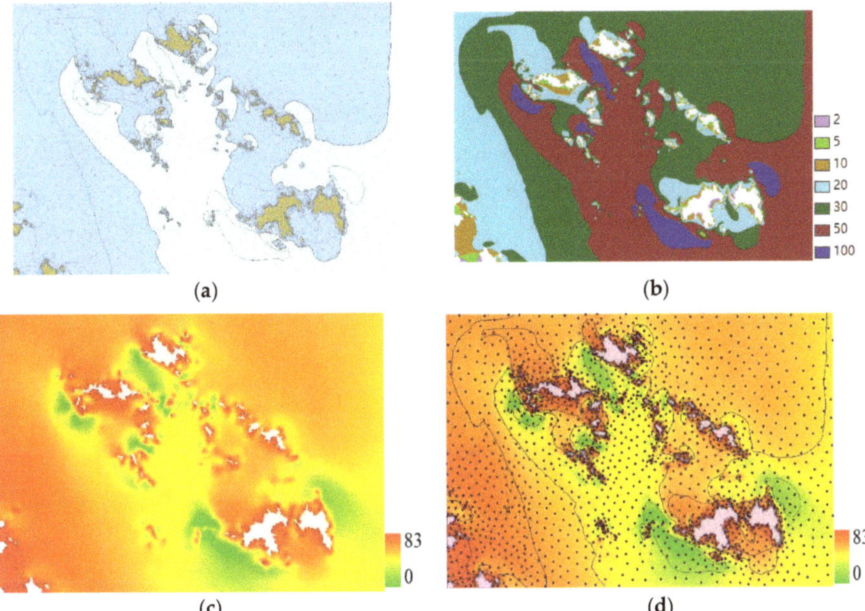

Figure 5. Electronic navigation Chart and Grid Depth Distribution Map. (**a**) S57 Vector Electronic navigation Chart displayed by S52 Standard in the study area; (**b**) Isobath Area Map; (**c**) Depth Distribution Map obtained by spline function interpolation approach with obstacles, in which the blank area is an obstacle; (**d**) Overlay with discrete depth points, contours, islands and reefs and other obstacles. In the depth distribution map, the magenta area is an obstacle.

3. Depth Safety Path Planning Method

3.1. Analysis of Hydrodynamic Properties of Unmanned Surface Vehicle

According to the hydrodynamic model established in Section 2.1, the navigation simulation of the USV is carried out, and the hydrodynamic properties of the USV under nine different working conditions are analyzed. The numerical simulation results are shown in Figures 6–8. The diagrams of navigation status, including heave and pitch motion in time domain under nine working conditions, are detailed in Supplementary Materials Figures S1 and S2. As shown in Figure 7a, the heave motion of the unmanned surface vehicle presents a gentle trend with the increase of speed. As shown in Figure 7b, the maximum heave amplitude is 0.7 m when the speed is 5 kn, and the significant wave height is 1.5 m, and the maximum downward settlement amplitude of the USV is 0.3 m (refer to Figure S1) when the speed is 8 kn, and the significant wave height is 1 m. It can be seen that when the USV navigates at a lower speed at a higher significant wave height, there will be a larger heave motion. In order to ensure the safety of navigation depth, the maximum downward settlement value is selected as the input parameter to calculate the minimum safe depth of navigation for unmanned surface vehicles.

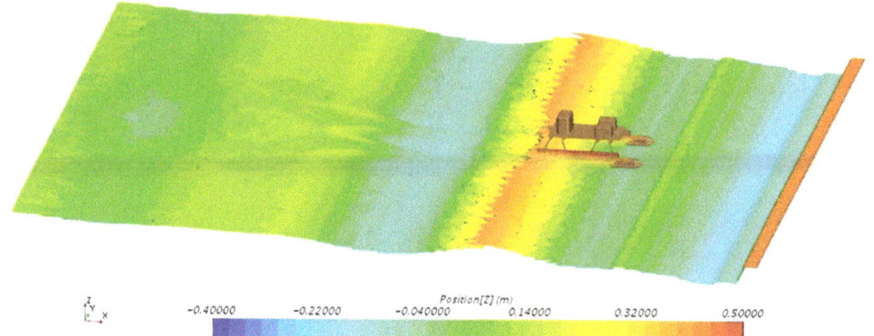

Figure 6. Numerical simulation results of the unmanned vehicle.

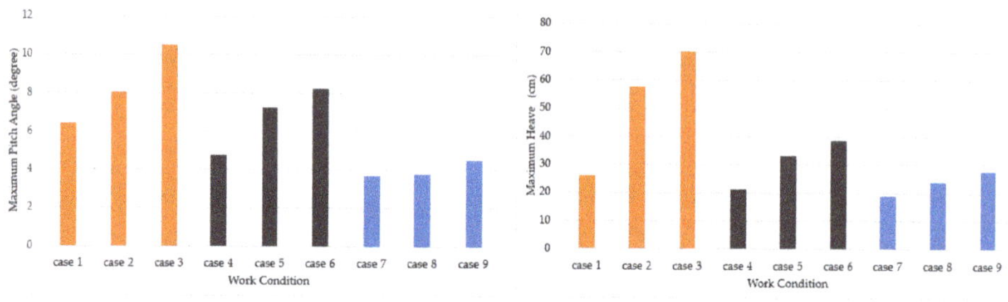

(**a**) Heave RMS value of nine cases　　　　(**b**) Maximum heave amplitude of nine cases

Figure 7. Analysis of heave of the USV.

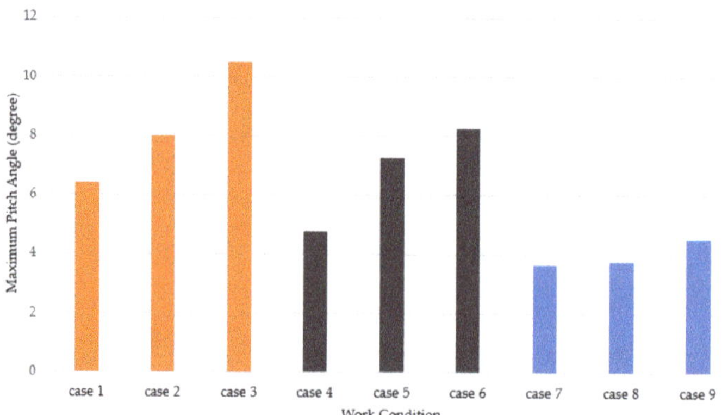

Figure 8. Maximum pitch Angle of nine cases.

The pitch of the ship, in waves, can also cause draft changes and induce danger of striking on the sea bed. As shown in Figure 8, the pitch angle of the USV decreases with the increase of speed and significant wave height. As shown in Figure 8, the maximum pitch angle occurs at the speed of 5 kn, and the significant wave height of 1.5 m, reaching 10.5 degrees. The minimum pitch angle occurs at 11 kn, and the significant wave height is 0.5 m, which is 3.7 degrees. The USV has a larger pitch angle at lower speed and higher significant wave height, which causes larger draft change and gives rise to potential hazards to navigation safety of the USV. In order to ensure the safety of navigation depth, the maximum pitch angle is used as the input parameter to calculate the minimum safe depth of navigation for USV.

3.2. Path Planning Algorithms Considering Depth Hazard

According to the results of the hydrodynamic analysis in Section 3.1, the minimum water depth required for safe navigation of unmanned surface vehicles is calculated by Equation (11):

$$S_{min} = z_{max} + 0.5L \tan \theta_{max} + \overline{T} + e_{enc}, \tag{11}$$

where, z_{max} is the largest downward settlement of the USV in several irregular waves at different speeds, L is the length of USV, θ_{max} is the maximum trim angle, $\overline{T}$ is the average draft of the USV, and e_{enc} is the depth error of ENC are considered. According to the minimum safe water depth for an unmanned surface vehicle, the restricted sea area of the unmanned surface vehicle is calculated, and the obstacle region is updated. The cost of water depth hazard in the obstacle region is set to 1.

Define the depth hazard of nodes N_i in feasible space as follows:

$$r(N_i) = \frac{S_{min}}{D(N_i)}, (0 < r(N_i) < 1), \tag{12}$$

$D(N_i)$ is the average water depth of each grid calculated by the interpolation algorithm in Section 2.2.

Total depth hazard of planned path is calculated by Formula (13):

$$R_s = \sum_{i=1}^{N} r(N_i), \tag{13}$$

where, i is the index of the grid through which the path passes, and N is the total number of all the grids through which the path passes. The depth hazard distribution map calculated by the combination of depth area and Formulas (11)–(13) is shown in Figure 9. The depth hazard map will be used as the input map of the path search algorithm.

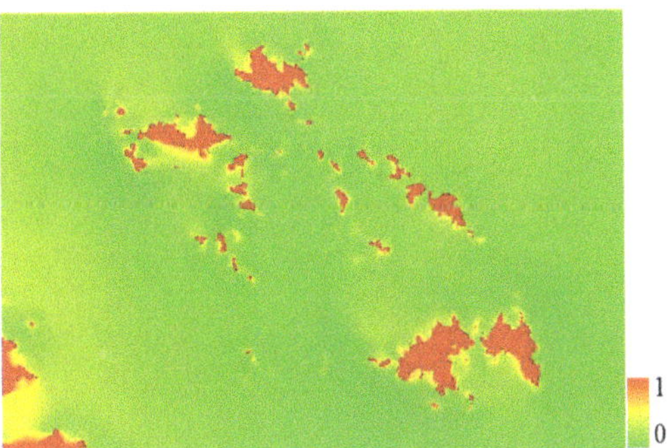

Figure 9. The depth hazard distribution map.

The depth risk map and A* algorithm are applied to plan the path. A* algorithm was proposed originally by Hart Peter et al. [9]. The algorithm searches the minimum cost path from the starting node to the terminating node through the minimum cost function (15),

$$h(N_i) = \sqrt{(x_{N_i} - x_G)^2 + (y_{N_i} - y_G)^2}, \tag{14}$$

$$f(N_i) = g(N_i) + h(N_i), \tag{15}$$

where $g(N_i)$ is the actual distance cost from the ith node N_i to the starting node S that has been paid in the raster map namely the distance traveling from the start node S to the *i*th node N_i. $h(N_i)$ is the heuristic distance from the start node to the goal node which has a great influence on the performance of A* algorithm, in which (x_{N_i}, y_{N_i}) is the coordinate the center of node N_i and (x_G, y_G) is the coordinate the center of goal node G. If it is always lower than (or equal to) real value, the shortest path can be found. The lower the number of extended nodes is, the slower the search speed. If it is equal to the actual value, the search only follows the best path and does not expand the redundant nodes, so the search speed is the fastest. If it is larger than the actual value, the shortest path cannot be guaranteed, but the search path is faster. In this paper, the Euclidean distance is chosen as a heuristic value when calculating the shortest distance path. Euclidean distance can ensure that the solution can be found if there is a shorter path. The shortest path planned by A* algorithm combined with the grid method is presented in Figure 10b.

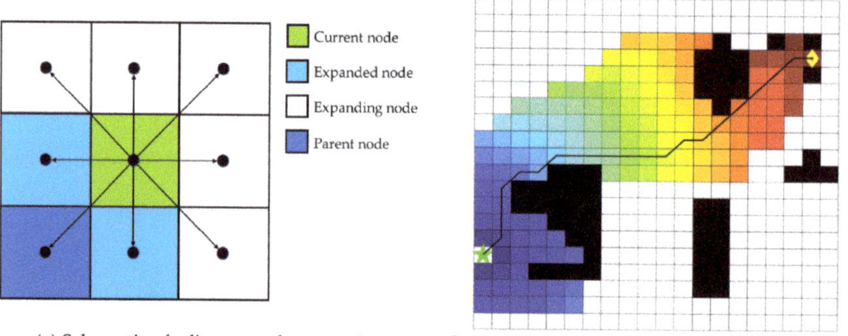

(**a**) Schematic of adjacent node expansion   (**b**) A* minimum distance algorithm search result

Figure 10. A* path planning algorithm schematic diagram. (**a**) Schematic of adjacent node expansion; (**b**) The shortest path of A* algorithm. The green star represents the starting point, and the yellow diamond represents the goal point. The shortest path is the black line between the starting point and the end point. The color grids are expanded nodes. The deeper the blue is, the closer the starting point is, and the deeper the red is, the closer the end point is. A set of white grids represents a feasible space, and black grids are obstacles.

The heuristic function of A* algorithm considering depth hazard degree is defined as:

$$f(N_i) = g(N_i) + h(N_i) + \alpha \sum_{i=1}^{N_c} r(N_i) \tag{16}$$

Not only the distance cost should be considered, but also the depth hazard, and the nodes with lower depth hazard degree should be selected to expand in neighbor nodes when expanding nodes. In Formula (14), $g(N_i)$. The distance cost from the N_i node to the starting point S. $h(N_i)$ is the heuristic distance cost from the N_i node to the goal node, $\sum_{i=1}^{N_c} r(N_i)$ is the water depth hazard cost sum from the starting node to the current node, and α is the constant greater than 0 used to control the weight of the risk cost in the total cost, thus controlling the impact of the water depth hazard on the final path. The pseudo code chart of the water depth risk level A*(WDRLA*) algorithm considering depth hazard is shown in Figure 11.

Algorithm 1. Traditional A* algorithm

Input: start node S and goal node G and water depth risk map

1: Initialize setOpen, setOpenCosts, setOpenHeuristics, setClosed, setPath
2: setOpen: = S, setClosed: =G
3: **while** setOpen $\neq \varnothing$ & G $\notin$ setOpen
4:    If $N_C = N_G$ **break**
5:    N_C: = Select the node from the setOpen whose $f(N_i)$ is smallest
6:    Expand the neighbor nodes, Calculate $f(N_i)$ of neighbor nodes
7:    Update setOpen, setOpenCosts, setOpenHeuristics,
8:    Update setClosed, setClosedCosts
9:    **for** each node of N_i of N_C **do**
10:      **if** $f(N_i) = 0 \| N_i \in setClosed$ **continue**
11:      **if** $N_i \notin$ setOpen
12:        add N_i into setOpen
13:        the parent node of N_i, $N_{i-parent} = N_C$
14:        calculate $f(N_i), g(N_i), h(N_i), \alpha \sum_{i=1}^{N_c} r(N_i, N_c)$
15:      **end if**
16:      **if** $N_i \in setOpen$
17:        $g'(N_i) = g(N_i, N_C) + d(N_i, N_C) + \alpha \sum_{i=1}^{N_c} r(N_i, N_c)$
18:        **if** $g'(N_i) \geq g(N_i)$ **continue**
19:      **else**
20:        $N_{ip} = N_C$
21:        $g(N_i) = g'(N_i)$
22:        $f(N_i) = g(N_i) + h(N_i) + \alpha \sum_{i=1}^{N_c} r(N_i)$
23:      **end if**
24:    $N_p = G$
25:    $setPath: = N_p$
26:    **while** $N_p \neq S$ **do**
27:      $N_p = N_{p-parent}$
28:      $setPath: = \{setPath: = N_p\}$
29: **Return:** $setPath$

Figure 11. WDRLA* algorithm pseudo code chart.

4. Discussions

The spline function interpolation method with obstacles is performed to get the raster depth area with the water depth point selected as input point feature and the smoothing coefficient chosen as 0 and the grid size set as 25 m. The feasible space is updated according to the calculated minimum depth of safe navigation (1.29 m). Related parameters are shown in Table 6. The simulation is also based on the following assumptions: Assuming the unmanned vehicle as a mass point, unmanned vehicles can safely cross the junction vertex of the blocked grid and free gird between the obstacle area and the non-obstacle area, and unmanned vehicles can complete the turnaround operation in the grid. The proposed approach is simulated using Matlab R2018b. All simulations are performed on a PC with Microsoft Windows 10 as OS with Intel i5 2.712 GHz quad core CPU and 8 GB RAM.

Table 6. Table of Related parameters.

Parameters	Value
Cell size	25 m × 25 m
L	3.2 m
$\overline{T}$	0.5 m
z_{max}	0.3 m
θ_{max}	11°
e_{enc}	0.2 m
α	10

The positions of USV are represented by the grid index value of row and column where the point is located. S(500, 500) is chosen as the starting node and G(650, 650) as the goal node. Three kinds of cost function, including only distance, cost seeing Equation (15), only risk level cost seeing Equation (13), cost of combining seeing equation (16), were used to calculate the paths of shortest distance path (SDP), the shorter and safer path (SSP), the safest path (SFP) The results are shown in Figure 12. Compared with Figure 12a–c, it can be seen that the path planning algorithm considering water depth risk level will lead to more expanded nodes, so the search speed will be slower than A* and the expanded nodes of WDRLA* is less than those of SFP seeing Figure 12b,c. Figure 12d,e indicate that there is little difference between SSP and SFP, which means approximate depth risk level, but SDP is closer to the obstacle, so the risk level is much higher. Water depth risk level A* algorithm (WDRLA*) can reduce the depth risk and improve the safety of path, but sacrificing a certain calculation time.

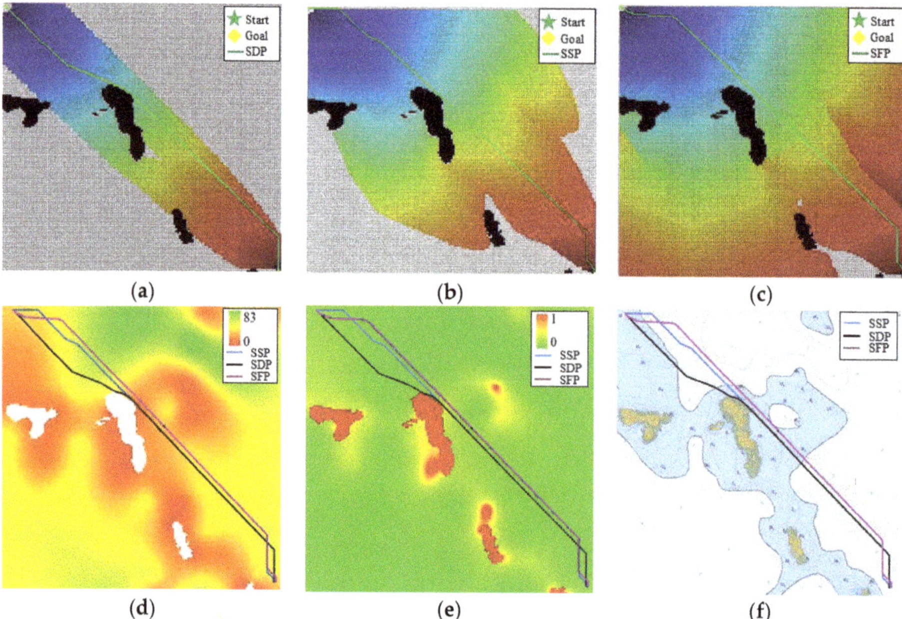

Figure 12. Comparison of three kinds path. (**a**) Shortest distance path (SDP) that ignores the depth risk; (**b**) Shorter and safer path (SSP) that considers the depth risk level and distance. (**c**) Safest path (SFP) that considers only the depth risk. The colorful grids in figure (**a**–**c**) are extended nodes. The space of white grids represents the feasible space, and the black grids are obstacles. (**d**) Three paths, shown on the depth distribution map. (**e**) Three paths, shown on the depth risk level distribution map. (**f**) Three paths, shown on S57 ENC. The black line is the SDP, the blue one is SSP, and the purple one is SFP.

Take S(710, 902) as the starting node, and G(990, 1408) as the goal node, Results of three kinds of paths are shown in Figure 13. From Figure 13, it can be drawn that there are marked differences between SDP and SSP in shallow coastal waters. SDP passes through a narrow channel with shallow water depth which is more dangerous, while SSP and SFP avoid narrow channel by choosing deeper water area, where SSP reduces the risk of the path by 39.61% (see Table 7). At the same time, the number of extended nodes increases greatly, so WDLRA* path search process is slower than A*. SSP increases 1652m and 10.54% in length compared with SDP. The length of SSP and SFP is approximate, but SSP with less expanded nodes.

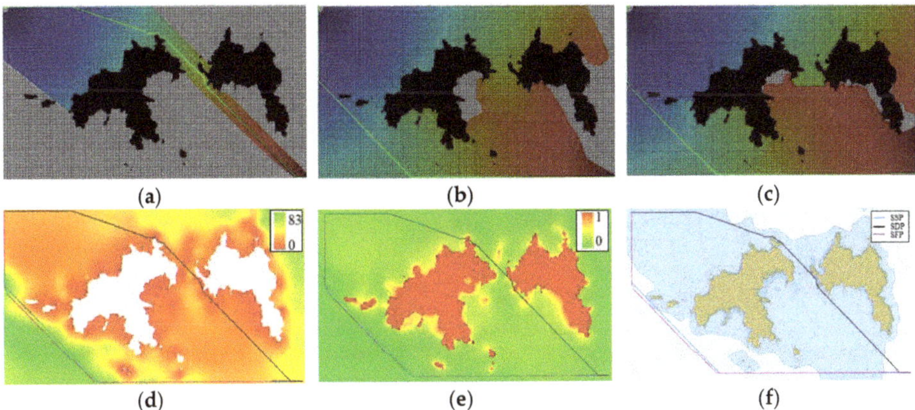

Figure 13. Comparison of three kinds path. (**a**) SDP that ignores the depth risk; (**b**) SSP that considers the depth risk level and distance. (**c**) SFP that considers only the depth risk. The colorful grids in figure (**a–c**) are extended nodes. The space of white grids represents the feasible space, and the black grids are obstacles. (**d**) Three paths, shown on the depth distribution map. (**e**) Three paths, shown on the depth risk level distribution map. (**f**) Three paths, shown on S57 ENC. The black line is the SDP, the blue one is SSP, and the purple one is SFP.

Table 7. Comparison of different starting and goal points.

Test	Start Point	Goal Point	Total Distance (m)		Risk Level		Nodes of Expanded	
			A*	WDLRA*	A*	WDLRA*	A*	WDLRA*
1	(710, 902)	(990, 1408)	15,667.5	17,319.5	38.86142	23.46830	37,682	102,283
2	(343, 975)	(417, 1066)	3063	3063.0	5.68736	5.47936	1363	4469
3	(408, 560)	(520, 680)	4457.75	4472.5	9.95472	7.03769	5851	7679
4	(200, 250)	(410, 530)	9270.5	9255.75	18.68029	17.92211	21,407	41,840
5	(500, 889)	(650, 1039)	5641.5	5405.5	14.43746	9.53375	2971	19,301
6	(500, 500)	(650, 650)	5567.75	5567.75	7.39605	7.31804	7761	15,533

In order to determine the reliability and consistency of the algorithm furtherly, disparate starting nodes and end nodes cases are simulated, and several parameters, such as the path length, the number of expanded nodes, and the risk level of the path, are compared simultaneously, in which the depth risk level is calculated by Equations (11)–(13). All of the results are shown in Table 7. From Table 7 combined with Figures 12–14, it can be seen that the depth risk level of the path can be reduced, and the path is relatively short, especially in the area with complicated underwater topography, such as Figure 13 with WDLRA* algorithm, but in the cost of decreasing computational efficiency with redundant expanded nodes (see Figures 12c and 13c). This method can ensure the safety of small USV navigating in complicated sea areas.

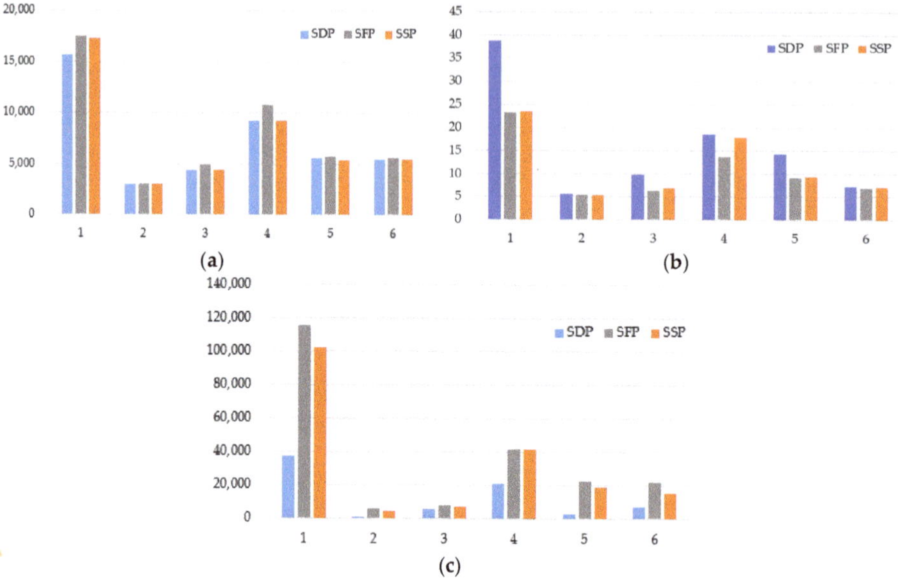

Figure 14. The comparison of key parameters of three different sorts of paths. (**a**) The total distance of the path. The unit of distance is m. (**b**) Water depth risk level of the path. (**c**) The number of expanded nodes of three sorts of cost functions.

5. Conclusions

In this paper, hydrodynamic numerical simulation of a new marine surveying and mapping USV platform developed by the author's research group is performed to evaluate its navigation status in irregular waves. It is found that the heave and pitch of the USV decrease with the increase of velocity and increase with the increase of significant wave height. The heave and pitch navigation status of the unmanned surface vehicle is evaluated through the analysis of 9 working conditions to ascertain minimum safe depth for navigation.

A grid environment modeling method considering water depth is carried out, in which the spline function interpolation algorithm with obstacles is employed to acquire the gridding prediction depth according to the discrete sparse depth points of ENC. The grid modeling method considering water depth can make the utmost of the water depth information provided by ENC to evaluate the depth risk level of path of unmanned surface vehicles.

The quantitative evaluation method of the depth risk level of the path is proposed, and the simulation of path planning is carried out based on A* algorithm using depth hazard degree as an index, named the water depth risk level A* algorithm (WDRLA*). The results show that WDRLA* can hit a safer and shorter path taking into account of the mean draft, the hydrodynamic property, including the maximum heave value and the maximum pitch angle of a USV sailing in an irregular wave, position errors, including ENC positioning errors, positioning system errors and chart depth errors. WDRLA* algorithm can reduce the risk of planned paths in shallow offshore waters to ensure the navigation safety of a USV. The method of path depth risk assessment is not only appropriate to USV path planning, but also can provide guidance for manned ship path planning.

The works are still not completed for the limitations existing. For example, we neglected the influence of tide to water depth, which determines dynamic depth, since the depth on ENC is defined as the vertical distance between the theoretically lowest tide level and seabed. In the future, dynamic water depth will be involved in to acquire bathymetric in better accuracy. The search efficiency of the safe path planning algorithm is expected to be improved in the future.

Appl. Sci. **2019**, *9*, 3228

Supplementary Materials: The following are available online at http://www.mdpi.com/2076-3417/9/16/3228/s1, Figure S1: Heave versus time, Figure S2: Pitch versus time.

Author Contributions: The manuscript was written S.L. and C.W. All authors discussed the original idea. conceptualization, S.L.; methodology, S.L. and C.W.; software, S.L.; validation, S.L.; formal analysis, C.W.; investigation, S.L.; resources, A.Z.; data curation, A.Z.; writing—original draft preparation, S.L. and C.W.; writing—review and editing, A.Z. and C.W.; visualization, S.L. and C.W.; project administration, A.Z.; funding acquisition, A.Z. and C.W.

Funding: This research was funded by National key research and development program of China, grant number 2018YFC1407400.

Conflicts of Interest: The authors declare no conflict of interest.

References

1. Liu, Z.X.; Zhang, Y.M.; Yu, X.; Yuan, C. Unmanned surface vehicles.: An overview of developments and challenges. *Annu. Rev. Control* **2016**, *41*, 71–93. [CrossRef]
2. Zheng, H.; Negenborn, R.R.; Lodewijks, G. Survey of approaches for improving the intelligence of marine Surface Vehicles. In Proceedings of the 16th International IEEE Conference on Intelligent Transportation Systems (ITSC 2013), The Hague, The Netherlands, 6–9 October 2013; pp. 1217–1223.
3. Mohanan, M.G.; Salgoankar, A. A survey of robotic motion planning in dynamic environments. *Robot. Auton. Syst.* **2018**, *100*, 171–185. [CrossRef]
4. Macharet, D.G.; Campos, M.F.M. A survey on routing problems and robotic systems. *Robotica* **2018**, *36*, 1781–1803. [CrossRef]
5. Campbell, S.; Naeem, W.; Irwin, G.W. A review on improving the autonomy of unmanned surface vehicles through intelligent collision avoidance manoeuvres. *Annu. Rev. Control* **2012**, *36*, 267–283. [CrossRef]
6. Liu, Y.C.; Bucknall, R. Path planning algorithm for unmanned surface vehicle formations in a practical maritime environment. *Ocean Eng.* **2015**, *97*, 126–144. [CrossRef]
7. Wu, P.P.; Campbell, D.; Merz, T. Multi-objective four-dimensional vehicle motion planning in large dynamic environments. *IEEE Trans. Syst. Man Cybern. B Cybern.* **2011**, *41*, 621–634. [CrossRef] [PubMed]
8. Ma, Y.; Hu, M.Q.; Yan, X.P. Multi-objective path planning for unmanned surface vehicle with currents effects. *ISA Trans.* **2018**, *75*, 137–156. [CrossRef] [PubMed]
9. Hart, P.E.; Nilsson, N.J.; Raphael, B. A formal basis for the heuristic determination of minimum cost paths. *IEEE Trans. Syst. Sci. Cybern.* **1968**, *4*, 100–107. [CrossRef]
10. Song, R.; Liu, Y.; Bucknall, R. Smoothed A* algorithm for practical unmanned surface vehicle path planning. *Appl. Ocean Res.* **2019**, *83*, 9–20. [CrossRef]
11. Nash, A.; Koenig, S. Any-Angle Path Planning. *AI Mag.* **2013**, *34*, 85–107. [CrossRef]
12. Choi, S.; Yu, W. Any-angle path planning on non-uniform costmaps. In Proceedings of the 2011 IEEE International Conference on Robotics and Automation ICRA, Shanghai, China, 9–13 May 2011; pp. 5615–5621.
13. Kim, H.; Lee, T.; Chung, H.; Son, N.; Myung, H. Any-angle Path Planning with Limit-Cycle Circle Set for Marine Surface Vehicle. In Proceedings of the 2012 IEEE International Conference on Robotics and Automation ICRA, Saint Paul, MN, USA, 14–18 May 2012; pp. 2275–2280.
14. Nash, A.; Koenig, S.; Likhachev, M. Incremental Phi*: Incremental Any-Angle Path Planning on Grids. In Proceedings of the International Joint Conference on Artificial Intelligence (IJCAI), Pasadena, CA, USA, 11–17 July 2009; pp. 1824–1830.
15. Nash, A.; Koenig, S.; Tovey, C. Lazy Theta*: Any-Angle Path Planning and Path Length Analysis in 3D. In Proceedings of the Third Annual Symposium on Combinatorial Search, SOCS 2010, Stone Mountain, Atlanta, GA, USA, 8–10 July 2010; pp. 147–154.
16. Daniel, K.; Nash, A.; Koenig, S.; Felner, A. Theta*: Any-Angle Path Planning on Grids. *J. Artif. Intell. Res.* **2010**, *39*, 533–579. [CrossRef]
17. Yang, J.M.; Tseng, C.M.; Tseng, P.S. Path planning on satellite images for unmanned surface vehicles. *Int. J. Nav. Arch. Ocean* **2015**, *7*, 87–99. [CrossRef]
18. Liu, C.G.; Mao, Q.Z.; Chu, X.M.; Xie, S. An Improved A-Star Algorithm Considering Water Current, Traffic Separation and Berthing for Vessel Path Planning. *Appl. Sci.* **2019**, *9*, 1057. [CrossRef]

19. Singh, Y.; Sharma, S.; Sutton, R.; Hatton, D.; Khan, A. A constrained A* approach towards optimal path planning for an unmanned surface vehicle in a maritime environment containing dynamic obstacles and ocean currents. *Ocean Eng.* **2018**, *169*, 187–201. [CrossRef]

20. Larson, J.; Bruch, M.; Ebken, J. Autonomous navigation and obstacle avoidance for unmanned surface vehicles. *Proc. SPIE* **2006**, *6230*. [CrossRef]

21. Svec, P.; Schwartz, M.; Thakur, A.; Gupta, S.K. Trajectory Planning with Look-Ahead for Unmanned Sea Surface Vehicles to Handle Environmental Disturbances. In Proceedings of the 2011 IEEE/RSJ International Conference on Intelligent Robots and Systems, San Francisco, CA, USA, 25–30 September 2011; pp. 1154–1159.

22. Naeem, W.; Irwin, G.W.; Yang, A.L. COLREGs-based collision avoidance strategies for unmanned surface vehicles. *Mechatronics* **2012**, *22*, 669–678. [CrossRef]

23. Svec, P.; Gupta, S.K. Automated synthesis of action selection policies for unmanned vehicles operating in adverse environments. *Auton. Robot.* **2012**, *32*, 149–164. [CrossRef]

24. Soulignac, M. Feasible and Optimal Path Planning in Strong Current Fields. *IEEE Trans. Robot.* **2011**, *27*, 89–98. [CrossRef]

25. Naus, K.; Wąż, M. The idea of using the A* algorithm for route planning an unmanned vehicle "Edredon". *Zesz. Nauk. Akad. Morska W Szczec.* **2013**, *36*, 143–147.

26. Kim, H.; Park, B.; Myung, H. Curvature Path Planning with High Resolution Graph for Unmanned Surface Vehicle. In *Robot Intelligence Technology and Applications 2012*; Springer: Berlin/Heidelberg, Germany, 2013; pp. 147–154.

27. Lee, T.; Kim, H.; Chung, H.; Bang, Y.; Myung, H. Energy efficient path planning for a marine surface vehicle considering heading angle. *Ocean Eng.* **2015**, *107*, 118–131. [CrossRef]

28. Fossen, T.I. *Handbook of Marine Craft Hydrodynamics and Motion Controli*, 1st ed.; John Wiley & Sons Ltd.: Chichester, UK, 2011.

29. Dhanak, M.R.; Xiros, N.I. *Springer Handbook of Ocean Engineering*; Springer: Berlin/Heidelberg, Germany, 2016.

30. Kim, H.; Kim, D.; Shin, J.U.; Kim, H.; Myung, H. Angular rate-constrained path planning algorithm for unmanned surface vehicles. *Ocean Eng.* **2014**, *84*, 37–44. [CrossRef]

Article

Multitask-Based Trajectory Planning for Redundant Space Robotics Using Improved Genetic Algorithm

Suping Zhao [1,2], **Zhanxia Zhu** [1,2,*] **and Jianjun Luo**[1,2]

[1] National Key Laboratory of Aerospace Flight Dynamics, Northwestern Polytechnical University, Xi'an 710072, China; suping36@mail.nwpu.edu.cn (S.Z.); jjluo@nwpu.edu.cn (J.L.)

[2] School of Astronautics, Northwestern Polytechnical University, Xi'an 710072, China

* Correspondence: zhuzhanxia@nwpu.edu.cn; Tel.: +86-138-9189-8260

Received: 29 April 2019; Accepted: 20 May 2019; Published: 30 May 2019

Abstract: This work addresses the multitask-based trajectory-planning problem (MTTP) for space robotics, which is an emerging application of successively executing tasks in assembly of the International Space Station. The MTTP is transformed into a parameter-optimization problem, where piecewise continuous-sine functions are employed to depict the joint trajectories. An improved genetic algorithm (IGA) is developed to optimize the unknown parameters. In the IGA, each chromosome consists of three parts, namely the waypoint sequence, the sequence of the joint configurations, and a special value for the depiction of the joint trajectories. Numerical simulations, including comparisons with two other approaches, are developed to test IGA validity.

Keywords: space robotics; redundant; free-floating base; multiple tasks; trajectory planning; genetic algorithm

---

1. Introduction

Space robotics are now increasingly employed in outer space for various tasks, such as the assembly and maintenance of the International Space Station (ISS) [1], and Lunar Base construction [2]. Examples of space robotics include the Japanese Experiment Module Remote Manipulator System (JEMRMS) of the Japan Aerospace Exploration Agency (JAXA) [3] and the Canadarm 2 of MacDonald Dettwiler and Associates Ltd. (MDA) [4], while examples of performing tasks include the Experimental Test Satellite VII (EST-VII) [5] and the Robot Technology Experiment [6]. The free-flying space robot (FFSR1) and the free-floating space robot (FFSR2) are two main types of space robotics. The base attitude of FFSR1 is controlled, contributing to establishing a good connection between the ground and the base spacecraft. This is because the attitude of the base spacecraft should be well-managed when sending signals to, or receiving signals from, the ground. The base spacecraft of FFSR2 is in free-floating mode, contributing to saving energy, which is part of the superiority of FFSR2 since energy is precious in outer-space environments.

The trajectory planning for space robotics, aiming at generating the time histories of joints (or end effector) and contributing to the desired motion of robots, attracts extensive attention from both scientists and practitioners. The study of trajectory planning is essential for robots before physical manipulation and has been well developed. A primary trajectory-planning approach is based on the inverse kinematics of space robotics. Concepts such as the generalized Jacobian matrix [7], the enhanced disturbance map [8], the path independent workspace [9] and the reaction null-space [10] have been successively proposed for trajectory planning. The disadvantage of this approach is that it leads to kinematic singularity in some cases. Aiming at avoiding this kinematic singularity, an approach based on the direct kinematics of space robotics has been well-received. To begin with, mathematical functions such as the polynomial, trigonometric [11], or the Bézier function [12] are employed to depict the joint trajectories. Then, according to predefined conditions, mathematical functions are depicted with a set of unknown coefficients. In the aforementioned mathematical functions, the polynomial function denotes the combination of constants and variables with limited addition and multiplication calculations, the trigonometric function is an elementary function with variables of angles, and the Bézier function contributes to the derivation of a smooth curve based on four random points. Next, the trajectory-planning problem is converted into a parameter-optimization problem, with the objective function of minimum maneuver time or maximum manipulability of the robotic system, or minimum attitude disturbance acting on the free-floating base spacecraft during the robotic maneuver. Finally, the unknown coefficients are optimized by the optimization algorithms, including the basic heuristic algorithms such as the particle-swarm optimization algorithm (PSO) [13] and genetic algorithm (GA) [14,15], and the improved optimization algorithms such as hybrid PSO [16] and the constrained differential evolution algorithm (DE) [17]. In the aforementioned algorithms, the PSO is a random search algorithm based on group collaboration, more specifically, simulating the foraging behavior of a flock of birds. The DE, based on population, is a self-adaptive optimization algorithm with global search capability. To the authors' knowledge, the approaches mentioned above aim at solving the point-to-point trajectory-planning problem, meaning that the space robot executes one task in each travel. If the robot executes two or more tasks in each travel, it could save much energy. Maneuvering time would also be reduced, which is a significant advantage in an emergency. Therefore, it is meaningful to study the multitask-based trajectory-planning problem (MTTP) for space robotics.

As a matter of fact, the MTTP for industrial robotics has been widely studied for its high productivity and low cost [18,19], and two categories of approaches were developed. For simplicity, the location of each task is usually considered to be a waypoint, which is also followed. One category of approaches aims at directly solving the MTTP. In [20], the MTTP is studied using the branch-and-bound method, where multiple inverse kinematic solutions of the robotic system are not considered. The branch-and-bound method aims at searching for solutions to solve optimization problems with constraints, where the feasible solution space is finite.

In [21], the MTTP is transformed into a mixed-integer nonlinear programming problem, where a solver is developed to improve calculation speed. The approach developed in [21] works well with a relatively small number of waypoints. In [22], an improved genetic algorithm (IGA) was developed where each chromosome consists of two parts. The first part represents the waypoint sequence, while the second part represents the sequence of the joint configurations at the waypoints, which corresponds to the first part. In [23], each chromosome consists of three parts, the waypoint sequence, the sequence of joint configurations, and the robot placement. The technique of dividing each chromosome into several parts was also employed in [24–26]. A common point of IGAs in [22–26] is that the parameter denoting the joint angular velocity is predefined as a constant. The other category of approaches aims at dividing MTTP into several subproblems which are then solved successively. In [27,28], MTTP is divided into two subproblems, the problem of the waypoint sequence, and the problem of joint trajectories. In [27], Tabu search was employed to optimize the waypoint sequence, after which joint trajectories were derived according to the inverse kinematics of industrial robotics. The Tabu search algorithm, a meta-heuristic algorithm, searches for the global optimal solution by constructing a Tabu table with the functions of cycling and memory. In [28], an improved lazy algorithm, according to the direct kinematics of industrial robotics, was developed to optimize the joint trajectories. Among MTTP documents for industrial robotics, manipulators are nonredundant, meaning that a point in the task space corresponds with finite points in the joint space. However, the redundant robot [29] provides the manipulator with high dexterity, contributing to the solution of problems such as obstacle avoidance, singularity avoidance, and joint limits. Unlike ground environments [30,31], there is microgravity in outer-space environments. Moreover, for FFSR2, strong coupling between manipulator and free-floating base is usually generated during a maneuver. Therefore, the aforementioned approaches, which were developed to solve MTTP for nonredundant industrial robotics, cannot be directly employed to solve MTTP for redundant space robotics.

In some cases, an urgent task should be performed in ISS, while the functional completeness of ISS should be guaranteed in advance. Component maintenance, assembly, and refueling of ISS should be finished successively, contributing to the functional completeness of ISS before performing the urgent task. Finishing a series of tasks successively also contributes to saving energy in outer space. Therefore, The MTTP for redundant space robotics is studied. In the MTTP, the location of each task is simplified as a task point, called waypoint. The position and attitude of each waypoint are predefined in the Cartesian space. First, the end effector is required to visit the waypoints with minimum time, thus the sequential order of visiting the waypoints should be optimized. Second, the joint configuration corresponding to each waypoint should be optimized, thus the feasible joint movements between adjacent waypoints can be guaranteed. Third, the joint movements should meet the predefined constraints. Piecewise continuous-sine functions with cubic polynomial arguments are employed to depict the joint trajectories along the waypoints, where each piece of sine function depicts one joint trajectory between adjacent waypoints. With predefined conditions, each piece of sine function is depicted with one unknown parameter which should be optimized. The MTTP is converted into a parameter-optimization problem. An IGA is developed to optimize the unknown parameters. In the IGA, each chromosome consists of three parts. The first part denotes the waypoint sequence, the second part denotes the sequence of joint configurations corresponding with the first part, and the third part denotes a special value corresponding with the depiction of the joint trajectories. Since the system is redundant, each waypoint corresponds with infinite joint configurations. Moreover, considering the sequence of joint configurations directly leads to combination explosion. An approach based on the concept of the dual-arm angle [32] was employed to formulate joint configurations. At each waypoint, eight joint configurations were derived with an assignment of the arm angle.

The rest is organized as follows. Preliminaries and adopted notation for space robotics are presented in Section 2. In the same section, the MTTP, the objective functions, the approach to depicting the joint configurations at each waypoint, and the approach of formulating the joint trajectories are explained. The IGA encoding and updating mechanisms, together with the IGA optimization process, are introduced in Section 3. In Section 4, numerical simulations are carried out to validate IGA, including comparisons with two other approaches. Section 5 concludes the work and gives an outlook for further research.

2. Preliminaries and Notation

2.1. Kinematic Models of Space Robotic Systems

The 2D representation of a space robotic system is outlined in Figure 1, where the system consists of $n + 1$ bodies $B_i (i = 0, ..., n)$. B_0 denotes the base spacecraft, usually called base, and $B_i (i = 1, ..., n)$ denotes the ith rigid link. The $n + 1$ bodies are connected by n revolute joints $J_i (i = 1, ..., n)$ and each joint provides the system with a single DOF (degree of freedom). Additional symbols in Figure 1 are explained as follows:

- $C_i \in R^1 (i = 0,...,n)$: mass center of body i;
- $a_i \in R^3 (i = 1,...,n)$: vector from joint i to mass center of body i with unit of [m; m; m], where m denotes the abbreviation of meter;
- $b_i \in R^3 (i = 0,...,n)$: vector from mass center of body i to joint $i + 1$ with unit of [m; m; m];
- $r_i \in R^3 (i = 0,...,n)$: position of mass center of body i with unit of [m; m; m];
- $p_i \in R^3 (i = 1,...,n)$: position of joint i with unit of [m; m; m];
- $r_g \in R^3$: position of system centroid with unit of [m; m; m];
- $p_e \in R^3$: position of end effector with unit of [m; m; m];
- $\Sigma_i (i = 0,...,n)$: coordinate frame fixed on B_i with origin O_i and axes X_i and Y_i;
- Σ_I: inertial coordinate frame with origin O_I and axes X_I and Y_I; and
- Σ_E: coordinate frame fixed on end effector with origin O_E and axes X_E and Y_E.

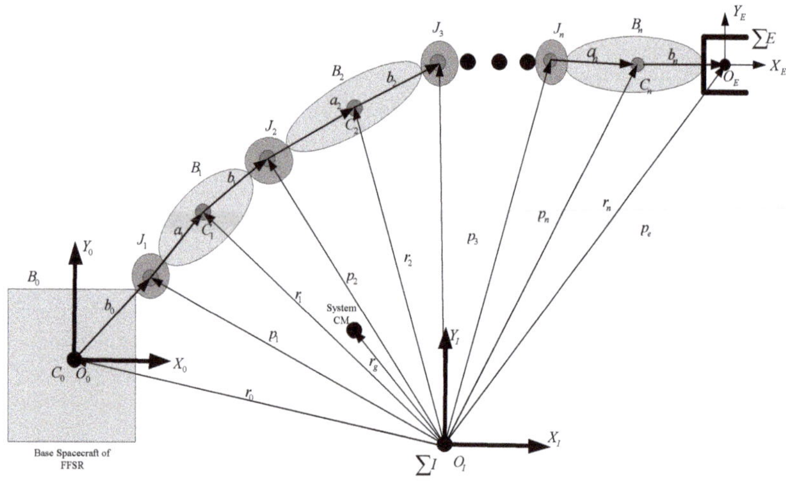

Figure 1. Two-dimensional sketch of a space robotic system.

The coordinate frame Σ_I coincides with the coordinate frame Σ_0 at initial time, and the coordinate frame Σ_i ($i = 1, ..., n$) is established using the D-H approach [33,34]. The D-H approach aims at establishing a coordinate frame fixed on each link and computing the coordinate transformation between adjacent coordinate frames by four D-H parameters. The coordinate frame Σ_i ($i = 1, ..., n$) is defined as follows: Z_i-axis is chosen along rotational direction of $(i + 1)$th joint, origin O_i is located at intersection of Z_i-axis with common perpendicular line of Z_{i-1}- and Z_i-axis, X_i-axis is chosen along common perpendicular line of Z_{i-1}- and Z_i-axis with positive direction from ith joint to $(i + 1)$th joint, and Y_i-axis is chosen to follow the right-handed rule. After the establishment of Σ_i ($i = 1, ..., n$), four D-H parameters a_i, α_i, d_i, and θ_i are specified as follows: a_i denotes the distance between Z_i- and Z_{i+1}-axis along X_i-axis; α_i denotes the angle from Z_i- to Z_{i+1}-axis around X_i-axis; d_i denotes the distance between X_{i-1}- and X_i-axis along Z_i-axis; and θ_i denotes the angle from X_{i-1}- to X_i-axis around Z_i-axis.

Two categories of space robotic systems are usually employed for manipulation tasks, the FFSR1 and the FFSR2, which are collectively called FFSR in Figure 1. The base attitude of FFSR1 is controlled, while the base of FFSR2 is in free-floating mode. If both the position and attitude of the base are fixed, the differential kinematic equation is formulated as:

$$\begin{bmatrix} v_e \\ \omega_e \end{bmatrix} = J_m \dot{q} \tag{1}$$

J_m in (1) is identical to the Jacobian matrix of industrial robotics. FFSR1 satisfies the linear momentum-conservation law, contributing to the following differential kinematic equation:

$$\begin{bmatrix} v_e \\ \omega_e \end{bmatrix} = \begin{bmatrix} -\dfrac{1}{M} J_{bv} J_{tw} + J_{mv} \\ J_{mw} \end{bmatrix} \dot{q} \tag{2}$$

However, FFSR2 satisfies the linear and angular momentum-conservation laws shown in Equation (3), where momentum value is usually set to 0 for simplicity.

$$H_b \dot{x}_b + H_{bm} \dot{q} = 0 \tag{3}$$

Since matrix H_b in Equation (3) is symmetric and positively definite,the velocity of the free-floating base can be expressed as:

$$\dot{x}_b = \begin{bmatrix} v_b \\ \omega_b \end{bmatrix} = -H_b^{-1} H_{bm}\dot{q} = J_{bm}\dot{q} \tag{4}$$

Therefore, the differential kinematic equation of FFSR2 can be expressed as:

$$\begin{bmatrix} v_e \\ \omega_e \end{bmatrix} = J_b \begin{bmatrix} v_b \\ \omega_b \end{bmatrix} + J_m\dot{q} = [J_m + J_b J_{bm}]\dot{q} = J_g\dot{q} \tag{5}$$

The symbols in Equations (1)–(5) are explained as follows:

- $q \in R^n$: joint configuration with unit of each element of q deg;
- $\dot{q} \in R^n$: angular velocity of joint configuration with unit of each element of $\dot{q}$ deg/s;
- $x_b \in R^6$: pose of base with unit of [m; m; m; deg; deg; deg];
- $\dot{x}_b \in R^6$: velocity of base with unit of [m/s; m/s; m/s; deg/s; deg/s; deg/s];
- $v_e \in R^3$: linear velocity of end effector with unit of [m/s; m/s; m/s];
- $\omega_e \in R^3$: angular velocity of end effector with unit of [deg/s; deg/s; deg/s];
- $J_b \in R^{6\times6}$: Jacobian matrix of base;
- $J_m \in R^{6\times n}$: Jacobian matrix of manipulator;
- $J_{bm} \in R^{6\times n}$: coupling Jacobian matrix;
- $H_b \in R^{6\times6}$: inertia matrix of base; and
- $H_{bm} \in R^{6\times n}$: coupling inertia matrix.

2.2. MTTP for Space Robotics

Space robots are required to execute a series of tasks, where the location of each task is simplified as a waypoint in the Cartesian space. The MTTP denotes that the end effector is constrained to visit a set of waypoints with minimum time. The position and attitude of the waypoints are predefined, while the optimal waypoint sequence is not given. Moreover, the space robot should meet the following constraints. First, the joint angular velocities need to be zero at each waypoint. Second, the joint angles and joint angular velocities are constrained to be within certain ranges. Third, for FFSR2, the attitude disturbance acting on the free-floating base should be minimized during the FFSR2 maneuver.

2.2.1. Objective Functions

A. Minimum Maneuver Time

In the MTTP for industrial robotics, minimum maneuver time [22,23] has been considered to improve the work efficiency of robots, which is also a factor for space robotics. Another key factor is to reduce the accident risk in outer space. In some cases, an urgent task should be performed in ISS, while the functional completeness of ISS should be guaranteed in advance. Component maintenance, assembly, and refueling of ISS should be finished in the shortest time, contributing to the functional completeness of ISS before performing the urgent task. Thus, urgent tasks are performed in time and the accident risk is reduced. One of the developed objectives is that the space robot is required to execute a series of tasks such as refueling with minimum maneuver time. Therefore, the end effector is required to visit a set of predefined waypoints with minimum time, which is expressed as:

$$\min F_1 := \sum_{j=1}^{N-1} T_j \tag{6}$$

In Equation (6), N denotes the number of waypoints and T_j denotes the time period during which the end effector moves along the jth subpath, i.e., from the jth waypoint to the $(j+1)$th waypoint. T_j is expressed as:

$$T_j = \max_i \left(t^i_{jf} - t^i_{js} \right) \tag{7}$$

In Equation (7), $i\ (i = 1, ..., n)$ denotes the ith joint. t^i_{js} and t^i_{jf} denote the initial and the final moving time instants of the ith joint, respectively, corresponding with the jth subpath. The n joints move for different time periods due to unequal angular distances and unequal angular velocities. The maximum of $\left(t^i_{jf} - t^i_{js} \right) (i = 1, ..., n)$ is defined as the time period employed by the space robot for the jth subpath.

B. Minimum Base Attitude Disturbance

During the FFSR2 maneuver, minimum attitude disturbance acting on the free-floating base is required, which is expressed as:

$$\min F_2 := \sqrt{\alpha_0^2 + \beta_0^2 + \gamma_0^2} \tag{8}$$

In Equation (8), α_0, β_0 and γ_0 are employed to depict the base attitude of FFSR2, denoting the Euler angles around the X-, Y-, and Z-axis of Σ_0, respectively. The order of rotation around the axes is $Z - Y - X$.

Remark 1. *The MTTP aims at searching for the optimal sequence of waypoints and the optimal joint movements. If not consider the optimal joint movements, the MTTP is the Traveling Salesman Problem (TSP) [18,20]. The TSP denotes that a salesman is required to visit a set of predefined cities with minimum time (or path length), while the time employed to stay at each city is not considered. This is because the time spent at each city has no influence on the optimal sequence of the cities, which also works in the MTTP. Therefore, the time period during which the end effector executes specific tasks at each waypoint is omitted.*

2.2.2. Depiction of Joint Configurations Corresponding to Each Waypoint

The inverse kinematics of redundant space robotics are complex, since one point in the task space corresponds to infinite points in the joint space. An approach based on the geometric construction of the robot, especially the robot called Spherical Revolute Spherical (SRS) [32,35–37], is well developed. Each component of the joint configuration is formulated as a function of the arm angle. By assigning a value to the arm angle, a finite number of joint configurations are obtained. Figure 2 presents the process for constructing the arm angle. The following legend is useful to understand the different subfigures. Figure 2a outlines the 7-DOF rotational robot SRS based on the D-H approach, together with the three main points S, E, and W. The shoulder point S denotes the intersection point of Z_i-axis ($i = 0, 1, 2$), the elbow point E denotes the origin O_3 of Σ_3, and the wrist point W denotes the intersection point of Z_i-axis ($i = 4, 5, 6$). Figure 2b outlines the reference plane constructed by Z_0-axis and ω, the arm plane SEW, and the arm angle ψ from the reference plane to the arm plane. Figure 2c coincides with Figure 2b, where ω denotes the vector from S to W. k and e denotes the vectors perpendicular to ω in the reference plane and the arm plane, respectively. Figure 2d outlines a case $\psi = 0$, meaning that the reference plane coincides with the arm plane. The following procedure is carried out for the derivation of the joint configuration $\mathbf{q} = [q_1, ..., q_7]^T$.

Step 1 Construction of the arm plane, the reference plane and the arm angle. First, the arm plane is constructed according to the points S, E, and W shown in Figure 2. Second, the reference plane is constructed based on ω and Z_0- or X_0-axis, since at least one of Z_0- and X_0-axes is not coincident with ω. The arm angle ψ is then depicted, shown in Figure 2.

Step 2 Derivation of the joint angle q_4. Project the arm plane SEW to the plane perpendicular to the rotation axis of the elbow joint, namely Z_3-axis. According to the geometric construction of SRS and the Pythagorean theorem, two values of q_4 are derived.

Step 3 Derivation of the shoulder-joint angles q_1, q_2, and q_3. First, the rotation transformation, rotating around ω with the rotation angle ψ, is derived. Second, the orientation of Σ_3 relative to Σ_0 using the concept of the shoulder reference attitude matrix [32]. Finally, two groups of shoulder-joint angles are derived.

Step 4 Derivation of the wrist-joint angles q_5, q_6, and q_7. Two groups of wrist-joint angles are derived according to the predefined attitude of the end effector, the rotation transformation and the should reference attitude matrix derived in **Step 3**, and the rotation matrix from Σ_3 to Σ_4.

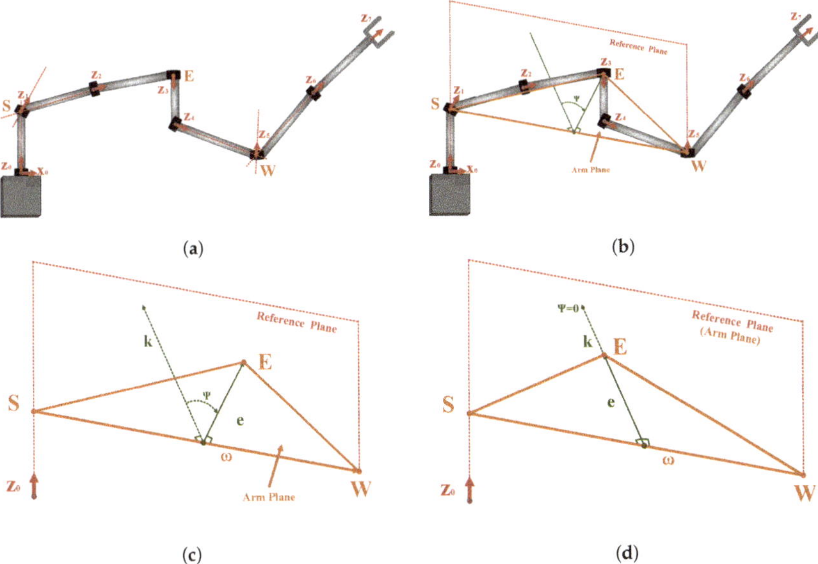

Figure 2. Process for constructing the arm angle ψ. (**a**) the 7-DOF robot SRS; (**b**) the arm angle ψ between the reference plane and the arm plane in SRS; (**c**) the arm angle ψ between the reference plane and the arm plane with vectors k and e, and (**d**) the case when the arm angle ψ is set to 0.

Remark 2. *A detailed explanation about the approach of computing* **q** *is expressed in [32]. According to the aforementioned procedure, eight groups of* **q** *are derived with an assignment of the arm angle ψ. ψ is set to zero considering the following aspects. First, the calculation accuracy of* **q** *corresponding to each waypoint declines as the absolute value of ψ increases. The result of* **q** *is derived with the highest accuracy in the case $\psi = 0$. Second, the assignment of $\psi = 0$ contributes to less computational cost [32]. Besides, the angular velocity of each joint is constrained to be zero at each waypoint, and is not affected by values of ψ.*

Remark 3. *Inspired by the work of [32], the joint configuration* **q** $= [q_1, ..., q_7]^T$ *denotes that the number of joints is seven, which is also the manipulator DOF meaning that the DOF of each joint is one. This category of robots has simple geometrical structure and contributes to convenient operation when performing tasks and less computational cost when verifying a developed approach. Study and application of this category of robots can also be found in [11–13,17,32].*

2.2.3. Formulation of Joint Trajectories among Waypoints

During the maneuver of the space robot when the end effector successively visits the waypoints, the joint trajectories are depicted with the piecewise-sine functions expressed in Equation (9). It can be seen that the sine function has the superiority of considering the physical limits on joint angles, and making the joint movements more practical.

$$q_j^i(t) = A \cdot \sin\left[a_{3j}^i(t - T_{j-1})^3 + a_{2j}^i(t - T_{j-1})^2 + a_{1j}^i(t - T_{j-1}) + a_{0j}^i\right] \tag{9}$$

In Equation (9), $i = 1, ..., n$, $j = 1, ..., N - 1$, and $T_0 = 0$. $q_j^i(\cdot)$ denotes an angular trajectory of the ith joint, corresponding with a jth subpath along which the end effector moves from the jth waypoint to the $(j + 1)$th waypoint. A denotes the physical limit on each joint angle, which is defined as 180deg. $a_{mj}^i (m = 0, 1, 2, 3)$ denotes the coefficients of the argument, where all a_{3j}^i share the same absolute value. Based on Equation (9), joint angular velocities are given by:

$$\dot{q}_j^i(t) = A \cdot \cos \left[a_{3j}^i (t - T_{j-1})^3 + a_{2j}^i (t - T_{j-1})^2 + a_{1j}^i (t - T_{j-1}) + a_{0j}^i \right] \cdot$$
$$\left[3a_{3j}^i (t - T_{j-1})^2 + 2a_{2j}^i (t - T_{j-1}) + a_{1j}^i \right] \tag{10}$$

Corresponding with the jth subpath of the end effector, the angle of the ith joint is denoted as q_{j0}^i at initial time t_{js}^i ($t_{js}^i = T_{j-1}$), and denoted as q_{jf}^i at final time t_{jf}^i, shown in Equations (11) and (12). To ensure the continuity of the joint movements during the whole travel, the space robot has the same joint configuration at the final time of jth subpath and at the initial time of $(j + 1)$th subpath, shown in Equation (13).

$$q_j^i(t_{js}^i) = q_{j0}^i \tag{11}$$

$$q_j^i(t_{jf}^i) = q_{jf}^i \tag{12}$$

$$q_{jf}^i = q_{(j+1)0}^i \tag{13}$$

In Equation (13), $q_{N0}^i = q_{(N-1)f}^i$. Moreover, joint angular velocities are limited to be zero at each waypoint, given by:

$$\dot{q}_{j0}^i = \dot{q}_{jf}^i = 0 \tag{14}$$

Substituting Equations (11)–(14) into Equations (9)–(10), the values of a_{0j}^i, a_{1j}^i, a_{2j}^i and t_{jf}^i are derived as follows:

$$a_{0j}^i = \arcsin\left(\frac{q_{j0}^i}{A}\right) \tag{15}$$

$$a_{1j}^i = 0 \tag{16}$$

$$a_{2j}^i = -\frac{3}{2} a_{3j}^i \left(t_{jf}^i - T_{j-1}\right) \tag{17}$$

$$t_{jf}^i = \left\{ \frac{2\left[a_{0j}^i - \arcsin\left(\frac{q_{jf}^i}{A}\right)\right]}{a_{3j}^i} \right\}^{\frac{1}{3}} + T_{j-1} \tag{18}$$

If the optimal value of $a_{3j}^i (i = 1, ..., n; j = 1, ..., N - 1)$ is derived, the joint trajectories corresponding with the jth subpath of the end effector and time period T_j are obtained. It should be noted that the case, $a_{3j}^i = 0$, is meaningless and not considered.

3. Improved Genetic Algorithm

3.1. Basic GA

The basic GA [38,39], inspired by natural selection, is a heuristic optimization algorithm with global search abilities. A certain number of individuals, usually called initial solutions, are randomly generated and contribute to the globality of the search space. During evolution, each individual, corresponding to a chromosome, is evolved toward the best position in the whole search space. Furthermore, chromosomes are updated according to fundamental genetic operators, including reproduction, crossover, and mutation. The reproduction generator contributes to the replication of chromosomes with better fitness values from parents to offspring. Crossover and mutation generators contribute to generating new chromosomes. On the basis of a crossover generator, the paired chromosomes exchange part of their genes with each other. On the basis of the mutation generator, one or several genes of a chromosome are updated. The flowchart of basic GA is shown in Figure 3.

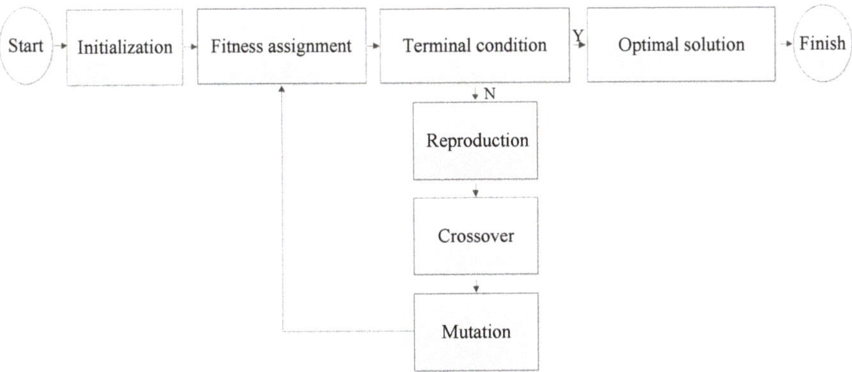

Figure 3. Flowchart of basic GA.

3.2. Improved GA

The basic GA is usually employed to optimize the polynomial coefficient or the path length in some problems, where chromosomes are assigned a category of meaning. Referring the composition mechanism of chromosomes studied in [22], an IGA was developed. Chromosomes were assigned three categories of meanings. Each IGA chromosome consists of three parts, as shown in Figure 4, where $\{W_1, W_2, ..., W_N\}$ denotes the waypoint sequence, $\{C_1, C_2, ..., C_N\}$ denotes the sequence of joint configurations, and OPC denotes a special value corresponding with the cubic coefficients of the arguments in the piecewise-sine functions. Joint configuration C_k $(k = 1, ..., N)$ corresponds with waypoint W_k.

Since the time period employed by the space robot for each subpath is positive, value

$$\left\{ \frac{2\left[a_{0j}^i - arcsin\left(\frac{q_{if}^i}{A} \right) \right]}{a_{3j}^i} \right\}^{\frac{1}{3}}$$

in (18) is positive. Therefore, a_{3j}^i is expressed as:

$$u_{3j}^i = sign\left(u_{0j}^i - arcsin\left(\frac{q_{if}^i}{A} \right) \right) \cdot abs\left(a_{3spe} \right) \tag{19}$$

In [24], a_{3spe} denotes the special value after decoding OPC, where $abs\left(\cdot \right)$ denotes the absolute value of $(\cdot)$.

The encoding mechanism is an essential component of the IGA, which is studied first. Then, the IGA updating mechanism, including initialization, reproduction, crossover, and mutation, is studied. Finally, the IGA work mechanism is developed.

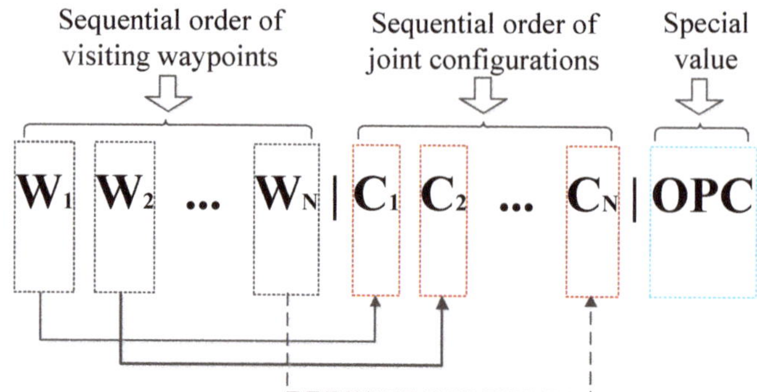

Figure 4. Composition of each chromosome in the improved genetic algorithm (IGA).

3.2.1. Encoding Mechanism

The encoding mechanism is essential, and the theoretical principle of IGA is implemented based on it. Traditional encoding mechanisms include the binary, the floating, the integer, and the symbol. The binary mechanism provides convenient encoding and decoding operations when solving discrete or continuous optimization problems. The floating mechanism provides results with high precision when solving continuous optimization problems with multidimensional functions. The integer mechanism is usually employed for solving combinatorial optimization problems such as the Traveling Salesman Problem [40]. The symbol mechanism contributes to the combination of GA and other algorithms. In IGA, the integer encoding mechanism is employed for the waypoint sequence, while the binary encoding mechanism is employed for the sequence of joint configurations and the special value corresponding with the cubic coefficients of the arguments, which is sketched in Figure 5.

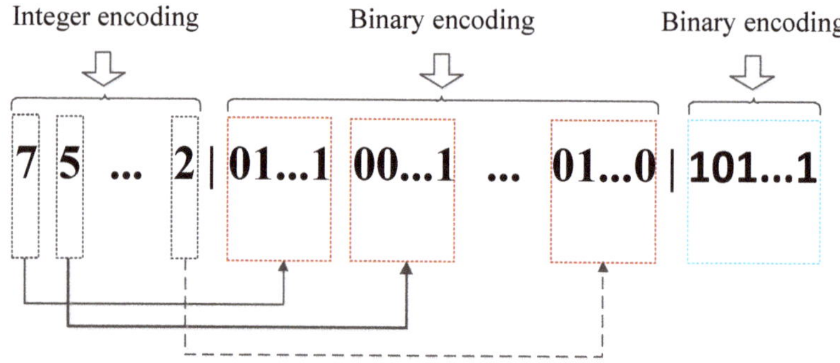

Figure 5. Encoding mechanisms of each chromosome.

A. First Part of Each Chromosome

In MTTP, the end effector is constrained to visit a set of predefined waypoints, and each waypoint is limited to being visited once. According to binary encoding, the random selection of the crossover and mutation mechanisms may lead to alphabet repetition, thus conflicting with the requirement of only visiting each waypoint once. Integer encoding is the best choice, where N positive integers correspond with N waypoints.

B. Second Part of Each Chromosome

The second part of each chromosome denotes the sequence of joint configurations corresponding to the first part of the chromosome, where binary encoding is implemented. Since eight special joint configurations corresponding to one waypoint are considered, each joint configuration is coded by one byte of three bits. Figure 6 represents the eight categories of bytes, where $CON_{im}(i = 1, ..., n; m = 1, ..., 8)$, also shown in Equation (20), and it denotes the mth joint configuration corresponding to the ith waypoint.

$$\mathscr{CON} = \begin{bmatrix} CON_{11} & CON_{11} & ... & CON_{18} \\ CON_{21} & CON_{21} & ... & CON_{28} \\ & ... & & \\ CON_{n1} & CON_{n1} & ... & CON_{n8} \end{bmatrix}. \tag{20}$$

C. Third Part of Each Chromosome

The third part of each chromosome denotes a special value corresponding to the cubic coefficients of the arguments in the piecewise-sine functions. According to the relation between a_{3j}^i and a_{3spe}, a_{3spe} is limited to be within $[-\pi, 0) \cup (0, \pi]$. Moreover, optimization a_{3spe} is constrained to be with an accuracy of $1e - 5$, where $524,288 = 2^{19} < 2\pi \cdot 10^5 < 2^{20} = 1,048,576$. Hence, the genes of this part of the chromosome consist of 20 bits.

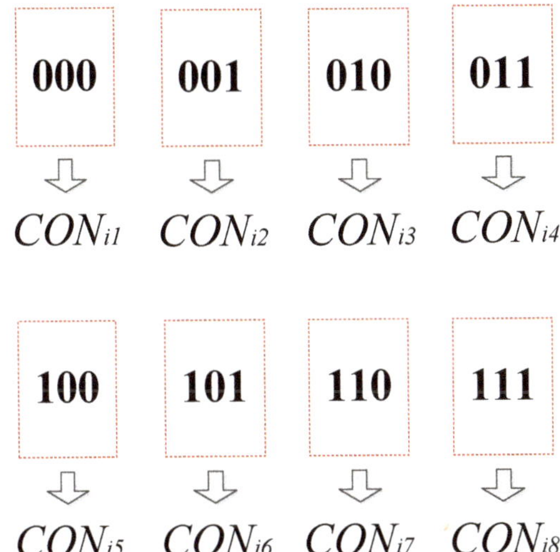

Figure 6. Eight categories of bytes and corresponding joint configurations.

D. Example of Chromosome Depiction

In this example, a space robot is constrained to visit five waypoints, and a feasible chromosome is shown in Figure 7, where the waypoint sequence is $\{5, 1, 2, 3, 4\}$ according to the first part of the chromosome. By the second part of the chromosome, the encoding of the joint configuration corresponding with the 5th waypoint is 001, namely CON_{52} in Equation (20) according to Figure 6. Similarly, encoding the joint configurations corresponding to the 1st, 2nd, 4th, and 3rd waypoints are CON_{14}, CON_{21}, CON_{42}, and CON_{35}, respectively. The decimal number of the third part of the chromosome is $325,588$, and the corresponding value within the range of $[-\pi, 0) \cup (0, \pi]$ is given by:

$$-\pi + 325,588 * \frac{\pi - (-\pi)}{2^{20} - 1} = -1.1906.$$

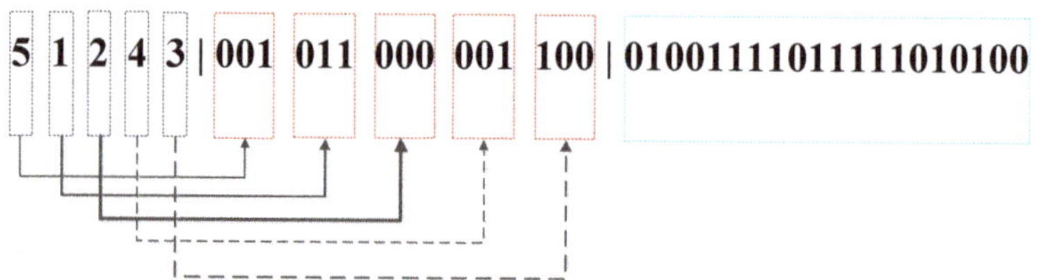

Figure 7. Feasible IGA chromosome in the example.

3.2.2. Updating Mechanism

A. Initialization

The initial population with 200 chromosomes is randomly generated. It should be noted that the first gene of each chromosome is 1, so that the first gene corresponds with the starting waypoint.

B. Reproduction

The reproduction operator aims at copying chromosomes from parents to offspring, where the roulette-wheel mechanism is employed in IGA. Each chromosome is assigned a proportion value, namely the fitness value of the chromosome to the sum of fitness values of the population. The chromosome with a large proportion of value is reproduced with high probability.

C. Crossover

The crossover operator is employed after the reproduction operator, where parent chromosomes are selected with a crossover probability defined as 0.6. The order, the two points, and the two-point crossover mechanisms are applied to the first, second, and third parts of every two selected and paired parent chromosomes, respectively. Figure 8 represents the crossover operation of two parent chromosomes and the offspring, where Par1 and Par2 denote two parent chromosomes, while Off1 and Off2 denote chromosomes after crossover operation.

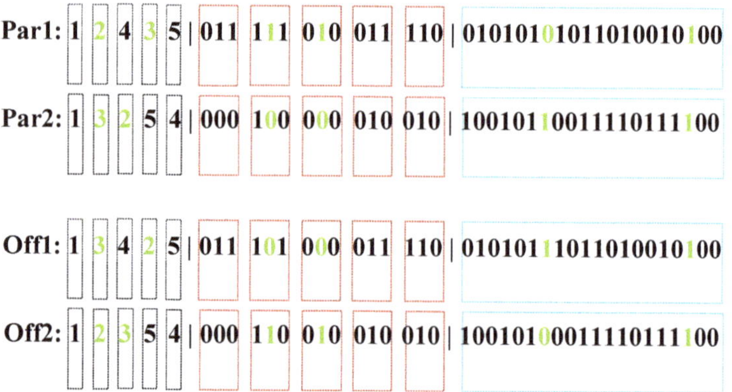

Figure 8. Two parent chromosomes and offspring after crossover operation.

D. Mutation

Mutation probability is defined as 0.15. For the first part of each chromosome, two genes are randomly selected to exchange values. Simultaneously, two random genes of the second part of the chromosome are selected, and the value of each selected gene is changed from ′1′ to ′0′ and vice versa. The mutation mechanism of the second part also works for the third part of the chromosome. Figure 9 shows the mutation operation of a chromosome, where Par denotes a selected parent chromosome, while Off denotes the chromosome after the mutation operation.

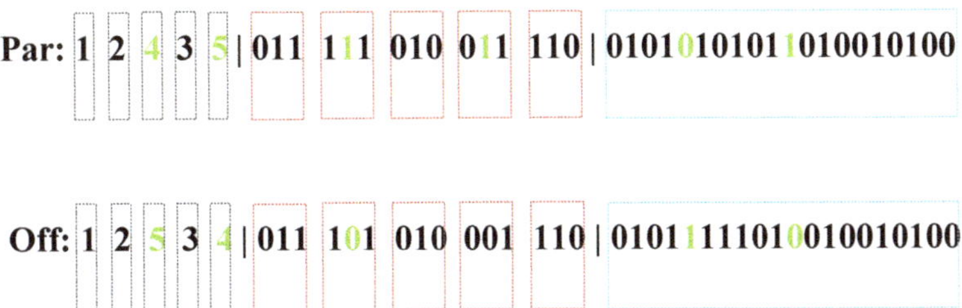

Figure 9. Parent chromosome and chromosome after mutation operation.

3.2.3. IGA Work Mechanism

Based on the above analysis, the IGA work mechanism is outlined in this part. Two-hundred individuals are randomly generated, where three parts of each chromosome are encoded according to Section 3.2.1. After this, the iterative loop starts. First, the fitness value of each chromosome, i.e., the value of the objective function, is computed after decoding. The chromosome with a minimum fitness value is denoted as the elite in each generation, which is saved directly for the next iteration. Second, the terminal condition is verified, where the maximum number of iterations is defined as 500. If the terminal condition is not followed, genetic operators, including reproduction, crossover, and mutation are successively executed as per Section 3.2.2. After this, the updated chromosomes, together with the saved elite, are denoted as the initial chromosomes for the next iteration. Once the terminal condition is satisfied, the loop stops, and the optimal solution is exported. Figure 10 shows the IGA flowchart.

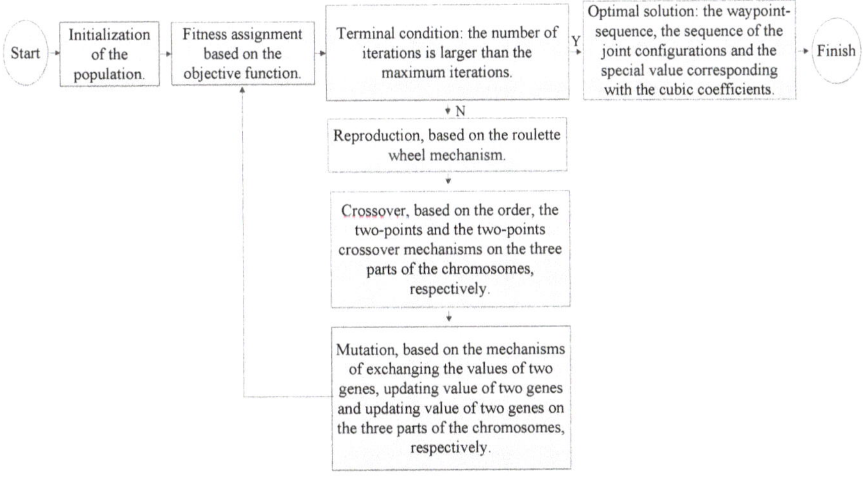

Figure 10. IGA flowchart.

4. Numerical Simulations

The space robotic system employed for simulation consists of a base spacecraft and a 7-DOF manipulator, and its parameters are shown in Table 1. Two categories of numerical simulations were developed by considering the state of the base spacecraft. For FFSR1, the base attitude is controlled, and only minimum maneuver time is required. For FFSR2, attitude disturbance acting on the free-floating base should also be minimized. Therefore, the objective function of MTTP for FFSR1 is $F : F = F_1$, while the objective function of MTTP for FFSR2 is $F : F = F_1 + \alpha F_2$. F_1 and F_2 are shown in Equations (6) and (8), respectively. α denotes the weight and is defined as 2.

Table 1. Physical parameters of the space robotic system.

Body	Mass	Length	Principle Moment of Inertial (kg $\cdot$m^2)		
	(kg)	(m)	I_{xx}	I_{yy}	I_{zz}
0	500	0.2	100	100	200
1	20	0.1	4	4	5
2	10	0.3	2	2	3
3	50	0.4	10	10	20
4	20	0.5	4	4	5
5	50	0.3	10	10	20
6	10	0.2	2	2	3
7	20	0.1	4	4	5

To validate the IGA, comparisons with two other algorithms were first developed. Afterward, two simulation cases on the detailed movements of space robotics using the IGA were developed.

4.1. Comparisons

Comparisons between IGA and two other algorithms, AL1 and AL2, were developed. In the IGA, the second part of each chromosome is encoded with the binary mechanism. In AL1, the integer mechanism is employed to encode the second part. This comparison refers to [22]. In both the IGA and AL1, the variation trend of each joint between adjacent waypoints is depicted with a sine function, whose argument is a cubic polynomial. The angular velocity of each joint, shown in Equation (10), varies during the maneuver of the space robot. Therefore, the third part of each chromosome is variable. However, joint angular velocities are defined as a constant in [20–26,41], leading to joint trajectories being depicted with the linear functions, and the third part of each chromosome being depicted with a constant. In AL2, joint angular velocity is defined as 0.8 rad/s (45.8 deg/s) [41].

Based on the IGA, AL1, and AL2, the MTTP for space robotics is solved successively. Two cases of space robotics are considered, where the MTTPs for FFSR1 and for FFSR2 are solved in Case 1 and in Case 2, respectively. To further validate the developed approach, different numbers of waypoints, 5, 7, 10, 15, 20, and 30, are considered in each case. With the same number of waypoints, each algorithm is carried out for 25 times. Tables 2–4 shows the average execution time (AET), the worst fitness (WF), the best fitness (BF) and the average fitness (AF) of 25 executions. The AET using AL2 is less than that using IGA or AL2. However, the optimization results using AL2 is much worse than results using IGA or AL1. Figure 11 represents the AF values. It should be noted that the optimal fitness value denotes the minimum value of F_1 in Figure 11a and denotes the minimum value of $F_1 + \alpha F_2$ in Figure 11b. First, the fitness values of AL2 are much larger than those of IGA and AL1, meaning that the joint movements with constant joint angular velocity do not work well. Second, the fitness values of AL1 grow larger than those of IGA as waypoints increase. It can also be seen that the difference between fitness values of IGA and AL1 in each case is not large. To further analyze the results using IGA and AL1, fitness values of IGA and AL1 plotted in Figure 11 are once again shown in Figure 12. In Case 1, the difference generated by IGA and AL1 is smaller than 1 when the number of waypoints is less than 10. The difference gradually becomes large as waypoints increase, and the difference is about 5 when the number of waypoints is 30. In Case 2, the difference is 0.9 when the number of waypoints is 5, and the difference is 3 when the number is 7. However, difference grows faster as waypoints increase, compared with the case of MTTP for FFSR1. It can be seen in Figure 12b that the difference is 8 when the number of waypoints is 30.

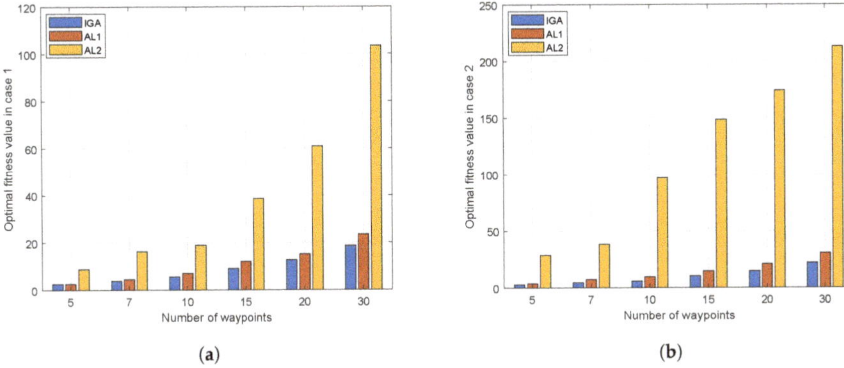

Figure 11. Optimal fitness values of IGA, AL1, and AL2 in two cases. (**a**) MTTP for FFSR1 and (**b**) MTTP for FFSR2.

Table 2. Average execution time and fitness values of IGA.

No.	C1AET	C1WF	C1BF	C1AF	C2AET	C2WF	C2BF	C2AF
5	29.6717	2.9690	2.4773	2.5280	87.2751	3.2721	2.6073	2.6871
7	41.7781	3.7754	3.7439	3.7680	102.3258	4.6112	4.1563	4.3424
10	62.6735	5.6198	5.3880	5.5059	130.2154	6.0007	5.5044	5.7854
15	88.4186	9.4253	8.6173	9.0766	165.2741	11.0144	9.5132	10.3125
20	118.4722	13.2953	12.0814	12.7159	201.4602	15.8477	14.0814	14.5411
30	168.7083	19.8873	16.9274	18.6520	268.4956	23.3607	21.0884	22.0983

Table 3. Average execution time and fitness values of AL1.

No.	C1AET	C1WF	C1BF	C1AF	C2AET	C2WF	C2BF	C2AF
5	30.2927	2.5802	2.5376	2.5734	87.7121	4.3226	2.9237	3.6392
7	43.4101	4.4033	3.9741	4.3858	105.3805	8.0266	6.9567	7.3691
10	63.5849	8.1634	6.4493	7.0059	132.4114	10.0733	9.1615	9.5277
15	89.5364	13.5983	10.9831	12.0514	170.8203	15.1020	13.8772	14.5412
20	123.4722	14.2953	16.0814	15.2159	206.5551	21.7545	19.0202	20.7822
30	175.4097	26.6370	21.8827	23.4604	275.3105	32.9934	28.1109	30.4985

Table 4. Average execution time and fitness values of AL2.

No.	C1AET	C1WF	C1BF	C1AF	C2AET	C2WF	C2BF	C2AF
5	17.5387	8.7288	8.7288	8.7288	40.7669	28.8014	28.8014	28.8014
7	19.4081	16.2850	16.2850	16.2850	45.7278	38.2449	38.2449	38.2449
10	24.9153	22.2850	16.4822	19.0011	53.4793	105.4628	90.2001	97.1588
15	35.6762	45.6080	34.0422	38.5686	70.6147	160.8218	138.5077	148.4189
20	46.5072	67.4052	58.3376	60.9117	86.0894	179.2049	167.3822	174.0670
30	68.3224	120.4871	98.7836	103.3427	130.1552	229.6452	200.9169	212.8989

According to the above analysis, the proposed approach is validated. It can be seen that depicting each joint angular trajectory with piecewise- and continuous-sine functions makes the joint angular velocity variable and contributes to a better solution. Generally, it is almost impossible for any optimization algorithm to find the best solution. However, compared with the integer mechanism of the second part in each chromosome, the binary encoding mechanism contributes to a preferable second-best solution.

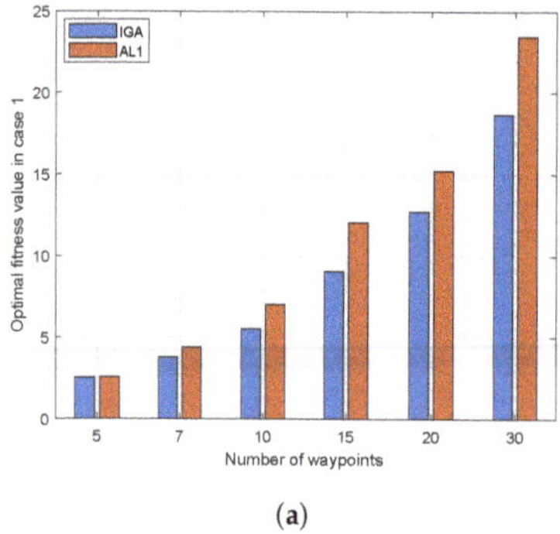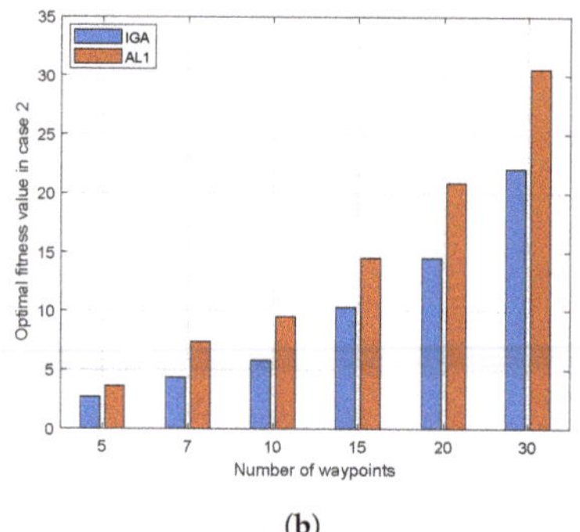

(a)   (b)

Figure 12. Optimal fitness values of IGA and AL1 in two cases. (**a**) MTTP for FFSR1 and (**b**) MTTP for FFSR2.

4.2. Simulation Cases

Two cases of numerical simulation results, based on the IGA, were developed for detailed motion analysis of FFSR1 and FFSR2. In both cases, the end effector is constrained to visit 10 waypoints in the 3D Cartesian space. The position and attitude of the 10 waypoints are expressed in Table 5.

Table 5. Position and attitude of ten waypoints.

No.	Position (m)	Attitude (deg)
1	$[1.61, 0.15, 0.15]$	$[148, -27, 75]$
2	$[1.00, -0.50, -0.45]$	$[150, -30, 80]$
3	$[1.50, 0.00, 0.00]$	$[100, 0, 100]$
4	$[1.00, -1.00, 0.50]$	$[80, 10, 100]$
5	$[0.00, 1.00, 1.00]$	$[60, 30, 120]$
6	$[-0.50, 0.50, 1.00]$	$[80, 80, 80]$
7	$[0.60, -0.50, 1.00]$	$[0, 80, 80]$
8	$[1.00, 0.50, 1.00]$	$[0, -10, 10]$
9	$[1.00, -0.50, 1.20]$	$[0, 0, 0]$
10	$[0.90, -0.50, 1.40]$	$[20, 0, -20]$

4.2.1. Case 1

In this case, the MTTP for FFSR1 is studied. Figure 13 represents the variation trends of fitness values using IGA, showing that the minimum maneuver time of FFSR1 is 5.196 s. Figure 14 represents the movements of FFSR1, where the red points in each subfigure denote the corresponding values at the waypoints. The following explanation is useful to understand the subfigures: Figure 14a,b represents the time histories of the position and attitude of the end effector, respectively; Figure 14c,d represents the time histories of the seven joint angles; Figure 14e,f represents the time histories of the seven joint angular velocities. Figure 14a,b shows that the optimal waypoint sequence is $1 \rightarrow 4 \rightarrow 7 \rightarrow 9 \rightarrow 2 \rightarrow 3 \rightarrow 10 \rightarrow 8 \rightarrow 5 \rightarrow 6$, and all waypoints are accurately visited. Figure 14c,d shows that the joints move smoothly, while Figure 14e,f shows that the joint angular velocities at each waypoint are zero. Moreover, Figure 14c–f shows that the joint angles and the joint angular velocities are within the predefined region. In Figure 14, each joint angle $q_i (i = 1, ..., 7)$ denotes one component of the joint configuration $\mathbf{q} = [q_1, ..., q_7]^T$, and each joint angular velocity dq_i denotes the 1st derivative of q_i.

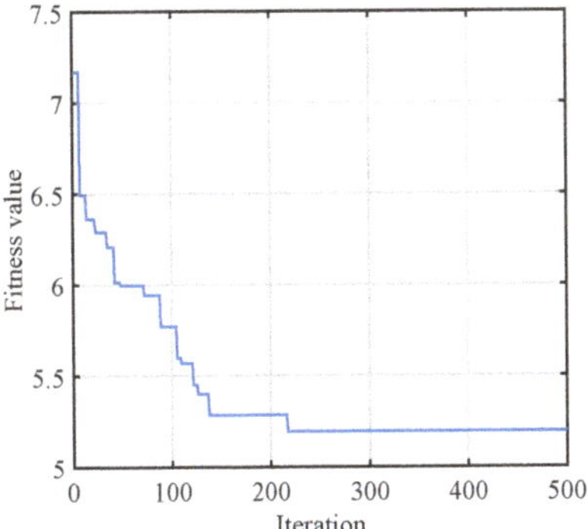

Figure 13. Variation trends of fitness values in the case of MTTP for FFSR1.

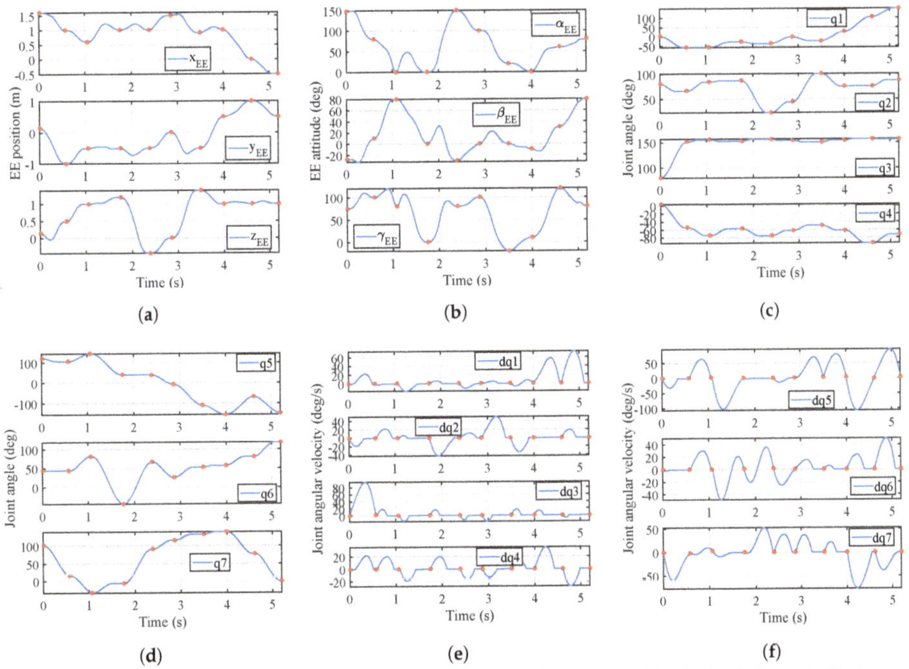

Figure 14. Case 1: FFSR1 movements. (**a**,**b**) Time histories of position and attitude of the end effector, respectively; (**c**,**d**) time histories of joint angles; (**e**,**f**) time histories of joint angular velocities.

4.2.2. Case 2

In this case, the MTTP for FFSR2 is studied. Figure 15 represents the variation trends of the fitness values using IGA, where the minimum maneuver time of FFSR2 is 6.056 s, and the module value of the base attitude in Equation (8) is 0.22 deg. FFSR2 movements are represented in Figure 16, where the red parts in each subfigure denote the values at the corresponding waypoints. The legend of Figure 16 is explained as follows: Figure 16a,b represents the time histories of the position and the attitude of the end effector, respectively; Figure 16c represents the time histories of the base attitude; Figure 16d,e represents the time histories of the seven joint angles; Figure 16f,g represents the time histories of the seven joint angular velocities. Figure 16a,b shows that the optimal waypoint sequence is $1 \rightarrow 2 \rightarrow 3 \rightarrow 7 \rightarrow 8 \rightarrow 10 \rightarrow 6 \rightarrow 5 \rightarrow 9 \rightarrow 4$, and all waypoints are accurately visited. It can be seen that the optimal waypoint sequence in this case is different from that obtained in Case 1, since one more objective function was considered in this case. Moreover, the randomness of IGA is a main factor. In Figure 16c, the variation amplitude of each Euler angle is less than 0.3 deg during the whole movement of FFSR2. Figure 16d,e shows that the joints move smoothly, while Figure 16f,g shows that the joint angular velocities at each waypoint are zero. Furthermore, Figure 16d–g shows that the joint angles and joint angular velocities are within the predefined region.

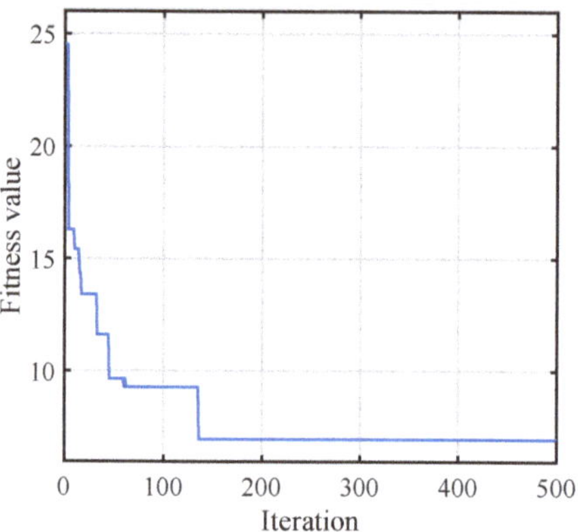

Figure 15. Variation trends of fitness values in the case of MTTP for FFSR2.

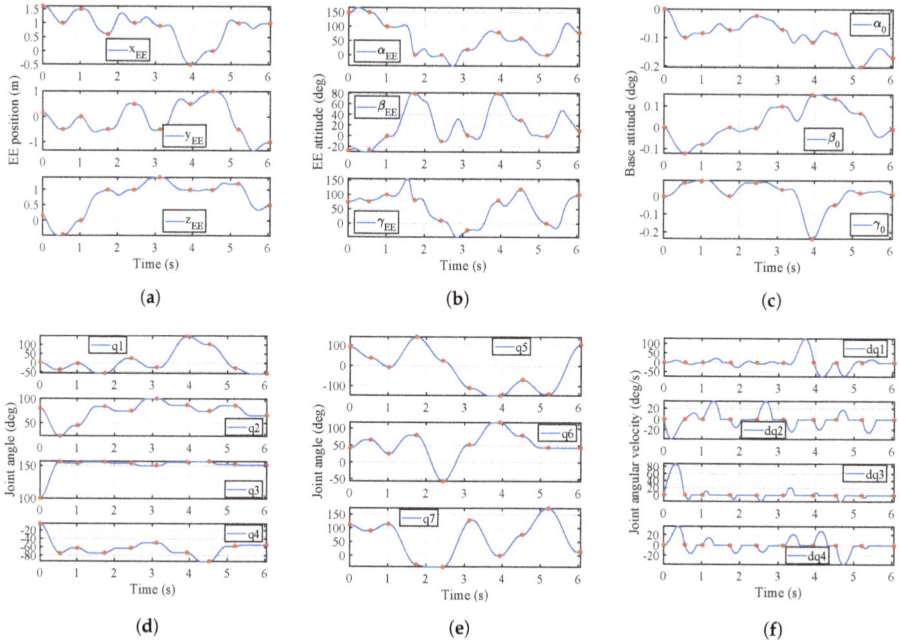

Figure 16. *Cont.*

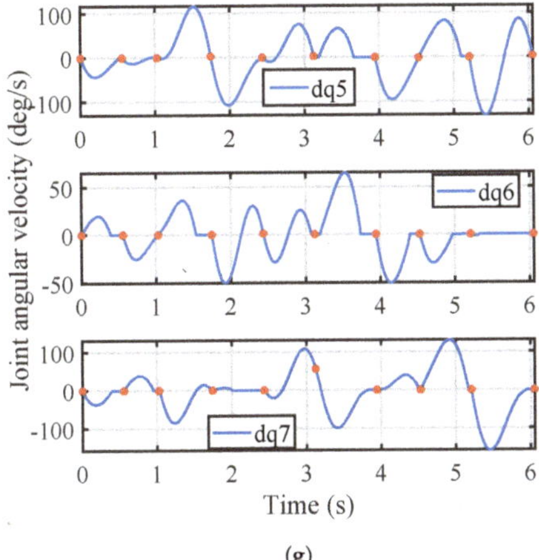

(g)

Figure 16. Case 2: FFSR2 movements. (**a,b**) Time histories of position and attitude of the end effector, respectively; (**c**) time histories of attitude of the free-floating base; (**d,e**) time histories of joint angles; (**f,g**) time histories of joint angular velocities.

5. Conclusions

This work studied the MTTP for space robotics, including the free-flying space robot (FFSR1) and the free-floating space robot (FFSR2). The trajectory-planning problem was converted into an optimization problem by depicting the joint movements with piecewise- and continuous-sine functions. An IGA was proposed to solve the optimization problem. The main contributions are expressed as follows:

(i) Cubic coefficients of arguments of piecewise-sine functions share the same absolute value, making the third part of each chromosome only consider one value after decoding. Therefore, computational cost is reduced.

(ii) In contrast to the setting on joint angular velocities in most recent studies about the MTTP for industrial robotics, each joint angular velocity varies based on Equation (10), contributing to the high-precision results.

For future research, the MTTP for space robotics will be further studied in the case that one or more obstacles appear, especially the case when obstacles move. Furthermore, the corresponding trajectory-tracking control problem will be studied, which is of great significance for practical applications.

Author Contributions: conceptualization, S.Z. and Z.Z.; methodology, S.Z.; writing—original-draft preparation, S.Z.; writing—review and editing, S.Z. and Z.Z.; supervision, Z.Z. and J.L.; funding acquisition, Z.Z.

Funding: This research was funded by the National Natural Science Foundation of China (Grant No. 11472213).

Acknowledgments: The authors sincerely acknowledge the financial support provided by the National Natural Science Foundation of China (Grant No. 11472213). The authors would also thank the members of the National Key Laboratory of Aerospace Flight Dynamics for their useful and valuable suggestions.

Abbreviations

The following abbreviations are used in this manuscript:

2D	Two-dimensional
AET	Average execution time
AF	Average fitness
AL1	Algorithm 1
AL2	Algorithm 2
BF	Best fitness
D-H	Denavit-Hartenberg
DE	Differential evolution algorithm
DOF	Degree of freedom
EST-VII	Experimental Test Satellite VII
FFSR1	Free-flying space robot
FFSR2	Free-floating space robot
GA	Genetic algorithm
IGA	Improved genetic algorithm
ISS	International Space Station
JAXA	Japan Aerospace Exploration Agency
JEMRMS	Japanese Experiment Module Remote Manipulator System
MDA	MacDonald Dettwiler and Associates Ltd.
MTTP	Multitask-based trajectory-planning problem
PSO	Particle-swarm optimization algorithm
SRS	Spherical revolute spherical
TSP	Traveling salesman problem
WF	Worst fitness

References

1. Bowman, L.M.; Belvin, W.K.; Komendera, E.E.; Dorsey, J.T.; Doggett, B.R. In-space assembly application and technology for NASA's future science observatory and platform missions. In Proceedings of the Space Telescopes and Instrumentation: Optical, Infrared, and Millimeter Wave, Austin, TX, USA, 10–15 June 2018. [CrossRef]
2. Bonneville, R. A truly international lunar base as the next logical step for human spaceflight. *Adv. Space Res.* **2018**, *61*, 2983–2988. [CrossRef]
3. Sato, N.; Doi, S. JEM remote manipulator system (JEMRMS) human-in-the-loop test. In Proceedings of the International Symposium on Space Technology and Science, Morioka, Japan, 28 May–4 June 2000.
4. Rembala, R.; Ower, C. Robotic assembly and maintenance of future space stations based on the ISS mission operations experience. *Acta Astronaut.* **2009**, *65*, 912–920. [CrossRef]
5. Oda, M.; Kibe, K.; Yamagata, F. ETS-VII, space robot in-orbit experiment satellite. In Proceedings of the IEEE International Conference on Robotics and Automation, Minneapolis, MN, USA, 22–28 April 1996; pp. 739–744. [CrossRef]
6. Hirzinger, G.; Brunner, B.; Dietrich, J.; Heindl, J. ROTEX-the first remotely controlled robot in space. In Proceedings of the IEEE International Conference on Robotics and Automation, San Diego, CA, USA, 8–13 May 1994; pp. 2604–2611. [CrossRef]
7. Umetani, Y.; Yoshida, K. Resolved motion rate control of space manipulators with generalized Jacobian matrix. *IEEE Trans. Robot. Autom.* **1989**, *5*, 303–314. [CrossRef]
8. Dubowsky, S.; Torres, M.A. Path planning for space manipulators to minimize spacecraft attitude disturbances. In Proceedings of the IEEE International Conference on Robotics and Automation, Sacramento, CA, USA, 9–11 April 1991; pp. 2522–2528. [CrossRef]
9. Papadopoulos, E.; Dubowsky, S. Dynamic singularities in free-floating space manipulators. In *Space Robotics: Dynamics and Control*; Xu, Y., Kanade, T., Eds.; Springer: Boston, MA, USA, 1993; pp. 77–100, ISBN 978-1-4615-3588-1.

10. Yoshida, K.; Hashizume, K.; Abiko, S. Zero reaction maneuver: Flight validation with ETS-VII space robot and extension to kinematically redundant arm. In Proceedings of the IEEE International Conference on Robotics and Automation, Seoul, Korea, 21–26 May 2001; pp. 441–446. [CrossRef]

11. Liu, X.; Baoyin, H.; Ma, X. Optimal path planning of redundant free-floating revolute-jointed space manipulators with seven links. *Multibody Syst. Dyn.* **2013**, *29*, 41–56. [CrossRef]

12. Wang, M.; Luo, J.; Walter, U. Trajectory planning of free-floating space robot using Particle Swarm Optimization (PSO). *Acta Astronaut.* **2015**, *112*, 77–88. [CrossRef]

13. Xu, W.; Li, C.; Liang, B.; Liu, Y.; Xu, Y. The Cartesian path planning of free-floating space robot using particle swarm optimization. *Int. J. Adv. Robot Syst.* **2008**, *5*, 301–310. [CrossRef]

14. Xu, W.; Liu, Y.; Liang, B.; Xu Y.; Li, C.; Qiang, W. Non-holonomic path planning of a free-floating space robotic system using genetic algorithms. *Adv. Robot.* **2008**, *22*, 451–476. [CrossRef]

15. Chen, Z.; Zhou, W. Path planning for a space-based manipulator system based on quantum genetic algorithm. *J. Robot.* **2017**, *2017*. [CrossRef]

16. Zhang, J.; Wei, X.; Zhou, D.; Zhang, Q. Trajectory planning of a redundant space manipulator based on improved hybrid PSO algorithm. In Proceedings of the IEEE International Conference on Robotics and Biomimetics (ROBIO), Qingdao, China, 3–7 December 2016; pp. 419–425. [CrossRef]

17. Wang, M.; Luo, J.; Fang, J.; Yuan, J. Optimal trajectory planning of free-floating space manipulator using differential evolution algorithm. *Adv. Space Res.* **2018**, *61*, 1525–1536. [CrossRef]

18. Alatartsev, S.; Stellmacher, S.; Ortmeier, F. Robotic task sequencing problem: A survey. *J. Intell. Robot. Syst.* **2015**, *80*, 279–298. [CrossRef]

19. Bonami, P.; Olivares, A.; Staffetti, E. Energy-optimal multi-goal motion planning for planar robot manipulators. *J. Optim. Theory Appl.* **2014**, *163*, 80–104. [CrossRef]

20. Little, J.D.; Murty, K.G.; Sweeney, D.W.; Karel, C. An algorithm for the traveling salesman problem. *Oper. Res.* **1963**, *11*, 972–989. [CrossRef]

21. Gentilini, I.; Margot, F.; Shimada, K. The travelling salesman problem with neighbourhoods: MINLP solution. *Optim. Method Softw.* **2013**, *28*, 364–378. [CrossRef]

22. Zacharia, P.T.; Aspragathos, N.A. Optimal robot task scheduling based on genetic algorithms. *Robot. Cim-Int. Manuf.* **2005**, *21*, 67–79. [CrossRef]

23. Baizid, K.; Chellali, R.; Yousnadj, A.; Meddahi, A.; Bentaleb, T. Genetic algorithms based method for time optimization in robotized site. In Proceedings of the IEEE/RSJ International Conference on Intelligent Robots and Systems, Taipei, Taiwan, 18–22 October 2010; pp. 1359–1364. [CrossRef]

24. Baizid, K.; Yousnadj, A.; Meddahi, A.; Chellali, R.; Iqbal, J. Time scheduling and optimization of industrial robotized tasks based on genetic algorithms. *Robot. Cim-Int. Manuf.* **2015**, *34*, 140–150. [CrossRef]

25. Suárez-Ruiz, F.; Lembono, T.S.; Pham, Q.C. RoboTSP-a fast solution to the robotic task sequencing problem. In Proceedings of the IEEE International Conference on Robotics and Automation, Brisbane, Australia, 21–25 May 2018; pp. 1611–1616. [CrossRef]

26. Bänziger, T.; Kunz, A.; Wegener, K. Optimizing human–robot task allocation using a simulation tool based on standardized work descriptions. *J. Intell. Manuf.* **2018**, 1–14. [CrossRef]

27. Kovács, A. Task sequencing for remote laser welding in the automotive industry. In Proceedings of the Twenty-Third International Conference on Automated Planning and Scheduling, Rome, Italy, 10–14 June 2013; pp. 457–461.

28. Saha, M.; Roughgarden, T.; Latombe, J.C.; Sánchez-Ante, G. Planning tours of robotic arms among partitioned goals. *Int. J. Robot. Res.* **2006**, *25*, 207–223. [CrossRef]

29. Gbenga, D.E.; Ramlan, E.I. Understanding the limitations of particle swarm algorithm for dynamic optimization tasks: A survey towards the singularity of PSO for swarm robotic applications. *ACM Comput. Surv.* **2016**, *49*. [CrossRef]

30. Li, G.; Zhang, F.; Fu, Y.; Wang, S. Joint stiffness identification and deformation compensation of serial robots based on dual quaternion algebra. *Appl. Sci.* **2019**, *9*, 65. [CrossRef]

31. Xue, Y. Mobile robot path planning with a non-dominated sorting genetic algorithm. *Appl. Sci.* **2018**, *8*, 2253. [CrossRef]

32. Xu, W.; Yan, L.; Mu, Z.; Wang, Z. Dual arm-angle parameterisation and its applications for analytical inverse kinematics of redundant manipulators. *Robotica* **2016**, *34*, 2669–2688. [CrossRef]

Appl. Sci. **2019**, *9*, 2226

33. Waldron, K.; Schmiedeler, J. Kinematics. In *Springer Handbook of Robotics*; Siciliano, B., Khatib, O., Eds.; Springer: Berlin, Germany, 2016; pp. 11–36, ISBN 978-3-319-32550-7.

34. Siciliano, B.; Sciavicco, L.; Villani, L.; Oriolo, G. *Robotics: Modelling, Planning and Control*; Springer: London, UK, 2010, pp. 39–104, ISBN 978-1-84628-641-4.

35. Zhou, D.; Ji, L.; Zhang, Q.; Wei, X. Practical analytical inverse kinematic approach for 7-DOF space manipulators with joint and attitude limits. *Intel. Serv. Robot.* **2015**, *8*, 215–224. [CrossRef]

36. Shimizu, M.; Kakuya, H.; Yoon, W.K.; Kitagaki, K.; Kosuge, K. Analytical inverse kinematic computation for 7-DOF redundant manipulators with joint limits and its application to redundancy resolution. *IEEE Trans. Robot.* **2008**, *24*, 1131–1142. [CrossRef]

37. Yu, C.; Jin, M.; Liu, H. An analytical solution for inverse kinematic of 7-DOF redundant manipulators with offset-wrist. In Proceedings of the IEEE International Conference on Mechatronics and Automation, Chengdu, China, 5–8 August 2012; pp. 92–97. [CrossRef]

38. Holland, J.H. *Adaptation in Natural and Artificial Systems Ann Arbor*; University of Michigan Press: Ann Arbor, MI, USA, 1975.

39. Goldberg, D.E. *Genetic Algorithm in Search, Optimization and Machine Learning*; Addison Wesley: Boston, MA, USA, 1989.

40. Grefenstette, J.; Gopal, R.; Rosmaita, B.; Van Gucht, D. Genetic algorithms for the traveling salesman problem. In Proceedings of the First International Conference on Genetic Algorithms and their Applications, Pittsburgh, PA, USA, 24–26 July 1985; pp. 160–168.

41. Abed, I.A.; Koh, S.P.; Sahari, K.S.M.; Tiong, S.K.; Tan, N.M. Optimization of task scheduling for single-robot manipulator using pendulum-like with attraction-repulsion mechanism algorithm and genetic algorithm. *Aust. J. Basic Appl. Sci.* **2013**, *7*, 426–445.

Article

A New Coverage Flight Path Planning Algorithm Based on Footprint Sweep Fitting for Unmanned Aerial Vehicle Navigation in Urban Environments

Abdul Majeed and Sungchang Lee *

School of Information and Electronics Engineering, Korea Aerospace University, Deogyang-gu, Goyang-si, Gyeonggi-do 412-791, Korea; abdulmajid09398@kau.kr
* Correspondence: sclee@kau.ac.kr; Tel.: +82-10-3232-5237; Fax: +82-2-3159-9969

Received: 11 March 2019; Accepted: 4 April 2019; Published: 8 April 2019

Abstract: This paper presents a new coverage flight path planning algorithm that finds collision-free, minimum length and flyable paths for unmanned aerial vehicle (UAV) navigation in three-dimensional (3D) urban environments with fixed obstacles for coverage missions. The proposed algorithm significantly reduces computational time, number of turns, and path overlapping while finding a path that passes over all reachable points of an area or volume of interest by using sensor footprints' sweeps fitting and a sparse waypoint graph in the pathfinding process. We devise a novel footprints' sweep fitting method considering UAV sensor footprint as coverage unit in the free spaces to achieve maximal coverage with fewer and longer footprints' sweeps. After footprints' sweeps fitting, the proposed algorithm determines the visiting sequence of footprints' sweeps by formulating it as travelling salesman problem (TSP), and ant colony optimization (ACO) algorithm is employed to solve the TSP. Furthermore, we generate a sparse waypoint graph by connecting footprints' sweeps' endpoints to obtain a complete coverage flight path. The simulation results obtained from various scenarios fortify the effectiveness of the proposed algorithm and verify the aforementioned claims.

Keywords: coverage flight path planning; footprints sweep; waypoint graph; navigation; urban environments; unmanned aerial vehicles

---

1. Introduction

Unmanned aerial vehicles (UAVs) are achieving ground-breaking success in many application areas such as temporary infrastructure, monitoring and tracking, data collection and surveying, and delivery of goods [1]. The modern UAVs have the ability to carry a variety of payloads' equipment to perform desired missions with realtime processing. The UAV market in experiencing an exponential growth due to the wide range of the practical applications in hazardous, uncertain and threatening areas [2]. Recently, due to the integration of internet of things and cloud technologies with the UAV the application scenarios of UAVs are becoming more and more advance [3]. The civilian real-life economic applications of UAVs such as traffic monitoring [4], air quality monitoring [5], disaster management [6], communication relays [7], and infrastructure inspection [8], among others, are more attractive applications. Despite having many promising applications in each sector, UAV usage without human control imposes several challenges that need to be resolved. Apart from the physical challenges, in many applications, a UAV needs the ability to compute a path between two pre-determined locations while avoiding various obstacles or to find a path which covers every reachable point of a certain area or volume of interest which is called coverage path planning (CPP). In this work, our focus is on the CPP for UAV's navigation in 3D urban environments with fixed obstacles for coverage missions.

CPP is classified as subtopic of the path planning in robotics where it is necessary to obtain a low cost path that covers the entire free space of a certain area or volume of interest with minimum overlapping [9]. CPP is a non-deterministic polynomial time hard (NP-hard) optimization problem. Due to the extensive use of UAVs in many fields for complex missions, the problem of CPP for single and multiple UAVs has been a very active area of research, especially during the last decade [10,11]. Depending upon the type of information available for the UAV workspace, the CPP are divided into two major categories global and local. Several CPP algorithms have been proposed for covering a regular or irregular shaped area of interest (AOI) with visual sensor and thermal sensor, etc. mounted on the UAV [12–14]. The basic approach adopted by most of the offline CPP algorithm is the area decomposition into non-overlapping subregions, determining the visiting sequence of the subregions, and covering decomposed regions individually in a back and forth manner to obtain a complete coverage path. Decomposition-based methods are promising in achieving the complete coverage of the target area. Classical exact cellular decomposition [15], morse based decomposition [16], landmark based topological coverage [17], grid-based methods [18], contact sensor-based coverage [19], and graph based methods [20] are well-known decomposition-based coverage methods.

Most of the existing CPP algorithms for UAVs do not provide thorough insight into complete coverage of the AOI by considering sensor footprints as coverage unit particularly regarding the efficient and complete coverage of the AOI with footprints' sweeps fitting in complex 3D urban environments. The authors of [21] have explained that most of the existing algorithms employ the same sweep direction in all subareas which may not be able to obtain optimal results. Recently, various approaches have been proposed to compute a low cost coverage path such as mirror mapping method [22], viewpoints sampling [23], in-field obstacles classification [24], optimal polygon decomposition [25], and context-aware UAV mobility [26]. Despite the success of such techniques, in most cases, either many locations of the target area are covered repeatedly, or computing time degrades [27]. In addition, most of the algorithms have limited applicability in relation with the shape of the target area [21,28]. When the shape of AOI is changed, the algorithm performance is no longer feasible. Furthermore, the majority of the algorithms do not consider the decomposition of the AOI with relation to the camera/sensor footprints which can significantly increase the number of turns in the path. To overcome the aforementioned limitations, this study proposes a new coverage flight path planning algorithm that fulfil the multiple objectives of CPP.

The rest of the paper is organized as follows; Section 2 explains the background and related work regarding well-known coverage path planning algorithms. Section 3 presents the proposed coverage flight path planning algorithm and explains its principal steps. Section 4 discusses the experiments and simulation results. Finally, conclusions and future directions are offered in Section 5.

2. Background and Related Work

This section presents the background and related work regarding the AOI decomposition techniques used for UAV CPP, coverage types, different types of the AOI used for coverage missions, geometric flight patterns, path optimization algorithms and application specific CPP methods. Cao et al. [29] defined the six requirements for coverage scenarios. In complex environments, it is not always possible to satisfy all six requirements. Therefore, in some cases, the priority is given to the mission completion with sacrifice on some of the requirements [30]. Prior research classified the CPP methods into two major categories: heuristic and cell decomposition-based methods. Heuristic based methods define procedures that should be followed during CPP [31,32]. In cell decomposition, the AOI is decomposed into non-overlapping cells of varying shapes and sizes. There exist several kinds of the AOI decomposition techniques in literature [30]. Boustrophedon decomposition [33] and trapezoidal decomposition [34] are two well-known exact cellular decomposition techniques that are widely used for coverage applications. Morse-based cellular decomposition is an advance form of the boustrophedon decomposition which is based on critical points of Morse functions [35]. The approximate cellular decomposition techniques divide the AOI into a set of regular cells.

Grid-based methods are used over approximate areas to generate coverage paths for the UAVs [36]. Convex decomposition transforms the irregular shaped AOI into regular shaped cells in order to reduce the number of turns [37]. Each decomposition method varies regarding the cell sizes, cell shapes, degree of computational complexity, and path quality. A pictorial overview of the most widely used AOI decomposition techniques is given in Figure 1.

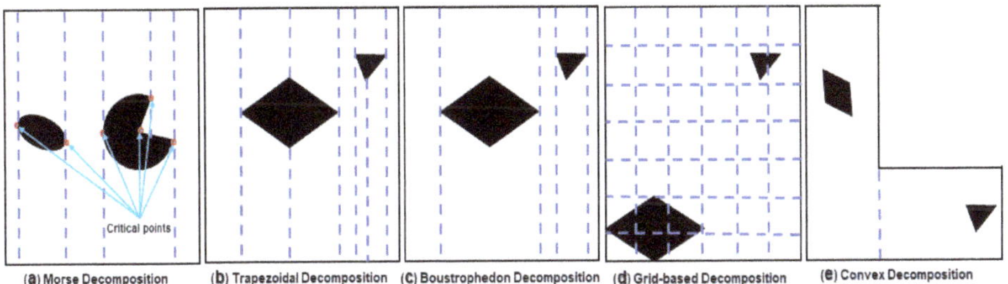

Figure 1. Most widely used areas of interest decomposition techniques for coverage path planning.

There are two types of coverage: simple and continuous. The shape of the AOI can be a rectangle, square, convex polygon, non-convex polygon, 3D cubes (i.e., buildings) or irregular for coverage missions. The most widely used geometric flight patterns for CPP are back and forth (BF), spiral (SP), zamboni, and hybrid patterns. Andersen et al. [38] compared different types of the geometric flight patterns in his work. The flight patterns are employed considering the UAV mobility constraints, shape of the AOI, coverage type, and application requirements. For example, the BF pattern is suitable when the AOI is rectangular and large in size. Jiao et al. [39] proposed an exact cellular decomposition based approach for concave area coverage. Initially, the concave shaped AOI is decomposed into the convex subareas using the minimum width sum approach [40], and then the BF motion pattern is applied to completely sweep the AOI. Xu et al. [41] proposed an optimal CPP approach for fixed wing UAVs with minimum overlapping and coverage guarantees. The proposed algorithm employs the boustrophedon decomposition to decompose the AOI into a set of cells. Later, the adjacency graph is constructed from the cells, and cells are swept using the BF motion patterns. The order of the cells visit follows the Eulerian circuit having start and end at the same vertex. The proposed approach is able to achieve the complete coverage of the AOI. However, in some cases, it will perform additional sweeps by not considering the sensor's footprint that will significantly affect the solution quality in the presence of complex geometry obstacles. Öst [42] proposed a CPP approach for concave shape AOI with sharp edges without sacrificing the guarantees on complete coverage.

In some scenarios, a UAV needs to visit several spatially distributed points of interests (POIs) in the target area to either collect data or provide wireless charging during the mission. In such cases, the CPP problem is divided into several small start-to-goal subproblems, where UAV needs to visit each POI only once while avoiding collision with the obstacles present in the AOI.Some authors considered the special cases of CPP in which AOI contains many interesting and non-interesting zones [43,44]. The interesting zones need careful coverage with higher resolution, so UAV covers such zones with low altitudes and vice versa.A new CPP approach for the aerial remote sensing in agriculture was proposed by Barrientos et al. [45]. The proposed approach decomposes the AOI based on the vehicle's capabilities. UAV has to fly at a certain constant height in order to ensure a specific resolution and CPP is performed using a wavefront planner [46]. The existing studies used several path optimization algorithms for UAV CPP. The ant colony optimization (ACO) [47], Genetic algorithm (GA) [48], wavefront algorithm [49], A* algorithm [50], theta* algorithm [51], particle swarm optimization (PSO) [52], and their improved versions, among other, are the most widely used path optimization algorithms for CPP.

A number of studies have explored a closely related method used for UAV CPP with coverage guarantees, the boustrophedon cellular decomposition based hierarchal CPP approach

(BCDH-CPP) [53], and the related coverage alternatives-based CPP (CA-CPP) algorithm [54]. The BCDH-CPP [53] is a cellular decomposition based CPP approach for the UAVs and it is a promising method for CPP in known environments inhabiting arbitrary obstacles. The proposed approach decomposes the AOI into cells, compute a Eulerian circuit traversing through these cells, and finally joins each cell in using seed spreader motion pattern to find a coverage path. The proposed approach determines the coverage path with excessive overlapping in some parts by not considering the UAV sensor/camera footprints as coverage unit. CA-CPP algorithm [54] introduced the four coverage alternatives to perform efficient coverage in both convex and non-convex areas. The proposed approach can reduce the number of turns and path length considering the appropriate directions and optimal line sweep. The proposed approach has very high computational cost. Meanwhile, the authors devised a method to reduce the overheads by enforcing adjacencies. However, due to extensive evaluation of various permutations, the proposed algorithm complexity varies exponentially with problem size. Furthermore, the decomposition-based CPP approaches discard the useful global information about the free spaces geometry which can lead to the excessive number of turns in the path.

The contributions of this research in the field of UAV global CPP can be summarized as follows: (i) it proposes a new coverage flight path planning algorithm that has potentials to obtain the minimum length path that ensures the perfect coverage of the target area with reduced computational time, number of turns, and path overlapping; (ii) it evaluates and selects the best coverage direction(s) by analyzing the span of the AOI, and exploiting free spaces geometry information to reduce the number of turns in the path; (iii) it introduces a novel footprints' sweeps fitting method in which an entire free space of a certain AOI can be swept with fewer and longer sensor/camera footprints' sweeps; (iv) it determines the visiting sequence of the footprints' sweeps in the form of closed path to reduce the global path length; (v) it generates a sparse waypoint graph by connecting the footprint sweeps' endpoints with neighbours' sweeps by taking into account their visiting sequence, UAV maneuverability constraints, and obstacles effect; (vi) it finds a coverage path that fully scan the desired area located in 3D urban environments.

3. The Proposed Coverage Flight Path Planning Algorithm

This section explains the conceptual overview of the proposed CPP algorithm and outlines its procedural steps. Figure 2 shows the conceptual overview of our proposed global CPP algorithm.

To find a path that covers an entire free space of a certain region of interest Q of a 3D urban environment of known geometry, we used the seven principle steps. Brief discussion about each principle step with the equations, procedures and examples are explained below.

3.1. Modelling of the UAV Operating Environment

After getting the coverage mission specification, the proposed algorithm models the UAV operating environment. Environment modelling usually refers to the classification of free spaces (Q_{free}) and obstacles regions ($Q_{obstacles}$). $Q_{obstacles}$ refers to the parts of the AOI Q where the UAV cannot fly due to the presence of obstacles. In contrast, Q_{free} refers to those parts of Q where UAV can operate without collision with obstacles. The obstacles' regions are represented with the geometrical shapes (i.e., squares, cubes, circles, and cylinder, etc.). In this work, we model the UAV operating environment from the raw environment containing the elevation data about the urban environment buildings and apartments, etc. with a set of 3D convex obstacles by calculating the convex hull. Each obstacle has random width, length and altitudes depending upon the real environment geometry. Each obstacle has eight vertices and six faces. The obstacle vertices u in the modelled environment can be represented with three co-ordinates values, $u = (x, y, z)$. All eight vertices of the ith obstacle along with their numerical values can be mathematically expressed as following matrix.

$$O_i = \begin{bmatrix} x_{min} & y_{min} & z_{min}; x_{min} & y_{min} & z_{max} \\ x_{min} & y_{max} & z_{min}; x_{min} & y_{max} & z_{max} \\ x_{max} & y_{min} & z_{min}; x_{max} & y_{min} & z_{max} \\ x_{max} & y_{max} & z_{min}; x_{max} & y_{max} & z_{max} \end{bmatrix} = \begin{bmatrix} 874 & 909 & 0; 874 & 909 & 252 \\ 874 & 1008 & 0; 874 & 1008 & 252 \\ 954 & 909 & 0; 954 & 909 & 252 \\ 954 & 1008 & 0; 954 & 1008 & 252 \end{bmatrix}.$$

The source location is represented with a 3D point p, where $p = (x_s, y_s, z_s)$ and destination location is represented with a point q, where $q = (x_t, y_t, z_t)$. In some coverage mission, the source p and destination q are the same because the UAV returns to the starting location after completing the mission. The objective of the proposed algorithm is to find a coverage path ζ that guarantees the perfect coverage of Q. The path ζ consists of nodes set which can be defined as, $\zeta \in Q_{free}$ and $\zeta \cap Q_{obstacles} = \varnothing$.

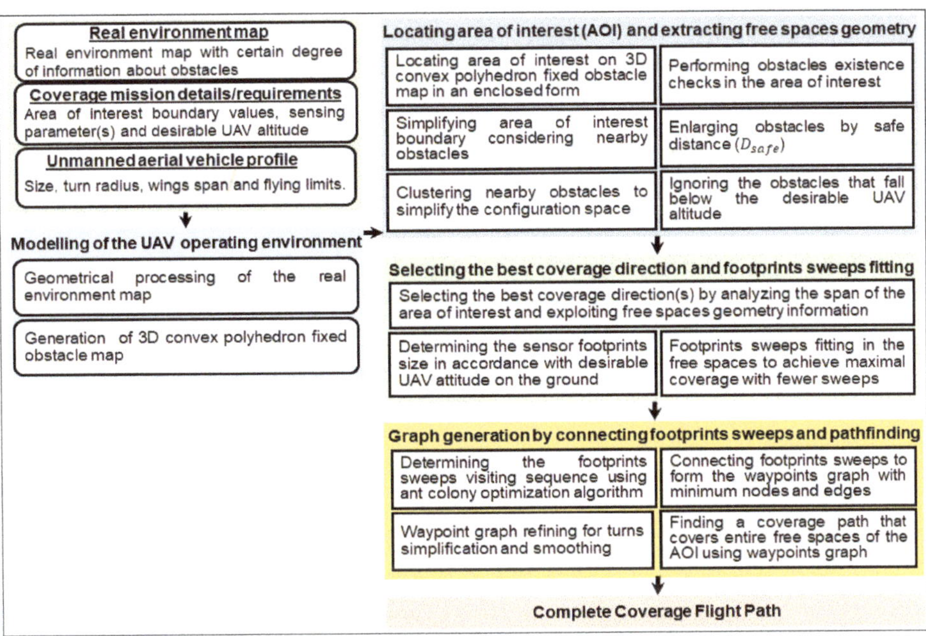

Figure 2. Conceptual overview of the proposed coverage flight path planning algorithm.

3.2. Locating Area of Interest on the Modelled Map and Extracting Free Spaces Geometry

This subsection briefly explains about locating the AOI on the modelled map and extracting free spaces geometry from the AOI for coverage mission.

3.2.1. Locating Area of Interest on the 3D Modelled Map

The AOI can be represented with the sequence of n vertices $\{v_1, v_2, v_3, \ldots, v_n\}$. Each vertex v_i has three co-ordinates values $(v_x(i), v_y(i), v_z(i))$. Considering v_i as the first vertex of the AOI, the next vertex to v_i will be $v_{next(i)}$, where $next(i) = i(mod\ n) + 1$. The edge e_i connecting two vertices v_i and $v_{next(i)}$ has length l_i, where $l_i = \left\| v_i - v_{next(i)} \right\|$. The AOI can contain many obstacles and no-Fly Zones (NFZ) of varying geometries. The AOI boundary is set of collision free vertices and edges. In some cases, the AOI vertices, edges or both can intersect with obstacles as shown in Figure 3.

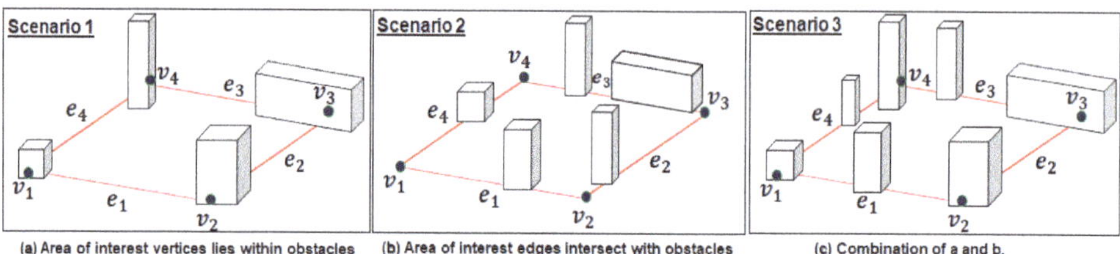

Figure 3. Different obstacles' intersections with the area of interest boundary.

3.2.2. Extracting Free Spaces Geometry

After locating the AOI in an enclosed form on the map, we extract free spaces geometry from the AOI. To do so, we first perform the obstacles' existence checks in the AOI. With the help of these checks, the AOI can be classified into three categories such as obstacles-free environment, obstacles inhibiting environment and area boundary obstacles only. We simplify the AOI boundary by considering the obstacles that intersect with the AOI edges or vertices and lying in close proximity of the AOI edges by shrinking the AOI inward considering the UAV size.

Those obstacles which are on the boundary of the AOI or in close proximity that can impact the UAV safety are removed and configuration space (CS) is simplified for mission. Later, we draw a line from one vertices of the AOI to its nearest adjacent vertices and rotate it to all n vertices to determine the obstacles' existence. If no obstacles intersect with the line, then the AOI can be regarded as the obstacle-free. Meanwhile, if the obstacles exist, the proposed algorithm determines whether the obstacles are part of the area boundary obstacles only or in-field obstacles. In the former case, the proposed algorithm performs the coverage in simplified CS where the AOI is obstacle-free. In the latter case, the obstacles' enlargement (i.e., pushing the intersections by a D_{safe} out of obstacles.) and clustering are carried out to extract the free spaces geometry from the AOI. We enlarge the obstacles by safe distance D_{safe} and cluster the nearby obstacles which overlap each other due to D_{safe} addition or become so close to each other that UAV can collide. D_{safe} is an integer number whose value can be adjusted considering the UAV size, operating environment and obstacles shape. A pictorial overview of the enlarged obstacle by D_{safe} addition is shown in Figure 4b.

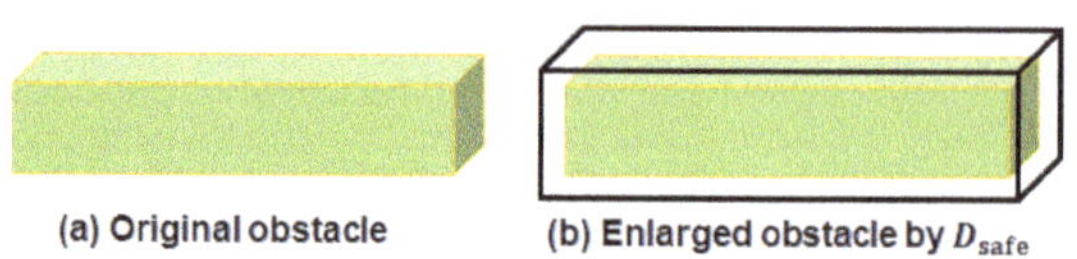

Figure 4. Overview of the enlarged obstacle by D_{safe} value.

We apply the minimum UAV altitude limits H_{min} to discard the obstacles that fall below the H_{min} and UAV can go over safely. Through the above-mentioned process, the AOI can be classified into traversable and non-traversable parts. In the traversal parts, we fit UAV sensor footprints' sweeps considering the appropriate coverage direction for the coverage missions.

3.3. Selecting the Best Coverage Direction (s) by Analyzing the Span of the Area of Interest and Exploiting Free Spaces Geometry Information

Selecting the best coverage direction is extremely important to reduce the number of turns in the flight path to preserve the UAV resources. The turns are costly in terms of the energy, mission time, and path length because the UAV has to reduce the speed, perform turn, and increase speed again. Figure 5 presents the overview of the coverage path obtained from the same AOI with two different directions. The path on the left has only three turns and the path on the right has eleven turns.

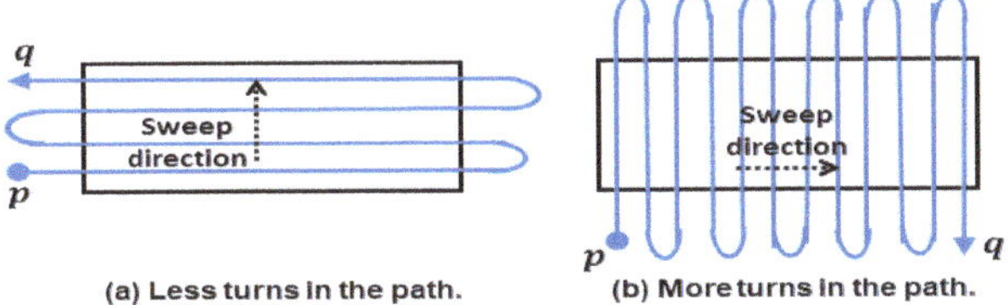

Figure 5. Overview of the coverage path obatined through two different sweep directions.

In this work, we select the best coverage direction (s) by analyzing the span D of the AOI and exploiting free spaces' geometry information. We compute the horizontal span H_s and vertical span V_s of the AOI for the best coverage direction selection for regular shaped AOI. Additionally, we exploit the free spaces' geometry knowledge to select the appropriate coverage direction (s). The coverage direction is chosen parallel to the maximal span axis and considering the free spaces tendency. We employ either one or multiple coverage directions depending upon the AOI complexity to reduce the number of turns. For the irregular shaped AOI, we compute the convex-hull and transform it to some approximate regular shape for the span calculation and best coverage direction selection. When both spans of the AOI are equal, the appropriate coverage direction is chosen considering the obstacles placement and free spaces tendency. By utilizing the global information about the AOI and following the procedure explained above, the best coverage direction can be selected which yields a smaller number of turns in the path. The proposed method utilizes the cumulative knowledge of the several parts of the AOI to find the best coverage direction which reduces the number of turns significantly.

3.4. Sensor Footprints Sweeps Fitting in Free Spaces of the Area of Interest

UAV carries a specific tool to cover the AOI in coverage missions. The tool can be a visual sensor, transmitter, digital camera, or spray tank depending upon the mission. In this work, we assume that UAV carries a visual sensor to cover the AOI during coverage. The attached tool has different sensing characteristics and varying footprint sizes depending upon the UAV height. If the UAV altitude is low, the footprint size is small and image resolution is high. In contrast, when the UAV altitude is high, the footprint size is bigger, and resolution is low. The sensing parameters are adjusted and usually taken at the start of the mission to achieve the desired objectives. UAV generally flies at constant altitude in coverage missions. However, in some cases, it pays attention to the priority of regions. The important regions are covered with low altitudes and vice versa. The sensor footprint F is of rectangular shape with fixed length F_l and width F_w, respectively. However, the size of the footprint keeps changing with the UAV altitude during the mission. After determining the sensor footprint size on the ground as shown in Figure 6a, we fit N sensor footprints' sweeps in such a way that maximal coverage can be achieved with fewer footprints' sweeps.

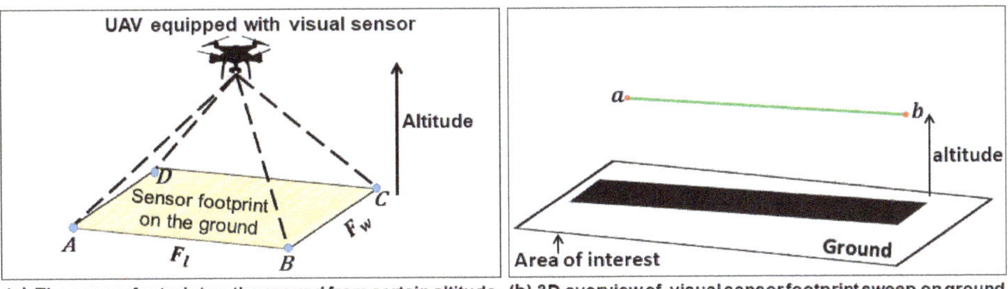

(a) The sensor footprint on the ground from certain altitude. (b) 3D overview of visual sensor footprint sweep on ground.

Figure 6. Overview of the sensor footprint on ground and footprint sweep in the AOI.

The N sensor footprints fitted in the AOI can be mathematically expressed as a set as shown in Equation (1):

$$R = \{F_1, F_2, F_3, \ldots, F_N\}. \tag{1}$$

Each footprint sweep has two parts: one is on the ground (i.e., 2D in rectangular form) and the other is from a certain height (i.e., 3D in a line segment form) similar to the sample shown in Figure 6b. The 3D one passes from the middle of the ground one and it is used as an actual path on which UAV moves during the mission. Considering the footprint as line segment, the start point a_i and endpoint b_i of footprint i are given as:

$$F_i = (a_i, b_i) = ((x_{a_i}, y_{a_i}, z_{a_i}), (x_{b_i}, y_{b_i}, z_{b_i})). \tag{2}$$

After fitting the footprints' sweeps, we find the midpoint of each footprint line segment part utilizing a and b values to find the visiting sequence of footprints' sweeps. The midpoint m_i of the ith footprint F_i can be computed using the a_i and b_i co-ordinates values. A pictorial overview of a footprint in rectangle form and sensor footprint sweep is shown in Figure 6b.

3.5. Determining the Footprints' Sweeps Visiting Sequence

After fitting the footprints' sweeps, the next step is to determine the sequence in which the footprints' sweeps will be visited to ensure the complete coverage. To find the low cost footprints' sweeps visiting sequence, we formulate it as a travelling salesman problem (TSP), and the ant colony optimization algorithm is employed to solve the TSP. Given M points set originally calculated from the footprint sweeps, endpoints are used to calculate a close path that visits each point exactly once. We specify the basic parameters of ACO according to problem size. After the parameter setting, the distance and sight matrix are computed between points. Then, l ants are deployed on random points and the computation is carried out for the pre-defined n_{itr} iterations. During the computation, each ant calculates the tour by remembering the vertices which have already been explored and choosing the nearby points from their current position based on action choice rule. The probability that point j will be visited by an ant k which is currently at point i can be computed from Equation (3):

$$p_{ij}^k = \begin{cases} \dfrac{[\tau_{ij}]^\alpha [\eta_{ij}]^\beta}{\sum\limits_{s \in allowed_k} [\tau_{ij}]^\alpha [\eta_{ij}]^\beta} & ,j \in allowed_k, \\ 0, & otherwise, \end{cases} \tag{3}$$

where τ_{ij} represents the intensity of the pheromones trial between point i and j, α is a parameter used to regulate the influence of τ_{ij}, the variable η_{ij} represents the visibility between point j and i, which is computed as $1/d_{ij}$ (where d_{ij} is the Euclidian distance between two points), β is a parameter used to regulate the influence of η_{ij} and $allowed_k$ represents the points that have not been visited by an ant k yet, respectively.

At the start, l ants are deployed to the m points randomly. Later, each ant makes the decision to choose the next point based on the transition probability p_{ij}^k given by Equation (3). After n_{itr} of this whole process, every ant computes a complete tour by visiting each point once. It is desirable to reinforce good solutions and the ant with the shortest tour should deposit more pheromones to find the low cost solution compared to other ants. Thus, the trail levels are updated, and each ant leaves a quantity of pheromones given by K/L_k, where K is a constant and L_k is the length of the tour. Meanwhile, the pheromones' quantity will decrease as time goes by. Therefore, the update rule of the τ_{ij} can be written as follows:

$$\tau_{ij}(t+1) = (1 - \rho)\tau_{ij}(t) + \Delta\tau_{ij}, \tag{4}$$

$$\Delta\tau_{ij} = \sum_{k=1}^{l} \Delta\tau_{ij}^k, \tag{5}$$

$$\Delta \tau_{ij}^k = \begin{cases} K/L_k, & if\ ant\ k\ travels\ on\ edge\ (i.j), \\ 0, & ,otherwise, \end{cases} \tag{6}$$

where variable t represents the iteration counter, ρ is the parameter to regulate the influence of τ_{ij}, $\Delta\tau_{ij}$ represents the total increase of trial level on a particular edge (i,j), and $\Delta\tau_{ij}^k$ represents the increase of trial level on the two respective edges (i,j) caused by an ant k. While determining the visiting sequence, we do not consider the obstacles' effect to find the minimum length global path. We consider the obstacles 'effect in the waypoint graph construction stage. With the help of ACO, we can get the set S, where $S = \{(F_1, n_1), (F_2, n_2), \ldots, (F_n, n_n)\}$ of footprint sweeps' visiting sequence that will be used for graph construction. In set S, the n_i refers to the visiting sequence of a particular sweep.

3.6. Waypoints Graph Generation by Connecting Footprints' Sweeps

The waypoint graph (WG) is constructed by connecting the footprint sweeps' endpoints for the pathfinding. Mathematically, WG is a double edge graph Γ of inter-reachable locations: $\Gamma = \{W, E\}$, where W represents the nodes set and E represents the edge set. The nodes of the Γ are the footprints' sweeps endpoints, while edges are straight lines for each sweep and connection from one sweep to another. Each footprint sweep is a line segment with the two endpoints a and b (i.e., $F_i = \overline{ab}$) as shown in Figure 6b. We connect the footprints' sweeps endpoint with the coincident (i.e., neighbours) sweeps endpoints to form a WG. Furthermore, in the WG construction process, we consider the obstacles' effect, UAV manoeuvrability constraints and footprints sweep visiting sequence to form the connected Γ. We devise a strategy to avoid obstacles by applying three avoidance options (i.e., left, right, top) in footprints' sweeps inter-connection. Given N footprints' sweeps in the form of line segments, we compute the collision-free connections between footprints' sweeps. While constructing the WG, the proposed algorithm evaluates and selects the best obstacles' avoidance option from the candidate options. For the appropriate options selection, the distance is computed between two sweeps by introducing the two points on three sides of the obstacles. The selection of the appropriate obstacle's avoidance option $AOAO$ for an obstacle O_i is carried out on the basis of following equation:

$$AOAO_i = min\{d_{left}, d_{right}, d_{top}\}, \tag{7}$$

where d_{left}, d_{right} and d_{top} represent the minimum distance required to avoid an obstacle i from left, right, and top, respectively. The value of d for each option can be computed between two sweeps with the help of two intermediate points. For example, the distance required to find the collision-free connection between two footprints' sweeps F_i and F_{i+1} is the sum of the three individual distances (i.e., $d\{(a,b),c), (c,d), (d,(a,b))\}$). The value of each option can be calculated with the help of following equation:

$$d_{left} = \sum_{j=1}^{3} d_j, \tag{8}$$

where d_j represents the minimum distance needed to connect two footprints' sweeps by bypassing an obstacle safely from left. Similarly, the distance values for the other two options are calculated and evaluated for the low cost connection formation. Meanwhile, if no obstacle exists between the footprints' sweeps, and collision can be avoided, then footprints' sweeps are connected with each other via straight lines. The waypoint graph Γ obtained from the AOI possess all properties of a roadmap. A pictorial overview of the waypoint graph Γ obtained by connecting endpoints of the N footprints' sweeps from the AOI having size $1500 \times 1500 \times 400$ is shown in Figure 7a.

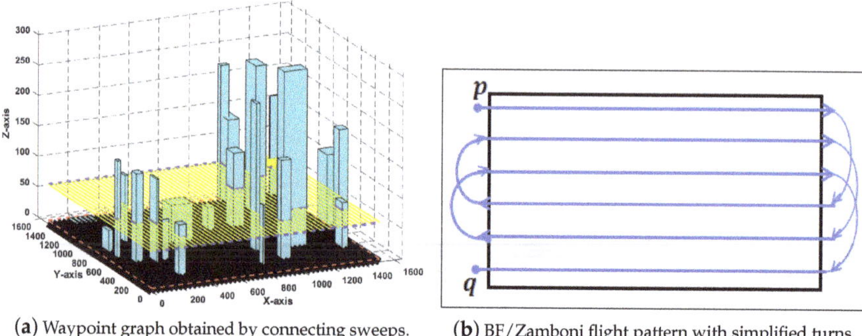

Figure 7. Pictorial overview of the waypoint graph and Back and forth/Zamboni flight pattern.

In Figure 7a, the yellow lines represent the footprints' sweeps and blue lines represent the collision-free connection between sweeps. Meanwhile, the black colour represents the area scanned by sensor footprints' sweeps. The proposed approach is applicable for both types of the UAVs, rotary-wing and fixed-wing. The rotary-wing UAVs present maneuverability advantages when making turns during the flight compared to the fixed wing UAVs. The WG produced by connecting the footprints' sweeps endpoints is directly usable for the rotary-wing UAV. Meanwhile, the fixed-wing UAV has strong maneuverability restrictions, demanding a bigger space to make turns. Therefore, the back-and-forth/zamboni flight pattern can be adopted to generate the WG, where the sweeps will be connected as shown in Figure 7b. In the BF/Zamboni flight pattern, the adjacent sweeps can be connected with the gap g of two or three depending upon the complexity of the AOI and sweeps placements. The suggested pattern can simplify the turn and UAV doesn't need to slow down while making the turn. However, the BF/Zamboni flight pattern can lead to the excessive path overlapping in the complex scenarios. The proposed approach simplifies the turns to the extent possible for the easier adoption using both types of the UAVs. The proposed approach generates a sparse WG with minimum nodes and edges to reduce the computational complexity of pathfinding process. The complete pseudo-code used to generate a waypoint graph from AOI is given in Algorithm 1. In Algorithm 1, the AOI Q containing N number of obstacles, footprint sweeps set R, footprint sweeps visiting sequence set S, and footprint connection gap g are provided as an input. Waypoint graph Γ, where ($\Gamma = \{W, E\}$) is obtained as an output. Line 2 implements the finding of appropriate footprint sweep to be connected with footprint sweep F_i out of $N - 1$ candidates based on optimized visiting sequence and sweeps connection gap. Line 3 implements the identification of the relevant endpoint's pairs through which two sweeps F_i and F_j will be connected with each other. Line 4 performs the check for the obstacle(s) existence between two sweeps to form the collision-free connection C_i. Lines 5–6 implement the connection formation in the presence of obstacle(s) between two sweeps while avoiding the obstacle with low cost option. Line 8 implements the connection formation if no obstacles exist between two sweeps. Furthermore, the same process continues until the complete waypoints is constructed. Finally, waypoint graph Γ of footprints' sweeps connection is returned as an output (line 11).

Algorithm 1: Waypoint graph generation from the AOI located in a 3D map.

Input	:(1) AOI Q containing N number of obstacles, where $N = \{o_1, o_2, o_3, \ldots, o_n\}$.
	(2) Set R of footprints' sweeps, where $R = \{F_1, F_2, \ldots, F_n\}$ and $F_i = (a_i, b_i)$.
	(3) Sweeps visitng sequence set S, where $S = \{(F_1, n_1), (F_2, n_2), \ldots, (F_n, n_n)\}$.
	(4) Sweeps connection gap (g).
Output	:Way point graph Γ

Procedure:

    for each footprint sweep F_i, where $F_i = F_1$ to $F_n \in R$ **do**

        Find next footprint sweep F_j considering the g value and optimized visiting sequence.

        Identify the relevant endpoint pair using (a_i, a_j, b_i, b_j) for connection C_i

        if INTERSECTS(C_i, o_i) **then**

            Evaluate and select the appropriate obstacle o_i avoidance option using Equation (7).

            Connect the footprints' sweeps F_i and F_j with the relevant endpoints.

        else

            Connect the footprints' sweeps F_i and F_j with the relevant endpoints.

        End if

    End for

    return Γ

3.7. Path Searching on the Waypoints Graph

After constructing the WG, the path searching is carried out to find the complete coverage path from the pre-determined location of the UAV. The source point p represents the point from where the UAV starts the mission and point q represents the endpoint of the mission. Once the complete graph is constructed, the pathfinding process starts from the point p and continues until the complete coverage is achieved. We used the ACO algorithm generated order for the collision-free pathfinding over WG with minimum overlapping and informed search. The proposed algorithm keeps track of the locations to be visited along the path by using the order generated for footprints' sweeps visits and their connections. We assume that UAV is able to take turns with sufficient accuracy while switching from one sweep to another during the mission. The proposed approach is able to find the minimum length path with less path overlapping, less number of turns, and reduced computation time compared to the existing methods in most scenarios.

4. Simulation Results and Discussion

This section presents the simulation results and key findings about the proposed concept. The improvements of the proposed algorithm were compared using four criteria; the improvements in computation time, path lengths, path overlapping, and number of turns with the existing closely related algorithms. To benchmark the proposed algorithm, we compared the proposed algorithm results with decomposition-based CPP methods, BCDH-CPP [53] and CA-CPP algorithm [54]. The simulation results were produced and compared on a PC running Windows 10, with a CPU Intel Core i5 of 2.6 GHz and 8.00 GB of RAM, using MATLAB version 9.4.0.81 (R2018a). In simulations of the proposed CPP algorithm, we consider a 25-kg UAV similar to our previous study [55]. We considered both global constraints that are related to the UAV operating environment and local constraints that are related to the UAV. The numerical values related to the local constraints are: maximum steering angle: p/6 radius and wing span: 1 m. We assumed a zero-wind scenario in our simulations and assumed that there exists no external inference that can impact the established path. We assumed that a UAV has sufficient power to finish the mission successfully in one round. The minimum and maximum UAV flight height limits are 25 m and 150 m (h_{min} = 25 m, h_{max} = 150 m). The safe distance value for collision avoidance with obstacles is set to 10 m (D_{safe} = 10 m). The ACO parameters were specified considering the problem size (i.e., no. of sweeps). The sensor footprint sweep width is set to 20 m

(F_w = 20 m) and footprint sweep length is set to 30 m (F_l = 30 m). We present the overview of the 3D maps used in the experiments and two exemplary coverage path results visually in Figure 8a,b. We compared the proposed algorithm results with the existing methods on two grounds: the varying obstacles densities and shape of the AOI. The relevant details about AOI sizes, obstacles' densities, and the shape of the AOI used in experiments are explained in Sections 4.1 and 4.2.

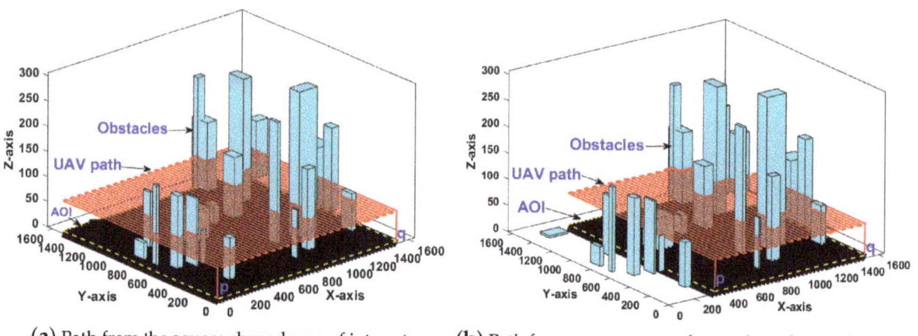

(**a**) Path from the square shaped area of interest.   (**b**) Path from non-convex polygon shaped area of interest.

Figure 8. Coverage path planning results from two different types of the area of interest.

4.1. Comparisons with the Existing Algorithms Based on Obstacles' Densities

The number of obstacles and their placement in the AOI significantly impact the performance of any CPP algorithm. To validate the proposed algorithm feasibility for coverage missions in 3D urban environments, we compared the proposed algorithm path lengths, computing time and path overlapping results with existing methods using five maps with varying obstacles' densities. For the evaluation, we compared the proposed algorithm results using three obstacles' density values (low, medium and high) on regular shaped AOI. All obstacles were placed randomly in the AOI in all cases. The low density obstacles' AOI has less number of obstacles and most parts of the area are traversable. In such areas, the most parts can be covered with fewer and longer sensor footprints' sweeps. In contrast, the high density obstacles AOI has less traversable parts and the path overlapping can increase due to complex obstacle geometry. The medium density obstacles areas have uniform distribution of obstacles and almost half of the spaces can be covered with the sweeps. The complete description about the AOI sizes used in experiments, and average running time and path length results' comparisons of the proposed CPP algorithm are shown in Table 1. The computing time, path length, and path overlapping results are the average of five runs in Tables 1–3.

Table 1. Proposed coverage path planning algorithm performance comparisons with the two existing algorithms.

Area of Intrest Size	BCDH-CPP Algorithm		CA-CPP Algorithm		Proposed-CPP Algorithm	
	Avg. Time (s)	Avg. Length (m)	Avg. Time (s)	Avg. Length (m)	Avg. Time (s)	Avg. Length (m)
500 × 600 × 300	13.53	8910.34	17.94	7901.12	10.54	8201.56
1000 × 1200 × 300	18.40	35,997.11	22.89	34,807.45	15.64	35,417.67
12,000 × 1400 × 300	23.19	55,920.19	28.81	55,429.12	18.29	53,120.09
1500 × 1500 × 400	27.13	72,995.22	33.09	72,810.21	24.54	71,030.42
2000 × 2000 × 400	31.21	124,991.10	39.19	123,951.10	26.75	122,367.15

From the results, it can be observed that both computing time and path length increase with an increase in size of the AOI and obstacles' densities. Meanwhile, the proposed algorithm shows 11.2% and 19.7% reduction in computing time compared to the closely related algorithms. From the path length point of view, the proposed algorithm shows 9.1% and 4% reduction as compared to BCDH-CPP and CA-CPP method, respectively. When the environment complexity is low and AOI size is small (i.e., the first two cases), the CA-CPP algorithm yields shortest length path compared to proposed algorithm. Meanwhile, as the AOI size and obstacle densities grow, the performance of the proposed algorithm improves on both metrics (i.e., path lengths and computing time). Apart from the path lengths and computing time comparisons, we compared the proposed algorithm path overlapping results with the two closely related algorithms in each map (listed in Table 1). The proposed algorithm path overlapping results in average and its comparison with the existing methods are shown in Table 2.

Table 2. Path overlapping comparison between three algorithms for the same area of interest.

Algorithms	Evaluation Criteria	Maps Id				
		1	2	3	4	5
CA-CPP	Avg. path overlapping (m)	129.59	289.37	990.45	1258.98	2039.51
BCDH-CPP	Avg. path overlapping (m)	209.21	330.89	1023.07	1455.45	2237.98
Proposed	Avg. path overlapping (m)	171.09	301.12	810.95	1125.12	1927.45

The proposed algorithm on average gives 9.98 % improvements compared to the existing algorithms on five different maps of varying AOI size and obstacles densities. Furthermore, in all cases, the proposed algorithm gives the coverage ratio C_{ratio} of 1.0 that represents the perfect coverage (i.e., 100%). The proposed approach can be applied in indoor and outdoor environments for a variety of applications specifically for the aerial inspection in urban environments. In the proposed algorithm, the footprints' sweeps fitting and final paths are calculated offline. Meanwhile, the proposed algorithm can be applied in two phases: offline and online. The sweeps can be fitted in an offline phase and path searching can be done in an online phase.

4.2. Comparisons with the Existing Algorithms Based on the Shape of the Area of Interest

Apart from the obstacle's densities, the shape of the area of intreat is also a relevant factor that can significantly impact the CPP algorithm performance. CPP over the regular shaped AOI is relatively easy compared to the irregular shaped AOI. The regular shaped AOI coverage can be obtained with the simple BF pattern. Meanwhile, the irregular shaped AOI coverage requires the concavities modification and hybrid motion patterns to achieve the full coverage. The CPP complexity varies with the shape of the target area and it requires obstacles' geometry knowledge to find the low cost solution for the coverage missions. To address this concern, we compared the proposed algorithm performance with the existing methods over five different types of the AOI to support generality. We performed rigorous experiments to verify the algorithm performance in relation with the shape of the AOI. The average computation time, path length and path overlapping are shown in Table 3.

Table 3. Proposed algorithm performance comparisons with varying shapes of the area of interest.

Algorithms	Evaluation Criteria	Shape of the Area of Interest				
		Square	Rectangle	Polygon (convex)	Polygon (non-convex)	Irregular
CA-CPP	Avg. computing time (s)	31.59	15.84	24.98	36.51	23.88
	Avg. path length (m)	70,810.21	8193.11	51,900.12	55,723.41	48,611.16
	Avg. path overlapping (m)	1153.40	169.19	671.65	755.35	976.51
BCDH-CPP	Avg. computing time (s)	25.13	11.85	19.08	27.91	18.53
	Avg. path length (m)	70,998.52	8710.24	52,100.32	55,925.41	48,914.09
	Avg. path overlapping (m)	1365.23	189.31	698.12	795.35	1001.31
Proposed	Avg. computing time (s)	19.55	8.17	14.78	18.91	13.43
	Avg. path length (m)	68,730.42	8001.96	50,300.32	54,605.21	47,410.89
	Avg. path overlapping (m)	1025.12	161.99	621.34	705.15	954.21

Through simulations and comparison with the two existing algorithms using five different shapes of the AOI, on average, our proposed algorithm reduces the computing time of pathfinding by 21.34%.

From a path length point of view, it reduces path length by 8.98%. Additionally, the proposed algorithm has less path overlapping compared to the existing algorithms. We further compare our approach results of turning maneuvers with two existing methods in ten representative scenarios. Figure 9 shows the number of turns' comparisons between proposed and two existing algorithms. The proposed method shows 12.1% and 7.03% improvements in the number of turns as compared to BCDH-CPP and CA-CPP algorithm, respectively.

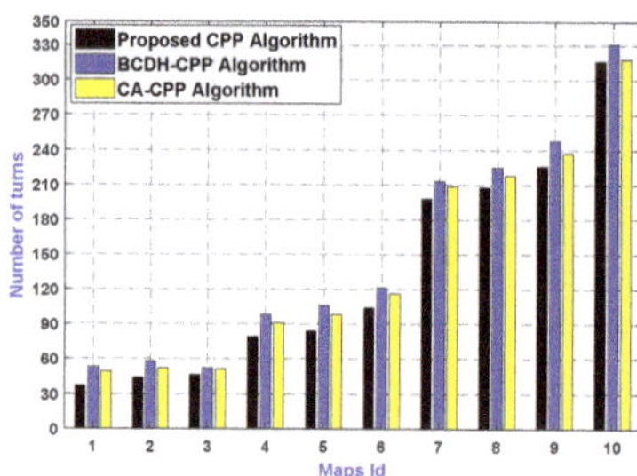

Figure 9. Turns comparisons: proposed algorithm versus BCDH-CPP and CA-CPP algorithms.

Apart from the numerical result, the proposed algorithm uses less number of sweeps compared to the decomposition based methods. Figure 10 shows the number of sweeps used by the proposed algorithm and decomposition based methods for the coverage of three cells (i.e., 5 to 7) of the AOI.

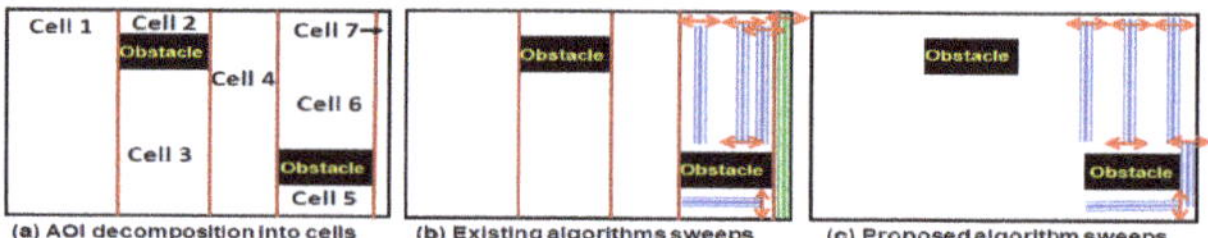

Figure 10. Footprints sweeps' comparisons with existing algorithms.

From the results, it can be observed that when the cell size is small and sufficient attention has not been paid to sensor/camera footprints size in area decomposition, the BCDH-CPP [53] and CA-CPP [54] algorithms (Figure 10b) need more sweeps compared to the proposed algorithm which can increase the path length in complex scenarios. In contrast, the proposed algorithm (Figure 10c) considers the sensor footprint as a coverage unit in the CPP and uses less sweeps to guarantee the perfect coverage.

5. Conclusions and Future Work

In this paper, we proposed a new coverage flight path planning algorithm based on footprints' sweeps fitting and a sparse waypoint graph for unmanned aerial vehicles' (UAVs) navigation in three-dimensional (3D) urban environments with fixed convex obstacles. The main goals of the proposed algorithm are to reduce the computational time, number of turns, and path overlapping while finding a minimum length path that passes over all reachable points of an area or volume of interest for UAVs flying at low-altitudes in 3D urban environments. We devised a novel footprints' sweeps fitting method by considering the UAV sensor footprint as a coverage unit that guarantees complete coverage of the AOI with fewer and longer sensor footprints' sweeps. Furthermore, we generate a sparse waypoint graph by connecting footprints' sweeps endpoints by considering footprints' sweeps visiting sequence, obstacles' effect, and maneuverability constraints to obtain a complete coverage flight path. Simulation results have shown that the proposed algorithm can achieve better performance compared to the closely related global coverage flight path planning algorithms. It finds a feasible path from the urban environments inhabiting substantial number of obstacles without sacrificing the guarantees on computing time, number of turns, perfect coverage, path overlapping, and path length.

The proposed CPP algorithm is practical, effective, feasible, and it has important significance in UAV practical application in urban environments. In addition, the proposed algorithm is directly inspired by the real-life applications of UAVs, and it is applicable for various coverage related applications in 3D urban environments with fixed obstacles. In future work, we are planning to enhance the proposed CPP algorithm for spatially distributed regions coverage.

Author Contributions: Conceptualization, S.L.; Data curation, A.M.; Investigation, A.M. and S.L.; Methodology, A.M.; Project administration, S.L.; Software, A.M.; Supervision, S.L.; Validation, A.M. and S.L.; Visualization, A.M.; Writing—original draft, A.M.; Writing—review & editing, S.L.

Funding: This work was supported by Institute for Information & Communications Technology Promotion (IITP) grant funded by the Korean Government (MSIT) (No. 2015-0-00893, Technology Development of DMM-based Obstacle Avoidance and Vehicle Control System for a Small UAV).

Conflicts of Interest: The authors declare no conflict of interest regarding the publication of this manuscript.

References

1. Kyrkou, C.; Timotheou, S.; Kolios, P.; Theocharides, T.; Panayiotou, C. Drones: Augmenting Our Quality of Life. *IEEE Potentials* **2019**, *38*, 30–36.
2. Wong, C.; Yang, E.; Yan, X.T.; Gu, D. Autonomous robots for harsh environments: A holistic overview of current solutions and ongoing challenges. *Syst. Sci. Control Eng.* **2018**, *6*, 213–219. [CrossRef]
3. Pinto, M.F.; Marcato, A.L.; Melo, A.G.; Honório, L.M.; Urdiales, C. A Framework for Analyzing Fog-Cloud Computing Cooperation Applied to Information Processing of UAVs. *Wirel. Commun. Mob. Comput.* **2019**. [CrossRef]
4. Karaduman, M.; Çınar, A.; Eren, H. UAV Traffic Patrolling via Road Detection and Tracking in Anonymous Aerial Video Frames. *J. Intell. Robot. Syst.* **2019**, 1–16. [CrossRef]
5. Wang, Y.C. Mobile Solutions to Air Quality Monitoring. In *Mobile Solutions and Their Usefulness in Everyday Life*; Springer: Cham, Switzerland, 2019; pp. 225–249.
6. Luo, C.; Miao, W.; Ullah, H.; McClean, S.; Parr, G.; Min, G. Unmanned Aerial Vehicles for Disaster Management. In *Geological Disaster Monitoring Based on Sensor Networks*; Springer: Singapore, 2019; pp. 83–107.
7. Ullah, S.; Kim, K.I.; Kim, K.H.; Imran, M.; Khan, P.; Tovar, E.; Ali, F. UAV-enabled healthcare architecture: Issues and challenges. *Future Gener. Comput. Syst.* **2019**, *97*, 425–432. [CrossRef]
8. Long, D.; Rehm, P.J.; Ferguson, S. Benefits and challenges of using unmanned aerial systems in the monitoring of electrical distribution systems. *Electr. J.* **2018**, *31*, 26–32. [CrossRef]
9. Choset, H. Coverage for robotics—A survey of recent results. *Ann. Math. Artif. Intell.* **2001**, *31*, 113–126. [CrossRef]
10. Dai, R.; Fotedar, S.; Radmanesh, M.; Kumar, M. Quality-aware UAV coverage and path planning in geometrically complex environments. *Ad Hoc Netw.* **2018**, *73*, 95–105. [CrossRef]
11. Yao, P.; Cai, Y.; Zhu, Q. Time-optimal trajectory generation for aerial coverage of urban building. *Aerosp. Sci. Technol.* **2019**, *84*, 387–398. [CrossRef]
12. Kim, D.; Liu, M.; Lee, S.; Kamat, V.R. Remote proximity monitoring between mobile construction resources using camera-mounted UAVs. *Autom. Constr.* **2019**, *99*, 168–182.
13. Oliveira, R.A.; Tommaselli, A.M.; Honkavaara, E. Generating a hyperspectral digital surface model using a hyperspectral 2D frame camera. *ISPRS J. Photogramm. Remote Sens.* **2019**, *147*, 345–360. [CrossRef]
14. Díaz-Delgado, R.; Ónodi, G.; Kröel-Dulay, G.; Kertész, M. Enhancement of Ecological Field Experimental Research by Means of UAV Multispectral Sensing. *Drones* **2019**, *3*, 7. [CrossRef]
15. Lingelbach, F. Path planning using probabilistic cell decomposition. In Proceedings of the 2004 IEEE International Conference on Robotics and Automation, ICRA'04, New Orleans, LA, USA, 26 April–1 May 2004; Volume 1, pp. 467–472.
16. Acar, E.U.; Choset, H.; Rizzi, A.A.; Atkar, P.N.; Hull, D. Morse decompositions for coverage tasks. *Int. J. Robot. Res.* **2002**, *21*, 331–344. [CrossRef]
17. Wong, S. Qualitative Topological Coverage of Unknown Environments by Mobile Robots. Doctoral Dissertation, The University of Auckland, Auckland, New Zealand, 2006.

18. Shivashankar, V.; Jain, R.; Kuter, U.; Nau, D.S. Real-Time Planning for Covering an Initially-Unknown Spatial Environment. In Proceedings of the Twenty-Fourth International Florida Artificial Intelligence Research Society Conference, Palm Beach, FL, USA, 18–20 May 2011.

19. Butler, Z.J.; Rizzi, A.A.; Hollis, R.L. Contact sensor-based coverage of rectilinear environments. In Proceedings of the 1999 IEEE International Symposium on Intelligent Control/Intelligent Systems and Semiotics, Cambridge, MA, USA, 17 September 1999; pp. 266–271.

20. Xu, L. Graph Planning for Environmental Coverage. Ph.D. Thesis, Robotics Institute, Carnegie Mellon University, Pittsburgh, PA, USA, August 2011.

21. Li, Y.; Chen, H.; Er, M.J.; Wang, X. Coverage path planning for UAVs based on enhanced exact cellular decomposition method. *Mechatronics* **2011**, *21*, 876–885. [CrossRef]

22. Nasr, S.; Mekki, H.; Bouallegue, K. A multi-scroll chaotic system for a higher coverage path planning of a mobile robot using flatness controller. *Chaos Solitons Fractals* **2019**, *118*, 366–375. [CrossRef]

23. Morgenthal, G.; Hallermann, N.; Kersten, J.; Taraben, J.; Debus, P.; Helmrich, M.; Rodehorst, V. Framework for automated UAS-based structural condition assessment of bridges. *Autom. Constr.* **2019**, *97*, 77–95. [CrossRef]

24. Zhou, K.; Jensen, A.L.; Sørensen, C.G.; Busato, P.; Bothtis, D.D. Agricultural operations planning in fields with multiple obstacle areas. *Comput. Electron. Agric.* **2014**, *109*, 12–22. [CrossRef]

25. Coombes, M.; Fletcher, T.; Chen, W.H.; Liu, C. Optimal polygon decomposition for UAV survey coverage path planning in wind. *Sensors* **2018**, *18*, 2132. [CrossRef]

26. Montanari, A.; Kringberg, F.; Valentini, A.; Mascolo, C.; Prorok, A. Surveying Areas in Developing Regions Through Context Aware Drone Mobility. In Proceedings of the 4th ACM Workshop on Micro Aerial Vehicle Networks, Systems, and Applications, Munich, Germany, 10–15 June 2018; ACM: New York, NY, USA, 2018; pp. 27–32.

27. Valente, J.; Del Cerro, J.; Barrientos, A.; Sanz, D. Aerial coverage optimization in precision agriculture management: A musical harmony inspired approach. *Comput. Electron. Agric.* **2013**, *99*, 153–159. [CrossRef]

28. Cabreira, T.; Brisolara, L.; Ferreira, P.R. Survey on Coverage Path Planning with Unmanned Aerial Vehicles. *Drones* **2019**, *3*, 4. [CrossRef]

29. Cao, Z.L.; Huang, Y.; Hall, E.L. Region filling operations with random obstacle avoidance for mobile robots. *J. Robot. Syst.* **1988**, *5*, 87–102. [CrossRef]

30. Galceran, E.; Carreras, M. A survey on coverage path planning for robotics. *Robot. Auton. Syst.* **2013**, *61*, 1258–1276. [CrossRef]

31. Lee, T.K.; Baek, S.H.; Choi, Y.H.; Oh, S.Y. Smooth coverage path planning and control of mobile robots based on high-resolution grid map representation. *Robot. Auton. Syst.* **2011**, *59*, 801–812. [CrossRef]

32. Lin, L.; Goodrich, M.A. Hierarchical heuristic search using a Gaussian Mixture Model for UAV coverage planning. *IEEE Trans. Cybern.* **2014**, *44*, 2532–2544. [CrossRef] [PubMed]

33. Choset, H. Coverage of known spaces: The boustrophedon cellular decomposition. *Auton. Robots* **2000**, *9*, 247–253. [CrossRef]

34. Choset, H.M.; Hutchinson, S.; Lynch, K.M.; Kantor, G.; Burgard, W.; Kavraki, L.E.; Thrun, S. *Principles of Robot Motion: Theory, Algorithms, and Implementation*; MIT Press: Cambridge, MA, USA, 2005.

35. Milnor, J.W.; Spivak, M.; Wells, R.; Wells, R. *Morse Theory*; Princeton University Press: Princeton, NJ, USA, 1963.

36. Nam, L.H.; Huang, L.; Li, X.J.; Xu, J.F. An approach for coverage path planning for UAVs. In Proceedings of the 2016 IEEE 14th International Workshop on Advanced Motion Control (AMC), Auckland, New Zealand, 22–24 April 2016; pp. 411–416.

37. Bochkarev, S.; Smith, S.L. On minimizing turns in robot coverage path planning. In Proceedings of the 2016 IEEE International Conference on Automation Science and Engineering (CASE), Fort Worth, TX, USA, 21–25 August 2016; pp. 1237–1242.

38. Andersen, H.L. Path Planning for Search and Rescue Mission Using Multicopters. Master's Thesis, Institutt for Teknisk Kybernetikk, Trondheim, Norway, 2014.

39. Jiao, Y.S.; Wang, X.M.; Chen, H.; Li, Y. Research on the coverage path planning of uavs for polygon areas. In Proceedings of the 2010 5th IEEE Conference on Industrial Electronics and Applications (ICIEA), Taichung, Taiwan, 15–17 June 2010; pp. 1467–1472.

40. Levcopoulos, C.; Krznaric, D. Quasi-Greedy Triangulations Approximating the Minimum Weight Triangulation. In Proceedings of the Seventh Annual ACM-SIAM Symposium on Discrete Algorithms, Atlanta, GA, USA, 28–30 January 1996; pp. 392–401.
41. Xu, A.; Viriyasuthee, C.; Rekleitis, I. Optimal complete terrain coverage using an unmanned aerial vehicle. In Proceedings of the 2011 IEEE International Conference On Robotics and automation (ICRA), Shanghai, China, 9–13 May 2011; pp. 2513–2519.
42. Öst, G. Search Path Generation With UAV Applications Using Approximate Convex Decomposition. Master Thesis, Linkoping University, Linkoping, Sweden, 2012.
43. Sadat, S.A.; Wawerla, J.; Vaughan, R.T. Recursive non-uniform coverage of unknown terrains for uavs. In Proceedings of the 2014 IEEE/RSJ International Conference on Intelligent Robots and Systems (IROS 2014), Chicago, IL, USA, 14–18 September 2014; pp. 1742–1747.
44. Sadat, S.A.; Wawerla, J.; Vaughan, R. Fractal trajectories for online non-uniform aerial coverage. In Proceedings of the 2015 IEEE International Conference on Robotics and Automation (ICRA), Seattle, WA, USA, 26–30 May 2015; pp. 2971–2976.
45. Barrientos, A.; Colorado, J.; Cerro, J.D.; Martinez, A.; Rossi, C.; Sanz, D.; Valente, J. Aerial remote sensing in agriculture: A practical approach to area coverage and path planning for fleets of mini aerial robots. *J. Field Robot.* **2011**, *28*, 667–689. [CrossRef]
46. LaValle, S.M. *Planning Algorithms*; Cambridge University Press: Cambridge, UK, 2006.
47. Dorigo, M.; Maniezzo, V.; Colorni, A. Ant system: Optimization by a colony of cooperating agents. *IEEE Trans. Syst. Man Cybern. Part B (Cybern.)* **1996**, *26*, 29–41. [CrossRef]
48. Larranaga, P.; Kuijpers, C.M.H.; Murga, R.H.; Inza, I.; Dizdarevic, S. Genetic algorithms for the travelling salesman problem: A review of representations and operators. *Artif. Intell. Rev.* **1999**, *13*, 129–170. [CrossRef]
49. Zelinsky, A.; Jarvis, R.A.; Byrne, J.C.; Yuta, S. Planning paths of complete coverage of an unstructured environment by a mobile robot. In Proceedings of the International Conference on Advanced Robotics, Tsukuba, Japan, 8–9 November 1993; Volume 13, pp. 533–538.
50. Hart, P.E.; Nilsson, N.J.; Raphael, B. A formal basis for the heuristic determination of minimum cost paths. *IEEE Trans. Syst. Sci. Cybern.* **1968**, *4*, 100–107. [CrossRef]
51. Nash, A.; Daniel, K.; Koenig, S.; Felner, A. Theta*: Any-Angle Path Planning on Grids. In Proceedings of the AAAI Conference on Artificial Intelligence, Vancouver, BC, Canada, 22–26 July 2007; pp. 1177–1183.
52. Fu, Y.; Ding, M.; Zhou, C.; Hu, H. Route planning for unmanned aerial vehicle (UAV) on the sea using hybrid differential evolution and quantum-behaved particle swarm optimization. *IEEE Trans. Syst. Man Cybern. Syst.* **2013**, *43*, 1451–1465. [CrossRef]
53. Xu, A.; Viriyasuthee, C.; Rekleitis, I. Efficient complete coverage of a known arbitrary environment with applications to aerial operations. *Auton. Robots* **2014**, *36*, 365–381. [CrossRef]
54. Torres, M.; Pelta, D.A.; Verdegay, J.L.; Torres, J.C. Coverage path planning with unmanned aerial vehicles for 3D terrain reconstruction. *Expert Syst. Appl.* **2016**, *55*, 441–451. [CrossRef]
55. Majeed, A.; Lee, S. A Fast Global Flight Path Planning Algorithm Based on Space Circumscription and Sparse Visibility Graph for Unmanned Aerial Vehicle. *Electronics* **2018**, *7*, 375. [CrossRef]

Article

Modeling and Analysis on Energy Consumption of Hydraulic Quadruped Robot for Optimal Trot Motion Control

Kun Yang, Xuewen Rong, Lelai Zhou * and Yibin Li

Center for Robotics, School of Control Science and Engineering, Shandong University, No. 17923, Jingshi Road, Jinan 250061, China; yangkundtc@gmail.com (K.Y.); rongxw@sdu.edu.cn (X.R.); liyb@sdu.edu.cn (Y.L.)
* Correspondence: zhoulelai@sdu.edu.cn; Tel.: +86-0531-8839-6083

Received: 19 March 2019; Accepted: 25 April 2019; Published: 28 April 2019

Abstract: Energy consumption is an important performance index of quadruped robots. In this paper, the energy consumptions of the quadruped robot SCalf with a trot gait under different gait parameters are analyzed. Firstly, the kinematics and dynamics models of the robot are established. Then, an energy model including the mechanical power and heat rate is proposed. To obtain the energy consumption, a cubic spline interpolation foot trajectory is used, and the feet forces are calculated by using the minimization of norm of the foot force method. Moreover, an energetic criterion measuring the energy cost is defined to evaluate the motion. Finally, the gait parameters such as step height, step length, standing height, gait cycle, and duty cycle that influence the energy consumption are studied, which could provide a theoretical basis for parameter optimization and motion control of quadruped robots.

Keywords: quadruped robot; energy model; foot force distribution; cubic spline interpolation

1. Introduction

Compared with wheeled or tracked robots, legged robots have certain advantages in the unstructured environment and have become one of the hotspots in the field of robotics. For legged robots, quadruped robots have simple structure and are easy to control; therefore, they're more widely used than biped and hexapod robots.

Foot trajectory planning is important for the study of quadruped robots. Sakakibara et al. [1] proposed a trot gait trajectory based on a sinusoidal curve and realized by He et al. [2]. This trajectory was simple and easy to be realized, but when the robot touched the ground, the bottom of the foot would slide and move. Semini et al. [3] used the cycloid foot trajectory to plan motions of HyQ—a hydraulically and electrically actuated quadruped robot. However, only a set of typical trot gait parameters including a 50% duty cycle was used. Zhang et al. [4] used a virtual model to study the torso motion control of a quadruped robot, where a flight toe trajectory generator was established. In the flight phase, a rectangular trajectory was used. However, the foot endpoint would have a big velocity before landing, which could cause an impact and could affect the control system. To deal with the above problem, a new foot trajectory needs to be studied. The foot trajectory can solve problems like the landing impact and more gait parameters need to be included in the trajectory.

Despite the fact that legged robots promise great mobility in rough and unstructured terrains, they still have some deficiencies in energy consumption when compared with tracked or wheeled robots. In order to improve the energy efficiency of the legged robots, many researchers studied the energy consumption of legged robots. In the last paragraph, the foot trajectories were studied in the view of motions not energy. In this part, energy models proposed by different researchers are introduced. Ikeda et al. [5] analyzed the energy flow of a quadruped robot with a flexible trunk joint. The energy consumption conditions were studied under different gaits. However, this paper did not establish any specific energy model of the robot. It only divided the energy flow into the energy input, the friction loss, and the collision loss. It lacked an analysis of the mechanical power change. Muraro et al. [6] and Silva et al. [7] both established their own energy model and studied the energy consumption by using different gait, and several criteria were proposed to evaluate the performances of different conditions. These two works focused on body forward velocity; therefore, only the step length and gait cycle were taken into consideration. Moreover, the energy model of References [6] and [7] only calculated the mechanical power, and the heat rate was ignored, which is a big part of the loss in practical application. Lei et al. [8] analyzed the energy consumption of a quadruped robot with three foot trajectories based on their trot gait. The energy model only took the mechanical power into account, and the influence on energy consumption caused by a duty cycle was not studied. Wang et al. [9] established the kinematics and dynamics models of a quadruped robot and studied the influences of different parameters on energy consumption. In this work, the feet forces were regarded as half of the robot total mass, which was not rigorous. Moreover, the heat rate was ignored in the energy model and only the situations with a 50% duty cycle were studied. Roy et al. [10] studied the effects of turning gait parameters on energy consumption of an electric actuated six-legged robot. In the paper, the joint positions were programmed by quintic polynomial. Moreover, the foot force distribution problem was discussed and a relative complete energy model was established. However, the heat rate of the hydraulic cylinder was different from the electric motor. In this work, a complete energy model that is suitable for a hydraulic actuated quadruped robot is studied.

In order to calculate the energy consumption correctly, the foot force distribution problem needs to be considered. Although the above studies have given a certain impetus to the high efficient trajectory planning and energy consumption research, the current energy models are either relatively simple or not suitable for hydraulic actuated quadruped robots. Therefore, a complete energy model as well as a universal foot trajectory for a hydraulic actuated quadruped robot needs to be studied. For hydraulic actuated quadruped robots, the hydraulic friction includes the Coulomb friction, static friction, and static decay friction. Therefore, the energy consumption due to friction is big and the heat rate must be considered in the energy model. In order to describe the robot motions better, a universal foot trajectory for a hydraulic actuated quadruped robot needs to be studied. Different gait parameters have to be included in the foot trajectory, and their influences on energy consumption should be studied.

This work expounds our study of the quadruped robot foot trajectory in planar motions. The main purpose of this work is to study the effects of different gait parameters on energy consumption. Firstly, the kinematics and dynamics model of the SCalf robot are established. An energy model including the mechanical power and heat rate is set up to calculate the energy consumption. A trot gait foot trajectory based on the cubic spline interpolation is proposed. The energy consumption variation caused by different gait parameters are studied through simulations.

2. Robot Modelling

The quadruped robot SCalf was build by the Center for Robotics of Shandong University, as shown in Figure 1. SCalf is a hydraulic actuated robot. The length and width of the robot are 1.4 m and 0.75 m respectively, and the overall weight is 200 kg. The robot consists of a trunk and four legs. The structure of the legs must be designed considering the fact that an improper design can limit the reachable space of the robot legs when operating in the normal quadruped gaits [11]. Therefore, the leg of SCalf is designed as a 3 DOFs (degree of freedom) structure to make the foot endpoint move in a three-dimension workspace with the hip joint as its origin. The three joints of the leg are defined as the rolling hip joint, pitching hip joint, and pitching knee joint. To make the robot easy to control, the SCalf is built symmetrically. The backward/forward configuration is used to improve the gradeability and load capacity.

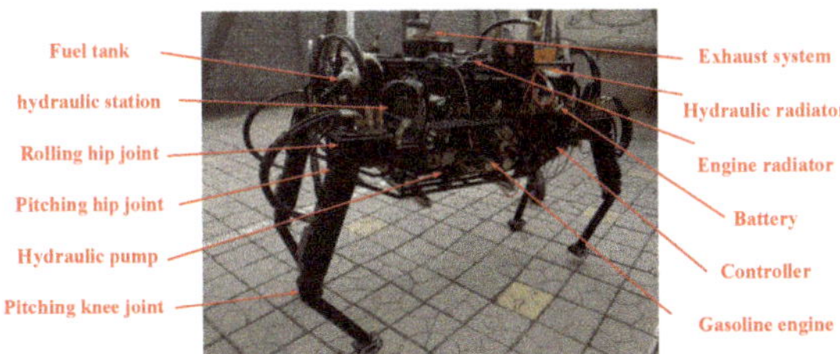

Figure 1. The overall structure of the SCalf robot.

A variable displacement piston pump is connect to the gasoline engine to provide power for the whole robot system. The onboard hydraulic system and the control system are integrated inside the trunk of the robot. The onboard hydraulic system can distribute a hydraulic flow to the joint actuators and can control the motions of the robot. The robot can be used to transport loads on rough terrains like jungle or mountain, and its load capacity is 200 kg.

To calculate the energy consumption of the robot, the following preconditions are made:

(a) Only the motions in the sagittal plane are studied. The robot is walking on a flat surface with the trot gait.

(b) When walking on a flat surface, the height change of the robot trunk's center of mass (COM) is ignored.

(c) The trunk of the robot can be regarded as a cuboid, and its COM is assumed to be at the geometric center of the body

2.1. The Kinematics Modelling

The legs of the robot have the same structure and mechanical parameters. Here, the right front (RF) leg is used as an example. The single leg model is shown in Figure 2. According to precondition (a), the motions of the rolling hip joint are not considered and the rolling hip joint position is set as 0.

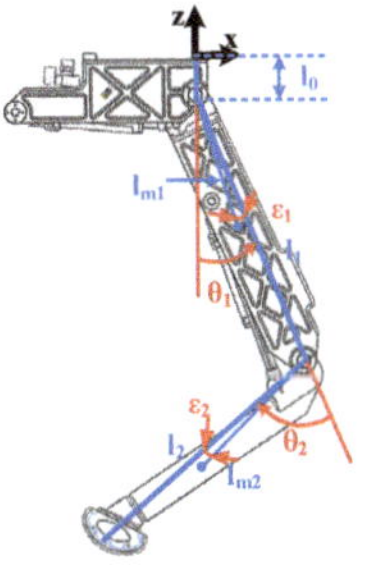

Figure 2. A simplified model of the right front (RF) leg.

In Figure 2, the RF leg can be divided into the hip, thigh, and shank parts. Since the motions of the rolling hip joint are neglected, the hip part can be regarded as the base. The origin of the coordinate frame $\{O_h\}$ is set at the rolling hip joint. The z-axis direction is vertical to the ground facing upward. The x-axis is pointing to the motion direction of the robot.

The mechanical parameters in Figure 2 are introduced as follows. The lengths of the hip, thigh, and shank parts are l_0, l_1, and l_2 respectively, where $l_0 = 45$ mm. m_1 and m_2 are the masses of the thigh and the shank parts, respectively. The joint positions of the pitching hip and pitching knee joints are denoted as θ_1 and θ_2, respectively. The COMs of the thigh and shank parts are defined by l_{m1}, l_{m2}, ε_1, and ε_2. When the thigh and shank parts rotate around their own joint shafts, the moments of inertia are I_1 and I_2, respectively. The values of these mechanical parameters are listed in Table 1 [12].

Table 1. The parameters of the thigh and shank parts.

Parameter (i = 1,2)	Thigh	Shank
Mass (m_i/Kg)	4.10	2.24
Length (l_i/mm)	451.00	405.00
COM position (l_{mi}/mm)	210.43	227.20
COM angle (ε_i/rad)	0.037	−0.083
Moment of inertia (I_i/Kg·m^2)	0.178	0.057

Based on the mechanical parameters, the location of the foot endpoint in coordinate frame$\{O_h\}$ can be written as follows, where $\theta_{12} = \theta_1 + \theta_2$

$$\begin{cases} x = -l_2 sin\theta_{12} - l_1 sin\theta_1 \\ z = -l_2 cos\theta_{12} - l_1 cos\theta_1 - l_0 \end{cases} \tag{1}$$

By solving Equation (1), the joint positions of the RF leg can be obtained as

$$\begin{cases} \theta_1 = -\phi - arctan(\dfrac{x}{-z - l_0}) \\ \theta_2 = \pi - arccos(\dfrac{l_1^2 + l_2^2 - \xi^2}{2l_1 l_2}) \end{cases} \tag{2}$$

where

$$\begin{cases} \xi = \sqrt{(-z - l_0)^2 + x^2} \\ \phi = arccos(\dfrac{l_1^2 + \xi^2 - l_2^2}{2l_1 \xi}) \end{cases} \tag{3}$$

The Jacobian matrix of the single leg model can also be calculated by Equation (1).

$$J = \begin{bmatrix} -l_1 cos\theta_1 - l_2 cos\theta_{12} & -l_2 cos\theta_{12} \\ l_1 sin\theta_1 + l_2 sin\theta_{12} & l_2 sin\theta_{12} \end{bmatrix} \tag{4}$$

2.2. The Dynamics Modelling

In order to obtain the energy consumptions of the robot, the joint torques must be acquired. The RF leg is used to build the dynamics model of the robot. The standard form of the dynamic equation can be written as Equation (5), where $D(\theta)$ is the inertia matrix, $H(\theta, \dot{\theta})$ is the Coriolis and centrifugal forces matrix, and $G(\theta)$ is the gravitational loading vector.

$$\tau_J = D(\theta)\ddot{\theta} + H(\theta, \dot{\theta})\dot{\theta} + G(\theta) \tag{5}$$

In the dynamics equation of Equation (5), the inertia matrix $D(\theta)$ is a 2×2 matrix and can be written as follows:

$$D(\theta) = \begin{bmatrix} D_{11}(\theta) & D_{12}(\theta) \\ D_{21}(\theta) & D_{22}(\theta) \end{bmatrix} \tag{6}$$

where

$$D_{11}(\theta) = I_1 + I_2 + m_1 l_{m1}^2 + m_2(l_1^2 + l_{m2}^2 + 2l_1 l_{m2} cos\theta_{\varepsilon 2}) \tag{7}$$

$$D_{12}(\theta) = D_{21}(\theta) = m_2 l_{m2}^2 + m_2 l_1 l_{m2} cos\theta_{\varepsilon 2} + I_2 \tag{8}$$

$$D_{22}(\theta) = m_2 l_{m2}^2 + I_2 \tag{9}$$

The Coriolis and centrifugal forces matrix $H(\theta, \dot{\theta})$ is shown in Equation (10).

$$H(\theta,\dot{\theta}) = \begin{bmatrix} H_{11}(\theta,\dot{\theta}) & H_{21}(\theta,\dot{\theta}) \\ H_{12}(\theta,\dot{\theta}) & H_{22}(\theta,\dot{\theta}) \end{bmatrix} \tag{10}$$

where

$$H_{11}(\theta,\dot{\theta}) = -\dot{\theta}_2 m_2 l_1 l_{m2} sin\theta_{\varepsilon2} \tag{11}$$

$$H_{12}(\theta,\dot{\theta}) = -(\dot{\theta}_1 + \dot{\theta}_2) m_2 l_1 l_{m2} sin\theta_{\varepsilon2} \tag{12}$$

$$H_{21}(\theta,\dot{\theta}) = \dot{\theta}_1 m_2 l_1 l_{m2} sin\theta_{\varepsilon2} \tag{13}$$

$$H_{22}(\theta,\dot{\theta}) = 0 \tag{14}$$

The gravitational loading vector $G(\theta)$ is

$$G(\theta) = \begin{bmatrix} m_1 g l_{m1} sin\theta_{\varepsilon1} + m_2 g(l_1 sin\theta_1 + l_{m2} sin\theta_{\varepsilon12}) \\ m_2 g l_{m2} cos\theta_{\varepsilon12} \end{bmatrix} \tag{15}$$

In the above equations, $\theta_{\varepsilon1} = \theta_1 + \varepsilon_1$, $\theta_{\varepsilon2} = \theta_2 + \varepsilon_2$, and $\theta_{\varepsilon12} = \theta_1 + \theta_2 + \varepsilon_2$.

Equation (5) shows the situation with no contact between the feet and the ground. When considering the ground reaction force, the dynamics model of the single leg changes to Equation (16), where F_G is the ground reaction force vector. The calculation of F_G will be introduced in the next part.

$$\tau - J_G^T F_G = D(\theta)\ddot{\theta} + H(\theta,\dot{\theta})\dot{\theta} + G(\theta) \tag{16}$$

2.3. The Foot Force Distribution

When analysing the foot force distribution, the slippage between the foot endpoint and the ground is neglected [13]. For the sagittal motions, the lateral forces of the foot endpoints can be ignored, and the ground reaction forces of each leg can be divided into a normal and a tangential component.

The ground reaction force can cause many negative effects like structural damages and control difficulties; therefore, the value of the ground reaction force must be minished. Here, the minimization of norm of the foot force method is used to minimize the norm solution of the foot force and to reduce the compact between the ground and the robot [14].

A coordinate frame $\{O_b\}$ is established at the geometric center of the body. The x axis points to the direction of the motion. The direction of the z axis is vertical to the ground and points upwards. The y axis is defined by the right-hand rule. The static force diagram is illustrated in Figure 3. Here, we assume the RF and LH (left-hind) legs are in the stance phase. $F_G = [f_{RF}, f_{LH}]^T$ is the ground reaction force vector which contains the ground reaction force of the RF and LH legs, where $f_{RF} = [f_{RF,x}, f_{RF,z}]^T$ and $f_{LH} = [f_{LH,x}, f_{LH,z}]^T$. The vector $V = [F_x, F_z, T_y]^T$ contains the forces and moment acting on the robot COM in sagittal plane. Under these conditions, the forces and moment balance equations can be written as follows:

$$F_x = f_{RF,x} + f_{LH,x} \tag{17}$$

$$F_z = f_{RF,z} + f_{LH,z} \tag{18}$$

$$T_y = f_{RF,x} z_{RF} - f_{RF,z} x_{RF} + f_{LH,x} z_{LH} - f_{LH,z} x_{LH} \tag{19}$$

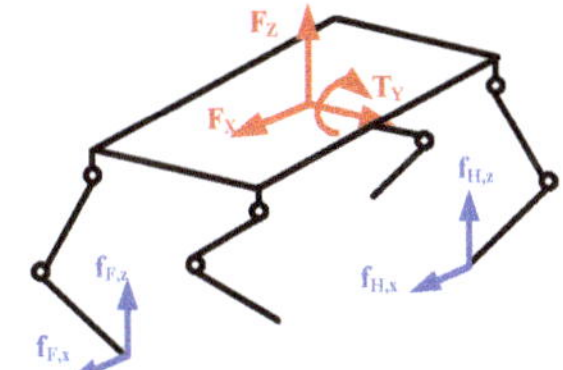

Figure 3. The static force diagram of SCalf.

The Equations (17) to (19) can be written in a matrix form as follows:

$$B \cdot F_G = V \tag{20}$$

where

$$B = \begin{bmatrix} 1 & 0 & 1 & 0 \\ 0 & 1 & 0 & 1 \\ z_{RF} & -x_{RF} & z_{LH} & -x_{LH} \end{bmatrix} \tag{21}$$

In Equation (21), $\{x_{RF}, z_{RF}\}$ and $\{x_{LH}, z_{LH}\}$ are the coordinates of the support foot endpoints in $\{O_b\}$. In Equation (20), there are 4 unknowns but only 3 equations. Here, we use the least squared method to realize the minimization of norm of the foot force method. The result is shown as follows.

$$F_G = B^T (BB^T)^{-1} V \tag{22}$$

In Equation (22), the matrix $B^+ = B^T (BB^T)^{-1}$ is also the pseudoinverse matrix of B. As the third component of V (T_y) is zero, the parameters in the last column of B^+ can be neglected. The value of B^+ is shown as follows.

$$B^+ = \begin{bmatrix} \dfrac{b_{11}^+}{2\Lambda} & \dfrac{b_{12}^+}{2\Lambda} & \times \\ \dfrac{b_{21}^+}{2\Lambda} & \dfrac{b_{22}^+}{2\Lambda} & \times \\ \dfrac{b_{31}^+}{2\Lambda} & \dfrac{b_{32}^+}{2\Lambda} & \times \\ \dfrac{b_{41}^+}{2\Lambda} & \dfrac{b_{42}^+}{2\Lambda} & \times \end{bmatrix} \tag{23}$$

where

$$b_{11}^+ = (x_{RF} - x_{LH})^2 + 2z_{LH}^2 - 2z_{LH}z_{RF} \tag{24}$$

$$b_{12}^+ = (x_{RF} + x_{LH})(z_{RF} - z_{LH}) \tag{25}$$

$$b_{21}^+ = (x_{RF} - x_{LH})(z_{RF} + z_{LH}) \tag{26}$$

$$b_{22}^+ = (z_{RF} - z_{LH})^2 + 2x_{LH}^2 - 2x_{LH}x_{RF} \tag{27}$$

$$b_{31}^+ = (x_{RF} - x_{LH})^2 + 2z_{RF}^2 - 2z_{LH}z_{RF} \tag{28}$$

$$b_{32}^+ = (x_{RF} + x_{LH})(z_{LH} - z_{RF}) \tag{29}$$

$$b_{41}^+ = (x_{LH} - x_{RF})(z_{RF} + z_{LH}) \tag{30}$$

$$b_{42}^+ = (z_{RF} - z_{LH})^2 + 2x_{RF}^2 - 2x_{LH}x_{RF} \tag{31}$$

$$\Lambda = (x_{RF} - x_{LH})^2 + (z_{RF} - z_{LH})^2 \tag{32}$$

The Jacobian matrix in Equation (16) can be written as follows:

$$J_G = \begin{bmatrix} J_F & 0 \\ 0 & J_H \end{bmatrix} \tag{33}$$

In Equation (33), J_F and J_H are the Jacobian of the RF and LH legs respectively.

3. The Energy Analysis of The Scalf

3.1. Energy Model

In order to analyze the energy consumption conditions of SCalf, the energy model must be established. For SCalf, the calculation of the energy consumption is obtained by the integration of the joint power. The joint power consists of the mechanical power and heat rate, as shown in Equation (34), where T is the gait cycle, E is the energy consumption in one gait cycle, and P is the joint power.

$$
\begin{aligned}
E &= \int_0^T P\,dt \\
&= \int_0^T \left(|\tau\dot{\theta}| + f_f\dot{x}_p \right)dt
\end{aligned}
\tag{34}
$$

In the second line of Equation (34), f_f and $\dot{x}_p$ are the friction and piston velocity of the hydraulic cylinder, respectively. The joint torque τ can be acquired by the dynamics analysis.

The structure of the hydraulic actuator is shown in Figure 4. The displacement and force of the piston can be measured by the position and force sensors. The motions of the piston are controlled by the servo valve.

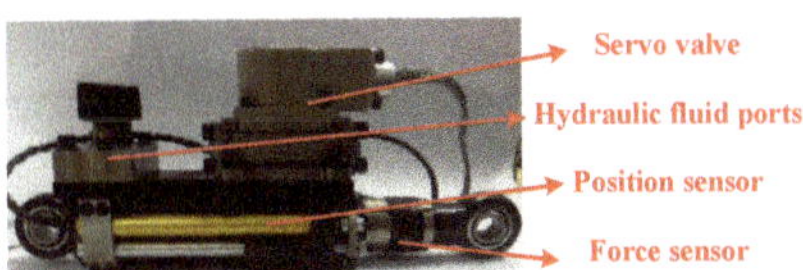

Figure 4. The structure of the hydraulic actuator.

The friction of hydraulic cylinders are strongly coupled by many parameters, and it has a high nonlinearity. It depends on many conditions like the cylinder velocity $\dot{x}_p$, the pressure difference Δp across the piston and possibly on the cylinder position x_p, and the temperature t [15]. The SCalf uses the asymmetry cylinders to realize the joint motions. The model parameters are different when the actuator are extending or shorting. According to the research made by Nissing, the model of the hydraulic frictions can be written in Equation (35) [16].

$$
f_f(\dot{x}_p) = \begin{cases} B^+\dot{x}_p + \left(F_{c0}^+ + F_{s0}^+ e^{-\frac{|\dot{x}_p|}{C_s^+}} \right), & \dot{x}_p > 0 \\[2ex] B^-\dot{x}_p - \left(F_{c0}^- + F_{s0}^- e^{-\frac{|\dot{x}_p|}{C_s^-}} \right), & \dot{x}_p < 0 \end{cases}
\tag{35}
$$

In Equation (35), B is the viscous friction coefficient (Ns/m), F_{c0} is the Coulomb friction coefficient (N), F_{s0} is the static friction coefficient (N), and C_S is the attenuation coefficient of static friction. Using the measured result made by Polzer et al. [17], the parameters in Equation (35) can be obtained as follows. When $\dot{x}_p \geq 0$, $B = 220$ Ns/m, $F_{c0} = 50$ N, $F_{s0} = 30$ N and $C_s = 0.015$ m/s. When $\dot{x}_p < 0$, $B = 180$ Ns/m, $F_{c0} = 50$ N, $F_{s0} = 20$ N, and $C_s = 0.007$ m/s.

The cylinder velocities can be obtained by the differential of cylinder lengths. The cylinder lengths can be calculated by the mechanical parameters in Figure 5 and Table 2.

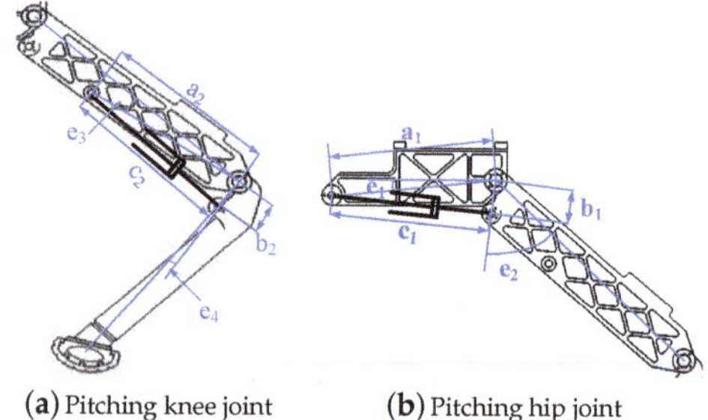

(**a**) Pitching knee joint   (**b**) Pitching hip joint

Figure 5. The mechanical structure parameters of the leg.

Table 2. The mechanical parameters used in Figure 5.

Joint	a_i (mm)	b_i (mm)	Angles (°)	
Pitching hip joint	289.26	56.87	$e_1 = 0.094$	$e_2 = 0.96$
Pitching knee joint	289.26	56.89	$e_3 = 0.14$	$e_4 = 0.041$

The relationship between the cylinder lengths and mechanical parameters are shown in Equation (36)

$$\begin{cases} c_1 = \sqrt{a_1^2 + b_1^2 - 2a_1 b_1 cos(-\theta_1 + \dfrac{\pi}{2} - e_1 - e_2)} \\ c_2 = \sqrt{a_2^2 + b_2^2 - 2a_2 b_2 cos(-\theta_2 + \pi - e_3 + e_4)} \end{cases} \tag{36}$$

3.2. Energetic Criterion

In order to evaluate the energy consumptions of robot motions, the corresponding energetic criterion should be designed.

In this paper, a criterion called energy cost is considered. The energy cost E_{cost} is defined as the energy consumption for a unit distance. It represents the energy consumption of the robot movements intuitively. It is defined by assuming that the negative work produced by the actuators are dissipated.

$$E_{cost} = \frac{1}{x(T) - x(0)} \int_0^T P dt \quad [J/m] \tag{37}$$

4. The Foot Trajectory Analysis

The SCalf team has already done some research on the gait trajectory planning [18–20], but they only considered the situations under the duty cycle as 50%. In this paper, a more general trot gait trajectory is proposed. In the trot gait, S is the step length, T is the gait cycle, H is the step height, and H_0 is the standing height, as illustrated in Figure 6. β is the duty cycle of the trot gait. The velocity of the robot v_r can be obtained by the step length S and the gait cycle T by using Equation (38).

$$v_r = 2S/T \tag{38}$$

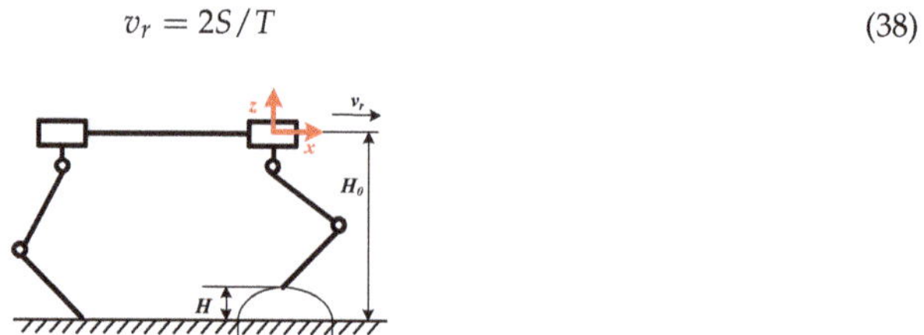

Figure 6. The illustration of the gait parameters.

During the movements of the robot, the legs will swing periodically with respect to the trunk. In order to make the quadruped robot walk at a steady speed, this work realizes the trajectory planning of the robot by simulating the trajectory of the quadruped animals' feet. Based on the motions of the horses, a foot trajectory including the cubic spline interpolation curve and the straight line is studied. The phases of the four legs are shown in Figure 7, where the two feet on the diagonal have the same movements.

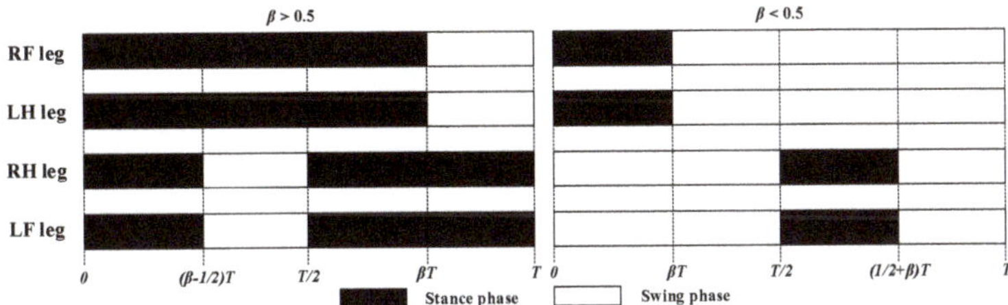

Figure 7. The leg phases of the trot gait.

Here, the trajectory of the RF leg is analyzed. During the stance phase, the trunk of the robot is moving forward. According to Equations (16), (22), and (34), in order to minimize the energy consumption, joint torques have to be as small as possible. Then, the first component of M, which is $F_x = Ma$, have to be small. Therefore, the trunk acceleration a is set as zero, and the motions of the stance phase become uniform motions in the x direction. The positions in the z direction are equal to $-H_0$. The trajectory equation of the stance phase in the coordinate frame $\{O_b\}$ can be obtained in Equation (39).

$$\begin{cases} x(t) = \dfrac{S}{2}\left(1 - \dfrac{2t}{\beta T}\right) \\ z(t) = -H_0 \end{cases}, \quad 0 < t \le \beta T \tag{39}$$

In order to obtain the parameters of the foot trajectory during the swing phase, the initial and terminated conditions of the stance phase are listed in Table 3, including the positions and velocities of the foot endpoint in the x and z directions.

Table 3. The initial and terminated conditions of the stance phase.

Items (time)	x Direction	z Direction
position (0)	$\dfrac{S}{2}$	H_0
position (βT)	$-\dfrac{S}{2}$	H_0
velocity (0)	$-\dfrac{S}{\beta T}$	0
velocity (βT)	$-\dfrac{S}{\beta T}$	0

In the swing phase, the cubic spline interpolation and the cosine curve are used in the x and z directions respectively. In the trajectory expression of the x direction, the coefficients of the cubic equation are obtained from the data in Table 3. For the cosine curve in the z direction, the foot endpoint reaches the curve's top in the middle of the swing phase. According to the above conditions, the foot trajectory can be written in Equation (40).

$$\begin{cases} x(t) = \dfrac{S}{2}\left(a_3 t^3 + a_2 t^2 + a_1 t + a_0\right) \\ z(t) = -H_0 + \dfrac{H}{2}\left\{1 - \cos\left[\dfrac{2\pi t}{(1 - \beta T)} - \dfrac{2\pi \beta}{(1 - \beta)}\right]\right\} \end{cases}, \quad \beta T < t \le T \tag{40}$$

where

$$\begin{cases} \gamma = (\beta - 1)^3 \\ a_0 = \dfrac{(\beta^3 - \beta^2 - 3\beta - 1)}{\gamma} \\ a_1 = \dfrac{-2(\beta^3 - 3\beta^2 - 3\beta - 1)}{\gamma \beta T} \\ a_2 = \dfrac{-6(\beta + 1)}{\gamma \beta T^2} \\ a_3 = \dfrac{4}{\gamma \beta T^3} \end{cases} \tag{41}$$

The curves of the foot trajectory are shown in Figure 8, including the curves of the x direction, of the z direction, and in coordinate frame $\{O_h\}$. Here, the gait parameters are chosen as follow: $S = 0.25$ m, $T = 0.5$ s, $H = 0.08$ m, $H_0 = 0.5$ m, and $\beta = 0.3$.

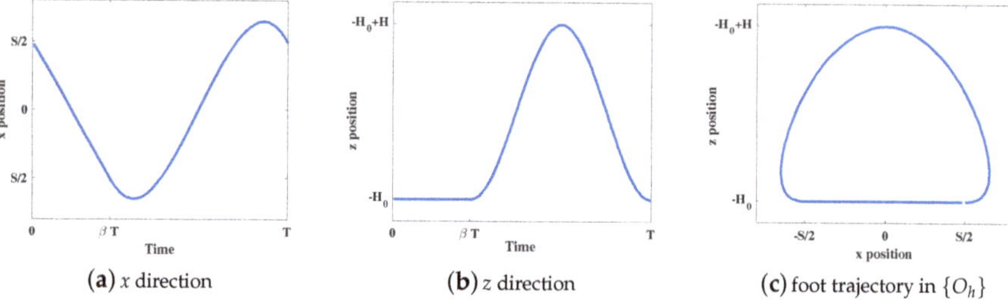

(**a**) x direction   (**b**) z direction   (**c**) foot trajectory in $\{O_h\}$

Figure 8. The foot trajectory curves.

The advantages of this foot trajectory are listed as follows:

(a) When gait phases switch, the velocities of the foot endpoint are zero, which can eliminate the collisions between the foot and the ground.

(b) The trunk velocity and the velocity of the swing leg are changing with no mutation.

(c) At the beginning and end of the swing phases, the foot endpoint will move back some distance, which can improve the ability of the robot to move over obstacles and to adapt to certain undulating grounds.

5. Simulations

In this section, we will study the influences of different gait parameters on the energy consumptions. The four legs of the SCalf have the same structure and mechanical parameters; therefore, the energy consumption of the RF leg can be regarded as a quarter of the total energy consumption.

A set of experiments are implementing in MATLAB. The total energy consumptions as well as the energy consumptions during the stance and swing phases using different gait parameters are calculated, and the energetic criterion is used to evaluate the robot motions. The flow chart of the simulation is illustrated in Figure 9.

In the simulations, the robot trunk can be regarded as a cuboid, and the mass of trunk is $M = 200$ kg. The distances between the origins of $\{O_b\}$ and $\{O_h\}$ in the x and z directions are 0.68 m and 0.15 m respectively, which are used to confirm the position of the foot endpoints in $\{O_b\}$.

Different gait parameters such as step length S, gait cycle T, step height H, standing height H_0, and duty cycle β are studied in this section. The default values of the gait parameters are listed in Table 4. Under the default condition, the velocity of the robot is 1 m/s.

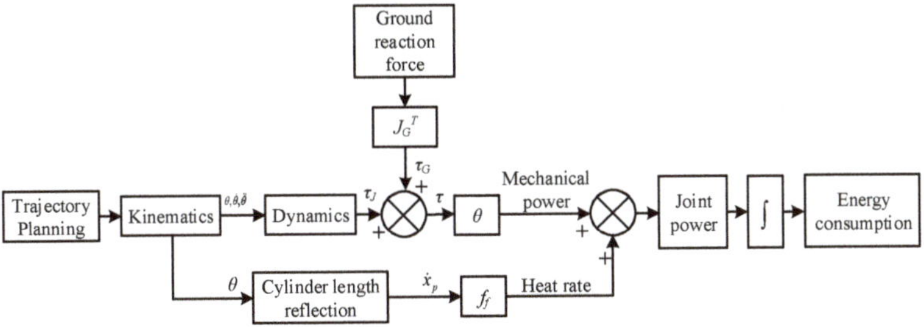

Figure 9. The flow chart of the simulation.

Table 4. The default values of the gait parameters.

Gait Parameter	Default Value
step length S	0.25 m
gait cycle T	0.5 s
step height H	0.08 m
standing height H_0	0.6 m
gait cycle β	0.5

5.1. Validation of the Energy Model

In order to verify the energy model proposed in this paper, some experiments are carried out. Different from the electrical actuated robots [21,22], the energy consumption of a hydraulic actuated robot is obtained by the data of joint sensors. For SCalf, the position and force sensors mounted on the actuator (Figure 4) are used to measure the cylinder lengths c_i and joint forces τ_i ($i = 1, 2$). Here, the equations of the RF leg are used. The mechanical parameters can be found in Table 2.

The joint angular velocities in Equation (34) can be obtained by the following equations.

$$\begin{cases} \theta_1 = \dfrac{\pi}{2} - arccos(\dfrac{a_1^2 + b_1^2 - c_1^2}{2a_1b_1}) - e_1 - e_2 \\ \theta_2 = \pi - arccos(\dfrac{a_2^2 + b_2^2 - c_2^2}{2a_2b_2}) - e_3 + e_4 \end{cases} \tag{42}$$

$$\dot{\theta}_i = \frac{d\theta}{dt}, \quad i = 1, 2 \tag{43}$$

The torques in Equation (34) are equal to the forces multiplied by their force arms, where $i = 1, 2$.

$$\tau_i = f_i \cdot a_i \sqrt{1 - cos(\dfrac{a_i^2 + c_i^2 - b_1^2}{2a_ic_i})} \tag{44}$$

In the experiment, the gait parameters are chosen from Table 4. The sensor data are recorded with a frequency of 80 Hz. In this work, data of 5 gait cycles are recorded. The curves of joint positions, joint torques, and the leg power of the RF leg are illustrated in Figures 10–12.

From the above experiment results, the joint positions, joint torque, and leg power have the same trends in the experiment and simulation. The energy consumptions during 5 cycles are 393.69 J and 421.80 J for the experiment and simulation, respectively. The mean powers of the RF leg in the experiment and simulation can be calculated as 157.48 W and 168.72 W respectively. The differences between the data are mainly caused by the foot force distribution. Based on the energy consumptions and average power, the proposed energy model is verified.

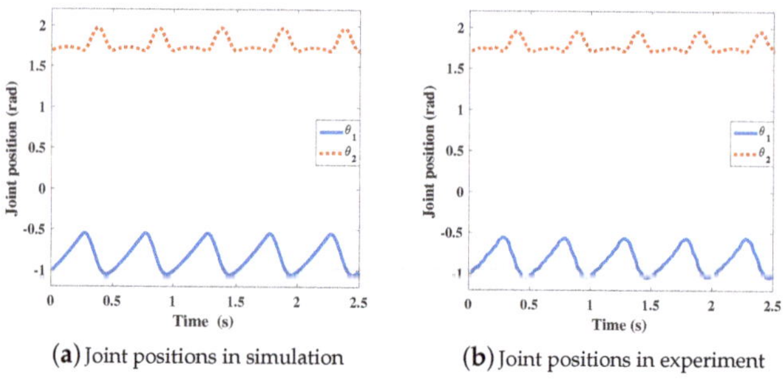

(**a**) Joint positions in simulation

(**b**) Joint positions in experiment

Figure 10. The joint positions of the RF leg.

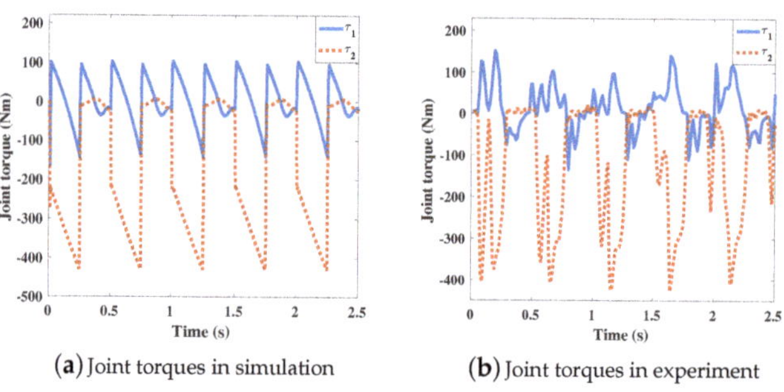

(**a**) Joint torques in simulation

(**b**) Joint torques in experiment

Figure 11. The joint torques of the RF leg.

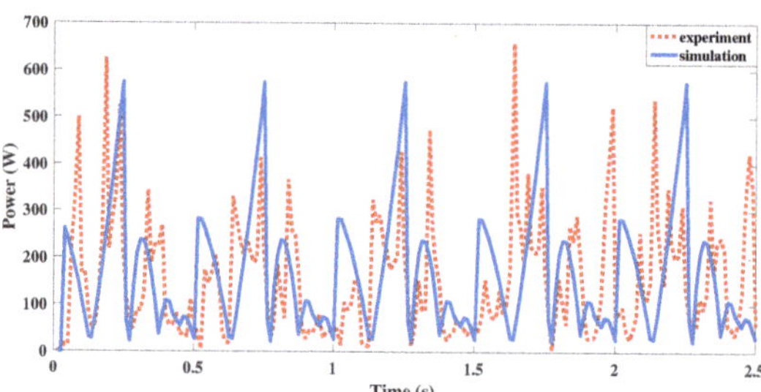

Figure 12. A comparison of the power variation between the experiment and simulation.

5.2. Simulation Results

5.2.1. Step Length

In this part, the step lengths are set as different values while the other gait parameters use default values. The step lengths ranges from 0.15 m to 0.5 m with an interval of 0.05 m. The energy consumption conditions and the E_{cost} are shown in Figure 13.

From Figure 13, we can know that with the increasement of the step length, the energy consumption during different phases and E_{cost} are both increasing but that the growth rate of E_{cost} is slower than the energy consumption.

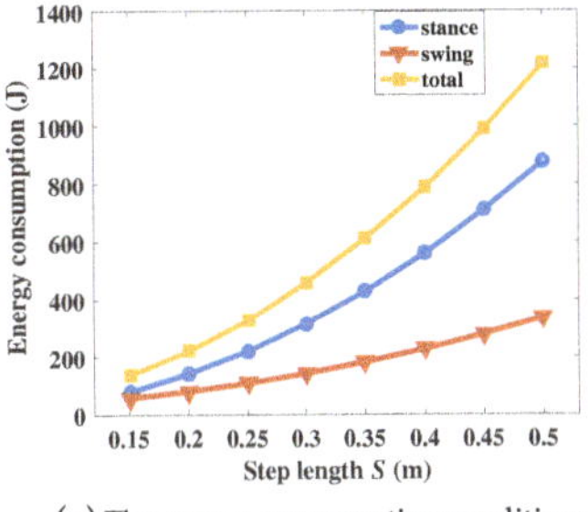 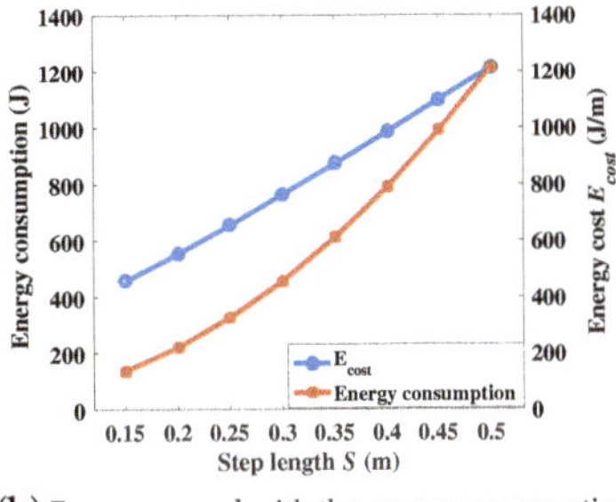

(**a**) The energy consumption conditions   (**b**) E_{cost} compared with the energy consumption

Figure 13. The energy consumptions with different step lengths.

5.2.2. Step Height

The energy consumption conditions with different step height are shown in Figure 14 and Table 5. The step heights H are chosen as 0.04, 0.08, 0.12, 0.16, and 0.20 m. According to Equation (37), when the step length S is set as a constant, the energy consumption is proportional to E_{cost}.

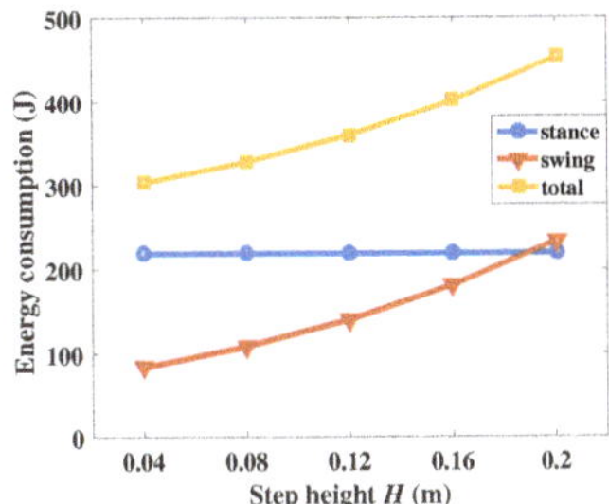

Figure 14. The energy consumptions with different step heights.

Table 5. E_{cost} with different step heights.

step height (m)	0.04	0.08	0.12	0.16	0.20
E_{cost} (J/m)	607.44	656.17	719.80	803.52	907.08

Based on the equations of the foot trajectory, the step height has no influence on the stance phase, which can be verified in Figure 14. With the augmentation of the step height, the movement distance of the leg endpoint as well as the joint energy consumptions will increase. Considering the obstacle crossing ability of the robot, a proper value of the step height needs to be chosen.

5.2.3. Standing Height

Considering the mechanical limitation and the ranges of the joint position, the standing heights H_0 are chosen between 0.5 to 0.7 m with a step of 0.05 m. The values of E_{cost} are proportional to the energy consumptions.

The results are illustrated in Figure 15 and Table 6. When the distance between the trunk and the ground is small, the joints have to sustain more weight, which will increase the joint torques, and the energy consumption will increase. However, if the distance between the trunk and the ground is too large, the COM of the robot will rise, which will reduce the movement stability of the robot.

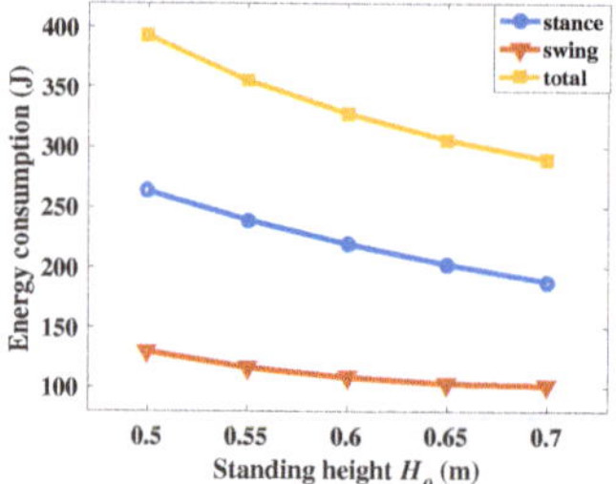

Figure 15. The energy consumptions with different standing heights.

Table 6. E_{cost} with different standing heights.

standing height (m)	0.5	0.55	0.6	0.65	0.7
E_{cost} (J/m)	787.17	712.88	656.17	612.50	580.59

5.2.4. Gait Cycle

The gait cycle is the reciprocal of the gait frequency (f). Here, we choose the gait cycles as 0.5, 0.6, 0.7, 0.8, 0.9, and 1.0 s to calculate the energy consumption. The results are listed in Figure 16 and Table 7.

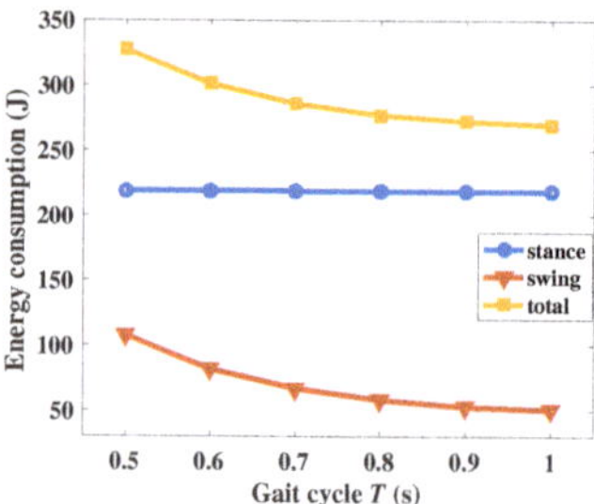

Figure 16. The energy consumptions with different gait cycle.

Table 7. E_{cost} with different gait cycles.

gait cycle (s)	0.5	0.6	0.7	0.8	0.9	1.0
E_{cost} (J/m)	656.17	603.20	572.72	555.15	545.33	539.83

With a large gait cycle, the velocity and acceleration of the joint angle will be decreased and the joint torque can be reduced according to Equation (5). From Figure 16, we can know that the energy consumption in the swing phase will decrease with the augmentation of the gait cycle but that the energy consumptions during the stance phase remain the same. For the motion of the SCalf, a gait with a large cycle can reduce the energy consumption but will make the motion unstable.

5.2.5. Duty Cycle

When the duty cycle is less or greater than 0.5, there will be conditions like a four-foot touchdown or a four-foot flight. The energy consumptions with different duty cycles are shown in Figure 17, and the E_{cost} are listed in Table 8.

For different duty cycles, the different energy costs E_{cost} are proportional to the energy consumptions, too. From Figure 17, we can know that the energy consumptions during the stance phases are nearly the same. When the duty cycle is large or small, the energy consumption during the swing phase will grow rapidly. The optimal duty cycle falls into the range of 0.3 to 0.5.

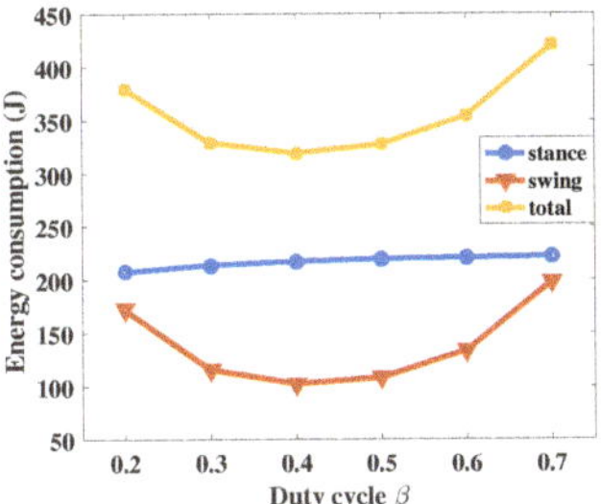

Figure 17. The energy consumptions with different duty cycles.

Table 8. E_{cost} with different duty cycles.

duty cycle	0.2	0.3	0.4	0.5	0.6	0.7
E_{cost} (J/m)	761.00	659.62	639.68	656.17	710.71	842.25

5.3. Discussion

In robot motions, the velocity of the robot is an important comprehensive index. According to Equation (38), the robot velocity is decided by the step length S and gait cycle T; both big S or short T can increase robot velocity. Based on the results in Figures 13 and 16, big S and short T will both increase the energy consumption. When choosing the values of S and T, the mechanical limitations and control stability should be considered. We notice that the growth rate of the energy consumption caused by S is higher than that caused by T. Therefore, when the robot velocity is determined, a gait with a small step length and a short gait cycle should be chosen.

For the step height H and standing height H_0, there are conflicts between low energy consumption and control effectiveness. Therefore, the selection of these parameters should consider the actual terrain and control situation.

Most researchers only study foot trajectories with the duty cycle as 50%. From the results in Figure 17, we can know that the optimal duty cycle falls into the range of 0.3–0.5, which would make the common trot gait become a flying trot gait.

According to the above discussions, for a typical situation with the robot velocity as 1 m/s, the gait parameters are chosen as follows: step length $S = 0.25$ m, gait cycle $T = 0.5$ s, step height $H = 0.08$ m, standing height $H_0 = 0.6$ m, and gait cycle $\beta = 0.4$. The energy consumptions under these gait parameters can be calculated as 217.45 J, 102.39 J, and 319.84 J for the stance phase, swing phase, and total consumption. The energy cost of this condition is 639.68 J/m.

6. Conclusions

In this work, the energy consumption conditions of the quadruped robot SCalf are studied. The kinematics model of the robot was studied, and the dynamics model was established based on the foot force distribution analysis. A complete energy model including the mechanical power and heat rate was derived. In order to describe the motions of the robot, a universal foot trajectory based on cubic spline interpolation is proposed. Different gait parameters including step length, gait cycle, step height, standing height, and duty cycle are considered in the trajectory. After obtaining the equations of the foot trajectory, the influences of different gait parameters on energy consumption are studied through a simulation. According to the results of the simulation, during the motion, a trot gait with a short step length and a small gait cycle should be chosen. The obstacle crossing ability and motion stability need to be considered in the selection of the standing height and step height. The duty cycle of the gait should be selected in the range of 0.3 to 0.5. For a typical situation with the robot velocity as 1 m/s, the gait parameters are chosen as follows: step length $S = 0.25$ m, gait cycle $T = 0.5$ s, step height $H = 0.08$ m, standing height $H_0 = 0.6$ m, and gait cycle $\beta = 0.4$.

Author Contributions: Conceptualization, K.Y. and L.Z.; software, K.Y.; writing, K.Y. and L.Z.; supervision, X.R. and Y.L.; funding acquisition, L.Z., X.R., and Y.L.

Funding: This work was supported by the National Key R&D Program of China (Grant No. 2017YFC0806505), the National Natural Science Foundation of China (Grant No. U1613223, 61603216, and 61773226), the National High-tech R&D Program of China (Grant No. 2015AA042201), and the Key R&D Program of Shandong (Grant No. 2017CXGC0901).

Conflicts of Interest: The authors declare no conflict of interest.

References

1. Sakakibara, Y.; Kan, K.; Hosoda, Y.; Hattori, M.; Fujie, M. Foot trajectory for a quadruped walking machine. In Proceedings of the IEEE International Workshop on Intelligent Robots and Systems, Towards a New Frontier of Applications, Ibaraki, Japan, 3–6 June 1990; pp. 315–322.

2. He, D.Q.; Ma, P.S. Simulation of dynamic walking of quadruped robot and analysis of walking stability. *Comput. Simul.* **2005**, *22*, 146–149.

3. Semini, C.; Tsagarakis, N.G.; Guglielmino, E.; Focchi, M.; Cannella, F.; Caldwell, D.G. Design of HyQ—A hydraulically and electrically actuated quadruped robot. *Proc. Inst. Mech. Eng. Part I J. Syst. Control Eng.* **2011**, *225*, 831–849. [CrossRef]

4. Zhang, G.; Rong, X.; Hui, C.; Li, Y.; Li, B. Torso motion control and toe trajectory generation of a trotting quadruped robot based on virtual model control. *Adv. Robot.* **2016**, *30*, 284–297. [CrossRef]

5. Ikeda, M.; Mizuuchi, I. Analysis of the energy flow on quadruped robot having a flexible trunk joint. In Proceedings of the 2015 IEEE International Conference on Robotics and Biomimetics (ROBIO), Zhuhai, China, 6–9 December 2015; pp. 1065–1071.

6. Muraro, A.; Chevallereau, C.; Aoustin, Y. Optimal trajectories for a quadruped robot with trot, amble and curvet gaits for two energetic criteria. *Multibody Syst. Dyn.* **2003**, *9*, 39–62. [CrossRef]

7. Silva, M.F.; Machado, J.T. Energy efficiency of quadruped gaits. In *Climbing and Walking Robots*; Springer: Berlin/Heidelberg, Germany, 2006; pp. 735–742.

8. Jingtao, L.; Feng, W.; Huangying, Y. Analysis on Trajectory Planning and Energy Consumption of Quadruped Robot. *Mach. Des. Res.* **2014**, *30*, 14.

9. Shoukun, W.; Xiaoyan, Z.; Guangrong, C. Research on energy consumption of hydraulic quadruped robot based on trot gait. *Trans. Beijing Inst. Technol.* **2016**, *36*, 399–404.

10. Roy, S.S.; Pratihar, D.K. Effects of turning gait parameters on energy consumption and stability of a six-legged walking robot. *Robot. Auton. Syst.* **2012**, *60*, 72–82. [CrossRef]

11. Chattunyakit, S.; Kobayashi, Y.; Emaru, T.; Ravankar, A.A. Bio-Inspired Structure and Behavior of Self-Recovery Quadruped Robot with a Limited Number of Functional Legs. *Appl. Sci.* **2019**, *9*, 799. [CrossRef]

12. Yang, K.; Zhou, L.; Rong, X.; Li, Y. Onboard hydraulic system controller design for quadruped robot driven by gasoline engine. *Mechatronics* **2018**, *52*, 36–48. [CrossRef]

13. Roy, S.S.; Singh, A.K.; Pratihar, D.K. Estimation of optimal feet forces and joint torques for on-line control of six-legged robot. *Robot. Comput. Integr. Manuf.* **2011**, *27*, 910–917. [CrossRef]

14. Roy, S.S.; Pratihar, D.K. Dynamic modeling and energy consumption analysis of crab walking of a six-legged robot. In Proceedings of the 2011 IEEE Conference on Technologies for Practical Robot Applications, Woburn, MA, USA, 11–12 April 2011; pp. 82–87.

15. Jelali, M.; Kroll, A. *Hydraulic Servo-Systems: Modelling, Identification and Sontrol*; Springer Science & Business Media: Berlin/Heidelberg, Germany, 2012.

16. Nissing, D. Identifikation, Regelung und Beobachterauslegung für elastische Großhandhabungssysteme. *Und Anwendungen Der Steuerungs- Regelungs-Und Informationstechnik* **2003**, *51*, 50–50. [CrossRef]

17. Polzer, J.; Nissing, D. Mechatronic design using flatness-based control to compensate for a lack of sensors. *IFAC Proc. Vol.* **2000**, *33*, 941–946. [CrossRef]

18. Rong, X.; Li, Y.; Meng, J.; Li, B. Design for several hydraulic parameters of a quadruped robot. *Appl. Math. Inf. Sci.* **2014**, *8*, 2465. [CrossRef]

19. Meng, J.; Li, Y.; Li, B. Control method and its implementation of quadruped robot in omni-directional trotting gait. *Robot* **2015**, *37*, 74–84.

20. Rong, X.; Li, Y.; Ruan, J.; Li, B. Design and simulation for a hydraulic actuated quadruped robot. *J. Mech. Sci. Technol.* **2012**, *26*, 1171–1177. [CrossRef]

21. Rönnau, A.; Sutter, F.; Kerscher, T.; Dillmann, R. On-board energy consumption estimation for a six-legged walking robot. In *Field Robotics*; World Scientific: Singapore, 2012; pp. 547–554.

22. Montes, H.; Mena, L.; Fernández, R.; Armada, M. Energy-efficiency hexapod walking robot for humanitarian demining. *Ind. Robot. Int. J.* **2017**, *44*, 457–466. [CrossRef]

Article

A Robotic Deburring Methodology for Tool Path Planning and Process Parameter Control of a Five-Degree-of-Freedom Robot Manipulator

Wanjin Guo [1,2,*], Ruifeng Li [2], Yaguang Zhu [1], Tong Yang [1], Rui Qin [1] and Zhixin Hu [1]

[1] Key Laboratory of Road Construction Technology and Equipment of MOE, Chang'an University, Xi'an 710064, China; zhuyaguang@chd.edu.cn (Y.Z.); tsshangzao@163.com (T.Y.); qinrui@chd.edu.cn (R.Q.); hzxin@chd.edu.cn (Z.H.)

[2] State Key Laboratory of Robotics and System, Harbin Institute of Technology, Harbin 150001, China; lrf100@hit.edu.cn

* Correspondence: guowanjin@chd.edu.cn; Tel.: +86-29-82334865

Received: 2 April 2019; Accepted: 14 May 2019; Published: 17 May 2019

Abstract: Industrial robotics is a continuously developing domain, as industrial robots have demonstrated to possess benefits with regard to robotic automation solutions in the industrial automation field. In this article, a new robotic deburring methodology for tool path planning and process parameter control is presented for a newly developed five-degree-of-freedom hybrid robot manipulator. A hybrid robot manipulator with dexterous manipulation and two experimental platforms of robot manipulators are presented. A robotic deburring tool path planning method is proposed for the robotic deburring tool position and orientation planning and the robotic layered deburring planning. Also, a robotic deburring process parameter control method is proposed based on fuzzy control. Furthermore, a dexterous manipulation verification experiment is conducted to demonstrate the dexterous manipulation and the orientation reachability of the robot manipulator. Additionally, two robotic deburring experiments are conducted to verify the effectiveness of the two proposed methods and demonstrate the highly efficient and dexterous manipulation and deburring capacity of the robot manipulator.

Keywords: industrial robotics; robotic deburring; tool path planning; process parameter control; dexterous manipulation

---

1. Introduction

Industrial robotics with advanced technology for the industrial automation field has received considerable attention in recent years [1–6]. Although the development and application of industrial robots are confined—to a certain extent—to the field of advanced manufacturing due to the lack of rigidity and precision of an industrial robot compared to a numerical control machine, the industrial robot, as a common and less expensive alternative to the latter, has great advantages and potentials especially in some applications that require low cutting forces, low precision requirements, large-sized complex shaped parts and/or multi-faces machining in one setup. Therefore, industrial robots were extensively invented and introduced into factories for robotic machining and other applications such as spray automation, spot welding and materials handling. Such robotic automation freed humans from heavy and tedious labor and such industries rapidly retooled their manufacturing lines into robot-integrated systems. Consequently, industrial robots offer a real gain of flexibility, modularity, and access for machining on production lines, and are viable solutions for enhanced productivity, quality, and safety. Industrial robotics are being widely adopted in the robotic machining applications and have become an area of significantly in-depth robotics research. More recent research on robotic machining and many diverse research domains have been investigated in recent years [7–14].

Several studies were concentrated on five-degree-of-freedom (five-DOF) hybrid robot manipulators for machining and fabrication applications. A five-DOF hybrid mechanism was developed, which included a synthesized two translational DOF and one rotational DOF (2T1R) parallel module [15]. A five-DOF hybrid mechanism which consists of a 3-DOF parallel platform and a X-Y table [16]. A parallel mechanism (3T1R) and a rotational table were integrated into a five-DOF hybrid robot manipulator [17]. A five-DOF hybrid mechanism was designed which includes a parallel manipulator (2T1R) with a rotational table [18]. A five-DOF hybrid mechanism including a 2-DOF redundant parallel manipulator was developed [19]. A five-DOF hybrid mechanism was investigated, consisting of a 3-DOF parallel manipulator with actuation redundancy and a 2-DOF worktable [20]. A five-DOF hybrid robot was introduced, which comprises of a parallel mechanism (1T2R) and two gantries [21]. Two five-DOF mechanisms were introduced, which are composed of a 3-DOF parallel mechanism connected in series with a 2-DOF wrist [22]. Some of the designs of these five-DOF hybrid robot manipulators are complex, since an application of the inferior-mobility robot manipulator demands a translatable table [15] or a rotatable one [16–18]. Also, the problem of achieving motion control is harder for a hybrid mechanism [19,20] with redundantly actuated joints than a non-redundantly actuated one. Moreover, a relatively small workspace for a robot manipulator is established to perform tasks [22].

Several research works are concentrated on the related technologies of robotic deburring. Practical mechanical deburring methods, including robotics for aluminum work parts and an overview of burr formation mechanism and morphology, were presented, and several deburring classifications were also proposed [23]. Also, a review of burr formation modeling and control, and factors governing milling burr formation were presented [24]. Robotic application of edge deburring with a controlled force progression pneumatic tool for a part of the manufacturing process of an aircraft engine detail was investigated [25]. A flexible robot-based cast iron deburring cell using a single-point laser displacement sensor was developed for small batch production of a robotized deburring task in a standard cast iron foundry scenario [26]. An adaptive robotic system using a custom-developed 3D laser-triangulation profilometer was developed for the robotic deburring of die-cast parts with position and orientation error correction [27]. Through discrete event simulation and 3D digital human modeling, a multi-purpose digital simulation approach was proposed for the sustainability enhancement of a real manufacturing cell of the aerospace industry, automated by robotic deburring [28].

The current research and applications of robotic deburring are mainly carried out by one or more traditional and commercial six-degree-of-freedom (six-DOF) serial industrial robot manipulators, which are not always suitable for the robotic deburring of complex shaped parts with highly dexterous orientations adjustment and efficient multi-faces deburring in one setup because of the defective nature of this kind of serial robot manipulator, e.g., the weakened structural rigidity of the extended robot manipulator when it is reaching far deburring position with a cantilever-beam-like structure, and the poor accessibility and small safety margins between robot manipulator and workpiece resulting from its difficulties in both obstacle avoidance adjustment and deburring orientation adjustment.

Several studies investigating tool path adaptation and process parameter control for robotic deburring were presented. A tool path adaptation for robotic deburring was presented in accordance with the registration using a custom 3D laser-triangulation profilometer to compensate the positioning errors of the currently processed workpiece, which is difficult to ensure repeatable clamping [27]. A tool path modification method based on a computer-aided design model and direct teaching was proposed to compensate for the position/orientation errors of the workpiece, in addition, impedance control was used to avoid applying an excessive contact force and a virtual wall was adopted to improve the force-control performance for the robotic deburring process [29]. An application of a human mimicking control strategy that mimics the human behavior during the manual deburring on the deburring of hard material items using an industrial robot was introduced [30]. By satisfying a set of constraints to properly perform the desired surface contact conditioning, a hybrid position/force control approach using task priority and sliding mode control was proposed for contact-driven robotic surface treatments such as deburring [31,32]. A set of optimal process parameter combination for robotic machining and the effect of process parameters such as spindle speed, feed rate and tool path strategies on the performance characteristics were investigated using the Taguchi–Grey relational approach and analysis of variance [33]. Edge robotic deburring with a controlled force progression pneumatic tool, as well as a methodology used to select and optimize the edge robotic deburring process, was presented [25]. A sliding mode control method based on radial basis function neural network was proposed for the deburring of industry robotic systems, without the requirements for strict constraints, an accurate model and exact parameters [34]. A fuzzy proportional–integral–derivative (PID) control method for deburring industrial robots was proposed and the PID controller parameters can be updated online at each sampling time to allow adaptive compensation for error and guarantee trajectory accuracy of the end-effector [35]. A vision-based approach, a Pythagorean hodograph quintic spline interpolator based on S curve acceleration/deceleration and an integrated process control structure consisting of an adaptive disturbance compensator, a sliding mode controller, and a friction compensator were investigated for force control and contour following in industrial applications such as deburring [36].

These studies of tool path adaptation and process parameter control for robotic deburring are usually conducted on one or more traditional and commercial six-DOF serial industrial robot manipulators. The tool path adaptation for robotic deburring implemented usually needs to supplement extra equipment or processes, e.g., using a vision system described in [27,36] or direct teaching [29]. Other studies on the process parameter control for robotic deburring have certain constraints, e.g., the deburring tool is an abrasive diamond disc in the control strategy for the process parameter control, and other deburring tools are not considered [30]; the designed control action and implementation are more intricate, such as [25,31,32,34,36]; the procedure of the proposed approach is more complicated, such as [33]; or detailed deburring process parameters such as robotic feed and spindle speed for the deburring industrial robot are not considered [35].

The deburring orientation adjustment of current industrial robots is usually not dexterous, especially at the far deburring position while simultaneously considering obstacle avoidance. There is an urgent need to develop a robot manipulator for robotic deburring of complex shaped parts with highly dexterous manipulation and very efficient five-face deburring in one setup. Moreover, there are very few related references for the tool path strategies which are implemented as proprietary software packages of robot manufacturers. It is particularly necessary to conduct the research on the tool path planning which is highly suitable for the deburring characteristics of the self-developed robot manipulator. Also, in order to perform an adaptive deburring process and finer deburring quality, it is very important to investigate an easy-to-implement and efficient process parameter control method with automatic-online errors correction.

The rest of this article is organized as follows. The structure of the robot manipulator and related research are described in Section 2. A robotic deburring tool path planning method and a robotic deburring process parameter control method are proposed for robotic deburring in Sections 3 and 4, respectively. A dexterous manipulation verification experiment is presented in Section 5. Two robotic deburring experiments, disc deburring experiment of an automobile hub and multifaceted edges deburring experiment of an automobile steering booster housing, are conducted in Section 6. Finally, conclusions are drawn in Section 7.

2. Structure Description and Related Research of the Robot Manipulator

This section outlines the structure, experimental platforms and previous theoretical research of a developed five-degree-of-freedom (five-DOF) hybrid robot manipulator. The structure of the robot manipulator is shown in Figure 1. The robot manipulator, consisting of a 3-DOF (3T) parallel module and a 2-DOF (2R) serial module, is a newly developed 3T2R hybrid robot manipulator. The first three translational motion axes J_1, J_2 and J_3 (consisting of ball screws) are driven by motor M_1, M_2 and M_3, respectively. The last two rotational motion axes J_4 and J_5 are driven by motor M_4 and M_5, respectively. The axis J_6 is the end-effector which contains a high-speed motor spindle. The designed structure parameter e is equal to zero, and the available range of the rotation angle is $\alpha \in [-\pi/4,\ \pi/4]$ because of the mechanical constraints.

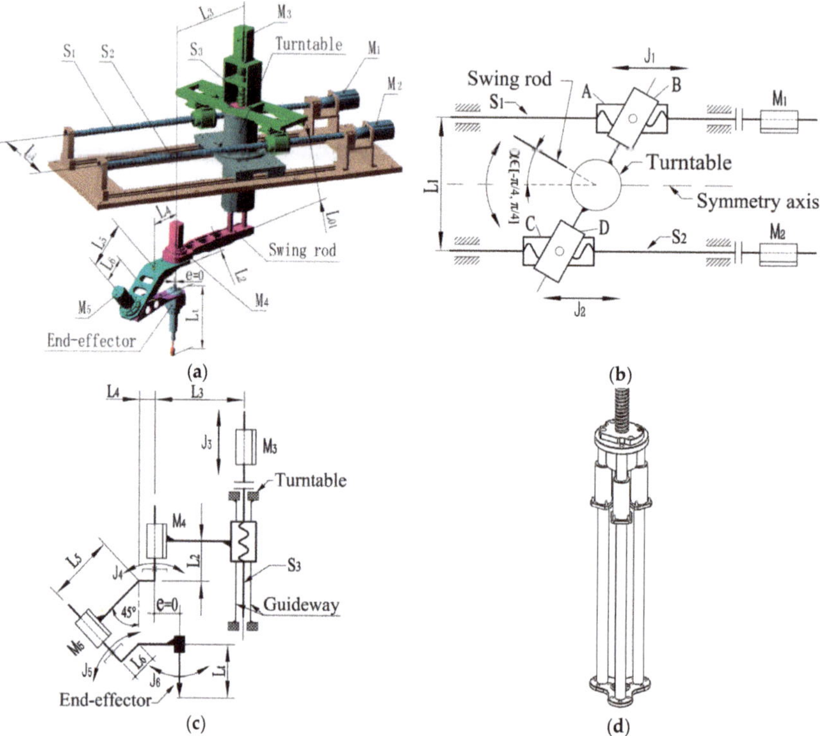

Figure 1. Robot manipulator: (**a**) 3D model; (**b**) top view; (**c**) front view; (**d**) guideway module.

One of the most significant advantages of the robot manipulator is its dexterity in manipulation and orientation reachability. The last two rotational motion axes J_4 and J_5 as shown in Figure 1 are connected with an angle of 45° in order to realize the dexterous manipulation and the orientation adjustment of the robot manipulator end-effector. The end-effector can reach all orientations in the upper half of the complete spherical surface. Hence, the robot manipulator has the five-face machining ability in one setup. The detailed realizations are described as follows.

When the processing point of the end-effector coincides with the intersection of axes J_4 and J_5, the position of the processing point of the end-effector can always keep at a same point (i.e., this point always coincides with the intersection), during the process of orientation adjustment in the upper half of the complete spherical surface only through the last two axes J_4 and J_5. And the position of the processing point can be adjusted when translating the first three axes J_1, J_2 and J_3. This means that the manipulation and the orientation reachability of the robot manipulator characterize a very high level of dexterity.

It is necessary to note that some nonnegligible dynamic effects which caused by backlash of ball screws and robot manipulator structural deformations. It is very meaningful to investigate these effects of the backlash and structural deformations and to compensate for them if necessary. Some related research can be referred to [37–40].

Experimental platforms of the robot manipulator are shown in Figure 2. Among them, the improved platform is superior to the original one in mechanical structure on translational vertical axis assembly. The cylinder guide pillar module (shown in Figure 3c) of the improved platform, and the guideway module (shown in Figure 1d) of the original one, are constituted by cylinder guide pillar and cylinder type support, and three sets of linear shafts and linear bearings, respectively. The former has greater torsional rigidity and strength of the vertical axis assembly around the vertical rotation axis than the latter. Finally, design structure and physical structure of the vertical axis assembly for the improved platform are shown in Figure 4.

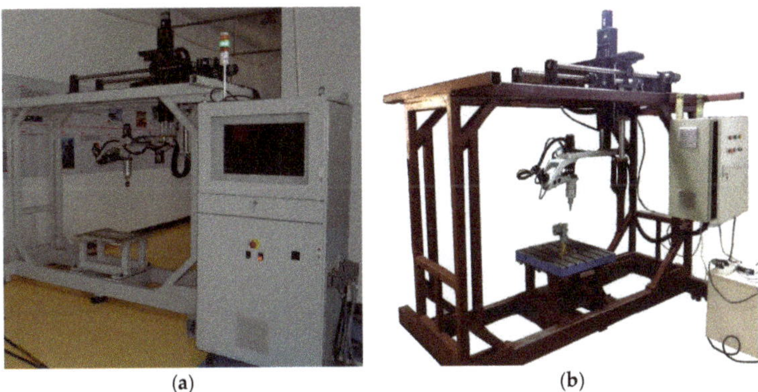

Figure 2. Experimental platforms of robot manipulator: (**a**) original platform; (**b**) improved platform.

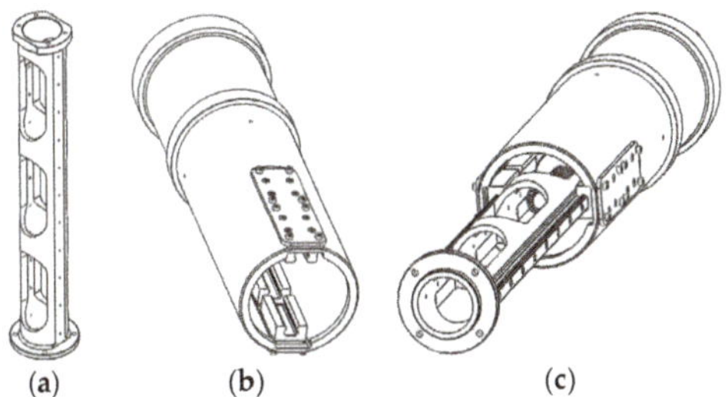

Figure 3. Improved cylinder guide pillar module: (**a**) cylinder guide pillar; (**b**) cylinder type support; (**c**) design structure diagram.

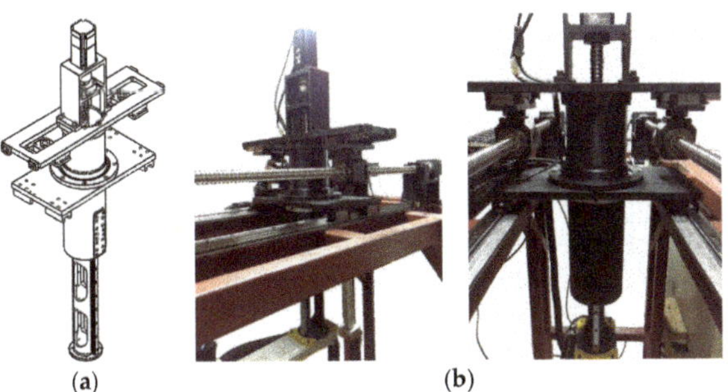

Figure 4. Vertical axis assembly of the improved platform: (**a**) design structure; (**b**) physical structure.

The reachable workspace of the robot manipulator based on structure parameters (Table 1) is shown in Figure 5. The top view and the front view of Figure 5 showed the range in the X, Y, and Z direction of the reachable workspace. Note that the shapes of the left end and the right end (in the top view of the Figure 5) are different. This is caused by the sway of the swing rod. The sway values are $L_3\cos\alpha$ and $L_3\sin\alpha$ in the X and Y directions, respectively. In the start phase, α increases from zero to the maximum attainable value ($\pm\pi/4$), and the sway value in the X and Y directions gradually decrease and increase, respectively. In the end phase, α decreases from the maximum attainable value ($\pm\pi/4$) to zero, and the sway values in the X and Y directions gradually increase and decrease, respectively. The mutual variation relationship is shown in Figure 5a.

In the previous research, kinematics analysis, dynamics analysis, dexterity analysis, multi-objective smooth trajectory planning and dynamic load-carrying capacity calculation, actual building of mechanical system and motion control system, and tests for the repeatability and accuracy of both the position and path for the five-DOF robot manipulator are conducted. For discussions of these domains and a more detailed listing of related research for the robot manipulator, see [41–44].

Table 1. Robot manipulator structure parameters (mm).

L_1	L_2	L_3	L_4	L_5	L_6	e	L_t	L_{01}	s_x/s_z
420	50	450	160	210	95	0	297	465	10

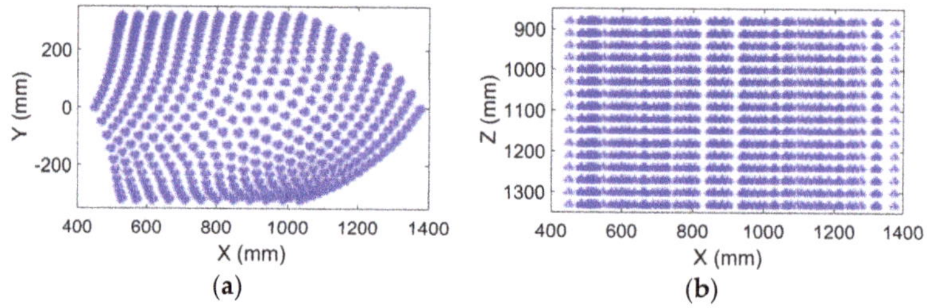

Figure 5. Reachable workspace of robot manipulator: (**a**) top view; (**b**) front view.

3. Robotic Deburring Tool Path Planning Method

In this section, a robotic deburring tool path planning method is proposed to handle the robotic deburring tool location (position and orientation) planning and the robotic layered deburring planning for the developed five-DOF robot manipulator.

In this article, the theoretical curve and theoretical surface, which are expected to be qualified after the final robotic deburring, are called the target curve and target surface, respectively. The deburring tool is a general term for various deburring tools, including the cutting tool with the end or side face edge, and the grinding tool such as grinding wheel head and rotary file.

3.1. Robotic Deburring Tool Planning Method

The proposed robotic deburring tool position planning method is discussed in two parts: surface deburring and curve deburring. In the first part (i.e., surface deburring), the tool contact path intersection line method [45] is used to generate the contact position of the deburring tool, i.e., the total locus of deburring tool contact points are derived from the intersection of a set of equidistant constraint planes and the target surface. Among them, the constraint plane spacing, the line spacing and step length of the tool path should be determined by some comprehensive factors such as deburring tool, deburring accuracy, deburring efficiency and deburring error. In the second part, i.e., curve deburring, the above comprehensive factors are also considered to directly generate the total locus of discrete deburring tool contact points. Upon obtaining the total locus of deburring tool contact points for surface deburring and curve deburring, considering the curvature variations of the surface and the curve, respectively, the total locus of deburring tool contact points is offset according to the layered deburring planning. After that, the deburring tool path of each layer is generated with some practical considerations of deburring tool length, blade edge and blade diameter, etc.

In this article, the orientation of the deburring tool is aligned with its side edge, defined as follows. Deburring tool orientations, the movement direction of the tool and the end edge of the tool are expressed as vectors **n**, **o** and **a**, respectively. A target curve is used as an example to illustrate the robotic deburring tool orientation planning method. The target curve is shown in Figure 6, and unit tangent vectors **τ** and unit inner normal vectors **f** of discrete positions are represented by blue arrows and green arrows, respectively.

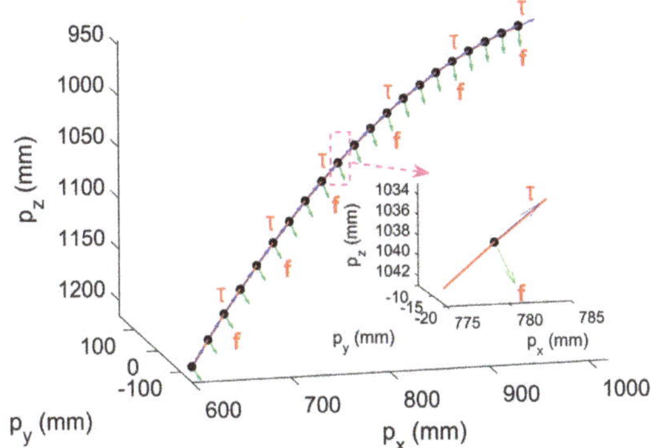

Figure 6. Example of unit tangent and normal vectors for discrete positions of target curve.

The proposed robotic deburring tool orientation planning method is proposed for deburring manners with the tool end edge and the tool side edge. The assignment of orientation vectors for these two deburring manners is described as follows. For the tool end edge (i.e., the deburring manner with the tool end edge): (a) the deburring direction of tool is planned along the tool end edge vector **a**, which is chosen as the unit inner normal vectors **f** of each discrete position of the target curve; (b) the movement direction of tool (vector **o**) is planned along the unit tangent vectors **τ** of each discrete position of the target curve and points to the next discrete position to be deburred; and (c) the tool vector **n** is obtained by **n** = **o** × **a**. For the tool side edge (i.e., the deburring manner with the tool side edge): (a) the deburring direction of tool is planned along the tool side edge vector **n**, which is chosen as the unit inner normal vectors **f** of each discrete position of the target curve; (b) the planned movement direction of tool (vector **o**) is the same as above deburring manner; and (c) the tool vector **a** is obtained by **a** = **n** × **o**.

In the above robotic deburring tool orientation planning, the feeding direction of the robot manipulator deburring corresponds to the deburring tool direction of the end edge or the side edge, that is, the feeding directions of the robot manipulator deburring along the unit inner normal vectors **f** for discrete positions of the target curve. The movement direction of the robot manipulator deburring corresponds to the movement direction of the tool, that is, the movement directions of the robot manipulator deburring along the unit tangent vectors **τ** for discrete positions of the target curve.

3.2. Robotic Layered Deburring Planning Method

The feature points on the object are selected based on the geometric characteristics of the deburring target, and the machining allowances are measured by the distance between each selected feature point and the target curve and/or surface. The detailed description of the proposed robotic layered deburring planning method for the layered deburring curve and/or surface is presented in the following steps:

Step 1: calculate the distance between each selected feature point and the target curve and/or surface, and find out the maximum one (which is denoted as D_{max}) as the maximum machining allowance for the layered deburring curve and/or surface.

Step 2: calculate the number of deburring layers n (i.e., the times of layered deburring) by the formula $n = \langle D_{max}/d_p + 1 \rangle$, where d_p and symbolic form $\langle \bullet \rangle$ represent the thickness of each selected deburring layer (i.e., cutting depth) and the rounding calculation, respectively. In order to improve the deburring quality, the deburring count is calculated based on different allowance d_l in the last layer: (a) set it to one when $d_l < d_p/2$ and the deburring allowance sets d_l, and (b) set it to two when $d_l \geq d_p/2$, here, the first deburring allowance sets $d_p/2$, and the second one sets the remains (i.e., $d_l - d_p/2$).

Step 3: plan the deburring tool path on each layer (i.e., the total locus of deburring tool contact points between the deburring tool and the object) based on the index of each layer n_j and its corresponding thickness d_{p_j}, that is, calculate the position offset relationship of each layer for points of the deburring tool path along the outward normal vector $\mathbf{n}_l = \begin{bmatrix} n_{lx} & n_{ly} & n_{lz} \end{bmatrix}^T$ of each discrete point derived from the target curve and/or surface (i.e., the deburring tool contact positions on the target curve and/or surface are obtained by the tool contact path intersection line method [45]), as follows:

$$
\begin{bmatrix} x \\ y \\ z \end{bmatrix}_{n_j} = \begin{bmatrix} x \\ y \\ z \end{bmatrix}_{n_{j-1}} + \begin{bmatrix} \Delta x \\ \Delta y \\ \Delta z \end{bmatrix}_{n_j} \quad j = 1, 2, \cdots, n, \tag{1}
$$

where, when $j = 1$, n_0 and $\begin{bmatrix} x & y & z \end{bmatrix}_{n_0}^T$ represent the number of located layer of the target curve and/or surface and positions of each discrete point of the target curve and/or surface, respectively. And the incremental position calculation is related by

$$
\begin{bmatrix} \Delta x \\ \Delta y \\ \Delta z \end{bmatrix}_{n_j} = d_{p\_j} \cdot \begin{bmatrix} n_{lx} \\ n_{ly} \\ n_{lz} \end{bmatrix}. \tag{2}
$$

Step 4: conduct the semi-finishing and finishing based on actual and special requirements according to Step 2 and Step 3.

Step 5: introduce the orthogonal overlapping and reciprocating manner, and the shape of the basic body surface that is connected with the target curve to conduct planning the deburring tool path for surface deburring and curve deburring at each layer.

In the above-mentioned deburring tool path planning for surface deburring and curve deburring, if the feature points of the object are unknown beforehand, the maximum machining allowance $D_{\max}$ can be estimated by the robot manipulator teaching according to the actual workpiece to be deburred, and the remaining planning steps are the same as above.

Usually, the loci of deburring tool contact points, which are obtained by only using the offset computed from the tool contact path intersection line method [45], may cause a large machining error due to the equal-distance offset. Compared with this tool contact path intersection line method [45], the proposed robotic deburring tool planning method for tool position and tool orientation in this section further extends to consider the curvature variations of the surface and the curve on the basis of this tool contact path intersection line method with the equal-distance offset, so as to ensure a more uniform deburring tool trajectory and repair these deficiencies caused by this method [45].

Please refer to the related documents for detailed explanation of machining mechanism and machining principles for topics such as deburring tool selection and its corresponding robotic deburring feed speed and tool spindle speed for different deburring workpieces with different curves or curves.

4. Robotic Deburring Process Parameter Control Method based on Fuzzy Control

A robotic deburring process parameter control method of the five-DOF robot manipulator for robotic deburring is proposed in this section. This proposed method is adopted in deburring process parameter adjustment control for the stable robotic deburring with a constant cutting speed and a constant cutting force via a designed fuzzy controller.

According to metal cutting principle, the cutting parameters include cutting speed, cutting feed and cutting depth. Taking carbide turning tools for carbon steel turning as an example, related researches showed that the biggest impact on tools is the cutting speed, followed by the cutting feed, and lastly the cutting depth [46–48]. More generally, the rough machining should select cutting parameters for the maximum productivity. Thus, a larger cutting depth (or cutting width) should be selected first, and after the majority of the machining allowance is removed, a suitable cutting feed (or a cutting thickness) and a cutting speed are selected in turn by the cutting condition. Finish machining generally uses a small cutting depth and cutting feed, and then a higher cutting speed to improve the machining accuracy and reduce the surface roughness.

In this article, robotic deburring process parameters—robotic spindle speed, robotic feed and robotic layered deburring thickness—are determined by cutting parameters—cutting speed, cutting feed and cutting depth—respectively. Among them, the robotic layered deburring thickness (i.e., cutting depth) planning and the robotic deburring tool location (position and orientation) planning are presented in the proposed robotic deburring tool path planning method in Section 3. In this section, issues to consider here pertain, mostly, to the adjustment control for robot manipulator deburring process parameters, that is robotic spindle speed and robotic feed, through a proposed robotic deburring process parameter control method.

Some classical control methods, such as proportional–integral–derivative (PID) control, which have been studied and practiced by broad researches, are very effective solutions for uncomplicated linear time-invariant systems [49–51]. In addition, modern control theory, such as sliding mode control, based on state variables has also been widely used to solve linear or nonlinear, and time-invariant or time-varying multi-input multi-output (MIMO) systems [52–54]. Although this kind of classical and modern control theory can overcome some internal disturbances of the control system, nevertheless, it is neither sufficient to ensure the stability of robotic deburring process in a robust enough way in real time, nor able to cope with environmental uncertainties. One goal of this article is to find a way of controlling the robotic deburring contact forces while maintaining constant cutting speed and cutting force of the robot manipulator presented earlier. In particular, this is critical when the interaction between the robot manipulator and deburring workpiece environment is of concern.

Because the fuzzy control with the ontological basis of fuzzy logic and fuzzy reasoning has the advantage of requiring neither the knowledge of the model structure nor the model parameters, it is strongly adaptive and highly robust to nonlinearity and variations of process parameters. Also, it can give fast response, effective noise suppression, and a better control effects when the model structure changes greatly. Therefore, related researches and applications based on fuzzy control have been favored by many scholars [55–58].

In this article, the process of the adjustment control for robot manipulator deburring process parameters (i.e., robotic spindle speed and robotic feed) has the characteristics of nonlinearity, noise and tight coupling between control loops, which cannot be solved sufficiently by classical and modern control methods such as PID-type controllers. In this section, a fuzzy controller with the above advantages of the fuzzy control is designed for a proposed robotic deburring process parameter control method to systematically accommodate robotic deburring.

The control objectives of the designed fuzzy control system in this section are the robot manipulator deburring process parameters, i.e., robotic spindle speed and robotic feed. The robotic spindle speed and the robotic spindle load of the robot manipulator deburring system can be controlled by adjusting the control voltage of the robotic spindle and the robotic feed, in order to realize the robotic deburring with a constant cutting speed and a constant cutting force. Hence, the designed fuzzy controller in this section is a two-input and two-output control system (shown in Figure 7). Here, input variables are robotic spindle speed and robotic spindle load, and output variables are control voltage of the robotic spindle and robotic feed.

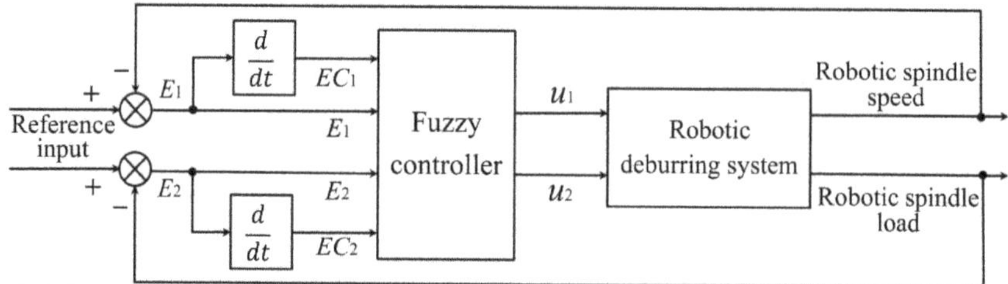

Figure 7. Schematic diagram for deburring process parameter control.

In this article, a parallel structure adopted is used by the fuzzy controller to independently control the control voltage of the robotic spindle and the robotic feed. There are two parts of representing fuzzy variables in the designed fuzzy controller. First, error of robotic spindle speed E_1 and change in error of robotic spindle speed EC_1, and error of robotic spindle load E_2 and change in error of robotic spindle load EC_2, are represented as input fuzzy variables for controlling the control voltage of the robotic spindle and the robot feed, respectively. Second, change in control voltage of the robotic spindle u_1 and change in robotic feed u_2 are represented as output fuzzy variables.

To achieve discrete membership functions, a discrete set with thirteen elements and a discrete set with seven elements are defined in the designed fuzzy controller for E_1, EC_1, and u_1, and for E_2, EC_2, and u_2, respectively, and corresponding sets are $U_1 = \left\{ \begin{array}{cccccccc} -6 & -5 & \cdots & -1 & 0 & +1 & +2 & \cdots & 6 \end{array} \right\}$ and $U_2 = \left\{ \begin{array}{ccccccc} -3 & -2 & -1 & 0 & +1 & +2 & +3 \end{array} \right\}$, respectively. Assume that variation ranges of E_1, EC_1, E_2 and EC_2 are $[-S_e, S_e]$, $[-S_{ec}, S_{ec}]$, $[-L_e, L_e]$ and $[-L_{ec}, L_{ec}]$, respectively. Combining these variation ranges and corresponding discrete sets, several quantizers corresponding to E_1, EC_1, E_2 and EC_2 are defined respectively as follows:

$$K_{Se} = 6/S_e, \tag{3}$$

$$K_{Sec} = 6/S_{ec}, \tag{4}$$

$$K_{Le} = 3/L_e, \tag{5}$$

$$K_{Lec} = 3/L_{ec}. \tag{6}$$

Here, each quantizer maps an inbound measurement within the variation range to a nearest integer element in the corresponding discrete set.

Similarly, scaling factors for the output variables (i.e., control variables) of u_1 and u_2, where corresponding variation ranges are $[-V_u, V_u]$ and $[-F_u, F_u]$, respectively, are defined as follows:

$$K_{Vu} = V_u/6, \tag{7}$$

$$K_{Fu} = F_u/3. \tag{8}$$

Assuming exact values obtained by the fuzzy reasoning for change in control voltage of the robotic spindle u_1 and change in robotic feed u_2 are u_{1i} and u_{2i}, respectively, then corresponding exact values of variation ranges V_{ui} and F_{ui} can be obtained by using above defined scaling factors. The detailed representations are given by

$$V_{ui} = K_{Vu} \cdot u_{1i}, \tag{9}$$

$$F_{ui} = K_{Fu} \cdot u_{2i}. \tag{10}$$

The fuzzy rule with automatic adjusting factors in the whole set using analytical expressions for fuzzy controller are usually designed as follows. Suppose that sets for error E, change in error EC and control variable u select as $E = EC = u = \{-N, \cdots, -2, -1, 0, 1, 2, \cdots, N\}$, the fuzzy rule can be expressed as

$$\begin{cases} u = -\langle \alpha E + (1 - \alpha)EC \rangle \\ \alpha = \frac{1}{N}(\alpha_s - \alpha_0)|E| + \alpha_0 \end{cases}, \qquad (11)$$

where $0 \leq \alpha_0 \leq \alpha_s \leq 1$ and adjustment factor $\alpha \in [\alpha_0, \alpha_s]$, and symbolic form $\langle \bullet \rangle$ represents the rounding calculation. Thus, the designed fuzzy rule can be automatically adjusted on the basis of the weight and magnitude of the error.

In this article, in order to optimize the designed fuzzy controller that provides good both static and dynamic stability properties, and robust performance, fuzzy rules are designed with automatic adjusting factors in the whole set, described by analytical expressions. Thus, in the whole set, weights of errors and changes in errors, which represent the effect on control results, can be automatically adjusted by errors. The detailed fuzzy rules for robotic spindle speed and robotic spindle load are designed respectively as follows:

$$\begin{cases} u_1 = -\langle \alpha_1 E_1 + (1 - \alpha_1)EC_1 \rangle \\ \alpha_1 = \frac{1}{6}(\alpha_{s1} - \alpha_{01})|E_1| + \alpha_{01} \end{cases}, \qquad (12)$$

where $0 \leq \alpha_{01} \leq \alpha_{s1} \leq 1$ and adjustment factor $\alpha_1 \in [\alpha_{01}, \alpha_{s1}]$, and symbolic form $\langle \bullet \rangle$ represents the rounding calculation; and

$$\begin{cases} u_2 = -\langle \alpha_2 E_2 + (1 - \alpha_2)EC_2 \rangle \\ \alpha_2 = \frac{1}{3}(\alpha_{s2} - \alpha_{02})|E_2| + \alpha_{02} \end{cases}, \qquad (13)$$

where $0 \leq \alpha_{02} \leq \alpha_{s2} \leq 1$ and adjustment factor $\alpha_2 \in [\alpha_{02}, \alpha_{s2}]$, and $\langle \bullet \rangle$ represents the same mentioned above.

It can be seen from the above that fuzzy rules are characterized naturally in that the adjustment factors α_1 and α_2 can be updated online and adjusted by the absolute value of the error $|E_1|$ and $|E_2|$, and there are six and three possible values, respectively, therefore, weights of errors and changes in errors can be automatically adjusted online by errors.

The automatic adjustment process of the above designed fuzzy rules conforms to the control characteristics of the human decision-making process, and these fuzzy rules described by the analytical forms exhibit their several own advantages over other representations in the optimization property, convenience, simplicity, and implementation in real time for fuzzy control algorithms.

5. Dexterous Manipulation Verification Experiment

In this section, an experiment is conducted on the improved experimental platform of the robot manipulator (shown in Figure 8a) to verify the dexterous manipulation of the robot manipulator, especially the dexterous manipulation of a certain processing point. An aluminum alloy casting automobile steering booster housing (shown in Figure 8b) was selected as the experimental object, and the dexterous manipulation verification experiment was carried out at the center hole on the cylindrical top of the booster housing (shown in Figure 8b).

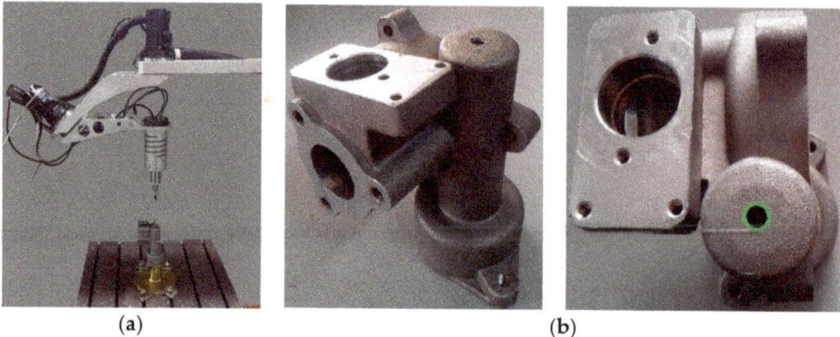

Figure 8. Experimental platform and experimental object: (**a**) experimental platform of improved robot manipulator; (**b**) overall appearance of booster housing.

Experimental verification of the dexterous manipulation was carried out on the platform of the improved robot manipulator. When the last two axes J_4 and J_5 both matched the actual initial state (shown in Figure 8a), the first three axes J_1, J_2 and J_3 were adjusted so that the tool tip of the end-effector reached the center hole on cylindrical top of the booster housing until the intersection of the last two axes J_4 and J_5 basically coincided with the center hole (shown in Figure 8b). Then the first three axes J_1, J_2 and J_3 remained stationary, and only the last two axes J_4 and J_5 rotated within the motion range. The characterizations of the dexterous manipulation of the robot manipulator are shown in Figure 9. The enveloping surface of the dexterous manipulation verification experiment is shown in Figure 10.

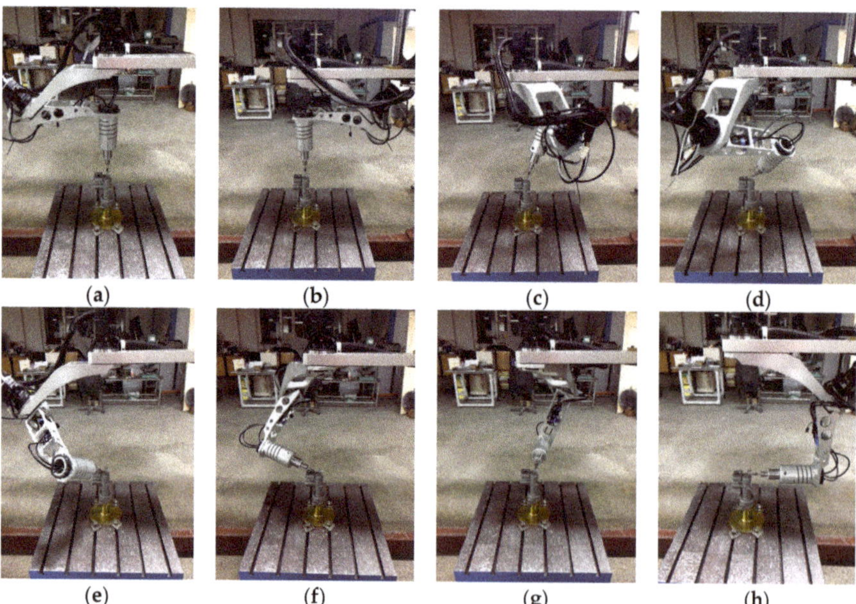

Figure 9. Experimental validation of the dexterous manipulation (unit is radian): (**a**) $\varphi_4 = 0$, $\varphi_5 = 0$; (**b**) $\varphi_4 = -\pi$; $\varphi_5 = 0$; (**c**) $\varphi_4 = -2\pi/3$; $\varphi_5 = \pi/6$; (**d**) $\varphi_4 = -\pi/6$; $\varphi_5 = \pi/3$; (**e**) $\varphi_4 = 0$; $\varphi_5 = \pi/2$; (**f**) $\varphi_4 = \pi/6$; $\varphi_5 = 2\pi/3$; (**g**) $\varphi_4 = 2\pi/3$; $\varphi_5 = 5\pi/6$; (**h**) $\varphi_4 = \pi$; $\varphi_5 = \pi$.

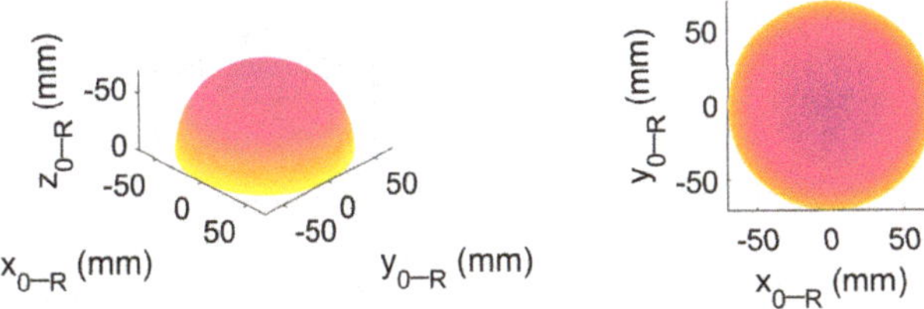

Figure 10. Enveloping surface of the experiment validation for the dexterous manipulation.

When the last two axes J_4 and J_5 traversed the motion ranges $[-\pi, \pi]$ and $[0, \pi]$, respectively, as shown in Figure 9, it can be found that the total locus of the tool tip of the end-effector appeared as an approximate enveloping half-spherical surface with a downward open, as shown in Figure 10, and the center of the half-spherical surface was the intersection of the last two axes J_4 and J_5. In addition, the total locus of the tool tip remained the same when the last two axes J_4 and J_5 traversed the motion ranges $[-\pi, \pi]$ and $[-\pi, 0]$, respectively. For the sake of simplicity, only the former case is illustrated in Figure 9. Note that the approximate enveloping half-spherical surface was not a strictly half-spherical surface, and the extremely small difference was caused by the structure parameter e, which was not strictly equal to zero due to assembly errors.

The above results of the dexterous manipulation experiment imply that the robot manipulator has a very high level of the dexterous manipulation and the orientation reachability. These available and excellent characterizations of the dexterous manipulation and the orientation reachability can provide appropriate and flexible manipulations and orientation adjustments with a very high level of dexterity for the robotic deburring of the robot manipulator.

6. Robotic Deburring Experiments

Two robotic deburring experiments were conducted on the experimental platforms of the robot manipulator in this section to show the effectiveness of the proposed robotic deburring tool path planning method and the proposed robotic deburring process parameter control method, and also to demonstrate the superiorly operational manipulation performance and the deburring ability of the robot manipulator.

6.1. Disc Robotic Deburring Experiment

In the first robotic deburring experiment, a disc deburring for an automobile hub was conducted on the original experimental platform of the robot manipulator (shown in Figure 2a). The experimental workpiece was an aluminum alloy casting blank of an automobile hub, as shown in Figure 11a, and the experimental deburring object were disc burrs of the automobile hub, as shown in Figure 11b.

It can be seen from Figure 11 that disc burrs of automobile hub were located on the inside of the mold cavity, so that these burrs were not suitable to be removed via the abrasive belt grinding. It was a good use case of a certain carbide rotary tool (i.e., a kind of high-speed machining file, shown in Figure 12) driven by the robot manipulator with a high level of dexterity of manipulation and orientation reachability. Furthermore, this kind of rotary tool with a relatively long side cutting edge is more suitable to execute the deburring for disc burrs of automobile hub. A double-cutting cylindrical round-head carbide rotary tool with the side cutting edge (shown in Figure 12b and its type is CX1020M06) was selected to carry out the deburring for disc burrs of automobile hub in the first experiment. The length of the side cutting edge and the diameter of the selected tool were 20 mm and 10 mm, respectively.

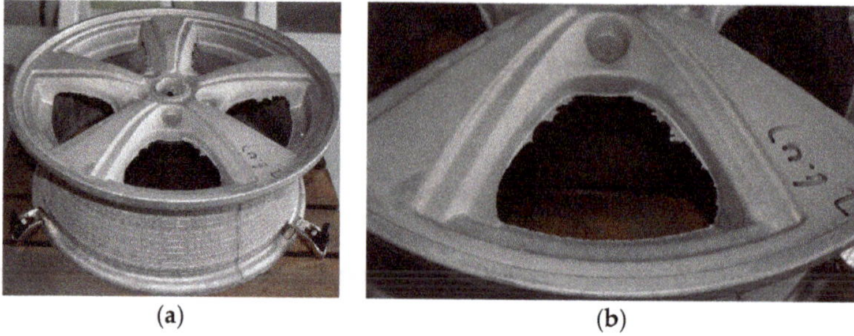

Figure 11. Casting blank of automobile hub: (**a**) overall appearance of automobile hub; (**b**) disc burrs of automobile hub.

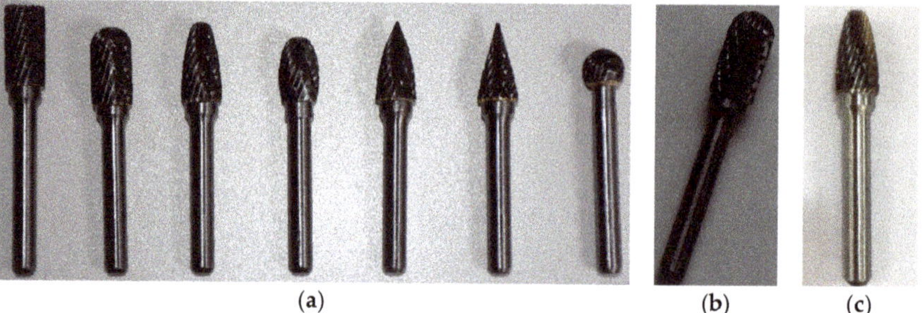

Figure 12. Carbide rotary tools: (**a**) series of tools; (**b**) cylindrical ball nose tool; (**c**) arch ball nose tool.

Based on the proposed robotic deburring tool path planning method, the detailed robotic deburring tool location (position and orientation) planning and the detailed robotic layered deburring planning are presented as follows. The selected tool can deburr once the entire thickness of burrs due to the length of the side cutting edge is significantly larger than the thickness of burrs with range between 0.37 mm and 0.72 mm. Thus, the disc burrs of automobile hub can be removed by executing robotic layered deburring several times through the proposed robotic deburring tool path planning method.

Disc burrs vary in size and exhibit a discontinuous distribution as can be seen in Figure 11b. Through the actual measurement, the maximum width dimension, i.e., the maximum machining allowance $D_{\max}$ was 7.36 mm. In the first experiment, the thickness of the single-layer deburring, i.e., the cutting depth d_p was set as about one third of the diameter of the side cutting edge, which was taken as 3 mm. The entire allowance can be deburred completely by repeating three times. Among them, the thickness of the last deburring layer was selected as 1.36 mm to improve the deburring quality.

Furthermore, a total of eighteen discrete points were selected on the entire target curve at the edge of disc body, as shown in Figure 13. The position of each contact point of the layered deburring tool is planned by the Equations (1) and (2) along the exterior normal $\mathbf{n}_l$ (red arrows are showed in Figure 14a) of each discrete point, as shown in Figure 14a. As mentioned above, disc burrs of the automobile hub were suitable for deburring with the tool side cutting edge and the contact orientations of the layered deburring tool were planned as shown in Figure 14b. The details were: (a) the deburring direction of tool, i.e., the tool side edge vector $\mathbf{n}$ was taken as the unit inner normal vector $\mathbf{f}$ of each discrete position (i.e., cyan arrows, as shown in Figure 14b, and they were in the opposite direction of the red arrows as shown in Figure 14a); (b) the movement direction of tool (vector $\mathbf{o}$) was planned along the unit tangent vectors $\boldsymbol{\tau}$ of each discrete position (i.e., green arrows, as shown in Figure 14b) and points to the next discrete position to be deburred; and (c) the tool vector $\mathbf{a}$ was derived as $\mathbf{a} = \mathbf{n} \times \mathbf{o}$. Finally, the disc deburring of the automobile hub was conducted in the clockwise direction based on the planned tool contact positions and tool contact orientations as shown in Figure 14 (symbols ①, ② and ③ represent the sequence of each layered deburring).

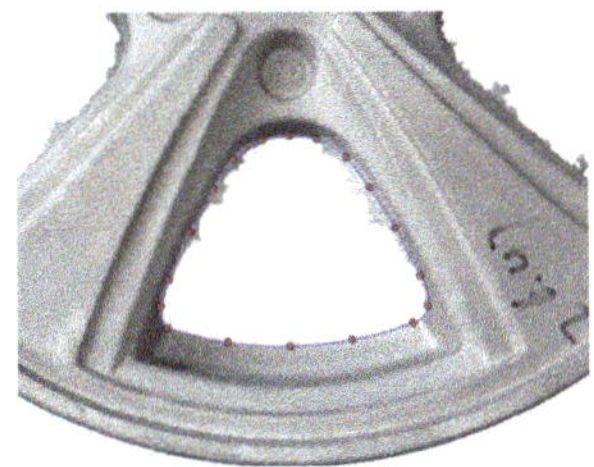

Figure 13. Target curve of disc and discrete points.

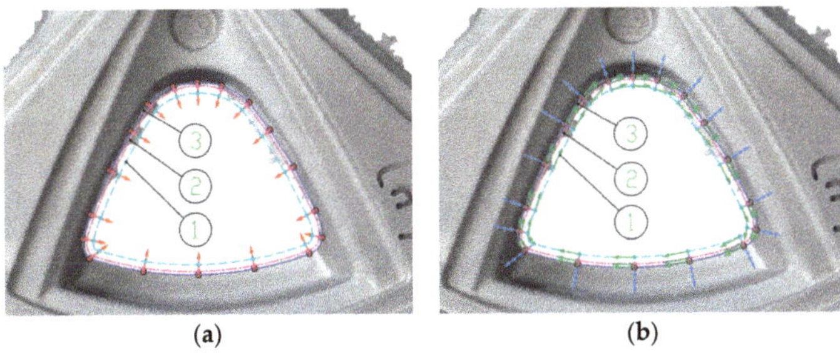

(a)   (b)

Figure 14. Tool locations planning of the layered deburring: (**a**) tool contact positions; (**b**) tool contact orientations.

Additionally, the proposed robotic deburring process parameter control method was applied to the robotic deburring of the first experiment in order to adjust robotic deburring process parameters with a constant cutting speed and a constant cutting force. In this experiment, the robotic spindle speed was selected to be 10,000 rpm, and the line speed of the robotic feed was set to 30 mm/s. Hence, the desired robotic spindle speed and the desired robotic spindle load were 10,000 rpm and 0.17N·m. Variation ranges E_1, EC_1, E_2 and EC_2 (corresponding to $[-S_e, S_e]$, $[-S_{ec}, S_{ec}]$, $[-L_e, L_e]$ and $[-L_{ec}, L_{ec}]$, respectively) were set to $[-3000, 3000]$, $[-800, 800]$, $[-0.05, 0.05]$ and $[-0.02, 0.02]$, respectively. Variation ranges of output variables, i.e., change in control voltage of the robotic spindle and change in robotic feed (corresponding to $[-V_u, V_u]$ and $[-F_u, F_u]$, respectively), were set to values such that $[-0.3, 0.3]$ and $[-3, 3]$, respectively. From (3)–(8), quantizers of E_1, EC_1, E_2 and EC_2, and scaling factors of u_1 and u_2, were obtained as $K_{Se} = 0.002$, $K_{Sec} = 0.0075$, $K_{Le} = 60$, and $K_{Lec} = 150$, and $K_{Vu} = 0.05$ and $K_{Fu} = 1$, respectively. Furthermore, adjustment factors α_1 and α_2 were set to $\alpha_1 \in [0.3, 0.7]$ and $\alpha_2 \in [0.2, 0.8]$, respectively. Hence, the fuzzy rules in the designed fuzzy controller for robotic spindle speed and robotic spindle load are expressed respectively as follows:

$$\begin{cases} u_1 = -\langle \alpha_1 E_1 + (1 - \alpha_1)EC_1 \rangle \\ \alpha_1 = \frac{1}{6}(0.7 - 0.3)|E_1| + 0.3 \end{cases} \tag{14}$$

$$\begin{cases} u_2 = -\langle \alpha_2 E_2 + (1 - \alpha_2)EC_2 \rangle \\ \alpha_2 = \frac{1}{3}(0.8 - 0.2)|E_2| + 0.2 \end{cases} \tag{15}$$

The experimental platform and the automobile hub workpiece of the first robotic experimental deburring are shown in Figure 15. The detailed robotic experimental deburring for disc burrs of the automobile hub on the original experimental platform is shown in Figure 16. Results of layered experimental deburring disc of the automobile hub workpiece are shown in Figure 17. Finally, the results of the target planning path and actual tool path of experimental deburring for the automobile hub workpiece are shown in Figure 18 (here, the blue line and the red line are the target planning path and the actual tool path, respectively). Deviation results of target planning path and actual tool path of experimental deburring for disc of hub are shown in Figure 19, the vertical axis indicates the magnitude of the deviation, and other two horizontal axes in the horizontal plane indicate the corresponding positions of the target planning path in the deburring experiment. In addition, a figure in the form of plane polar coordinates illustrating deviation results of this experiment deburring is shown in Figure 20. The maximum deviation was 1.23 mm, and these robotic deburring results can meet the experimental deburring requirements.

The effectiveness of the proposed robotic deburring tool path planning method and the proposed robotic deburring process parameter control method were verified in the first deburring experiment. Also, it can be seen that the robotic deburring orientations were adjusted dexterously, especially in the place where the local curvature changes greatly, and the dexterous deburring ability of the robot manipulator was fully demonstrated in the first deburring experiment.

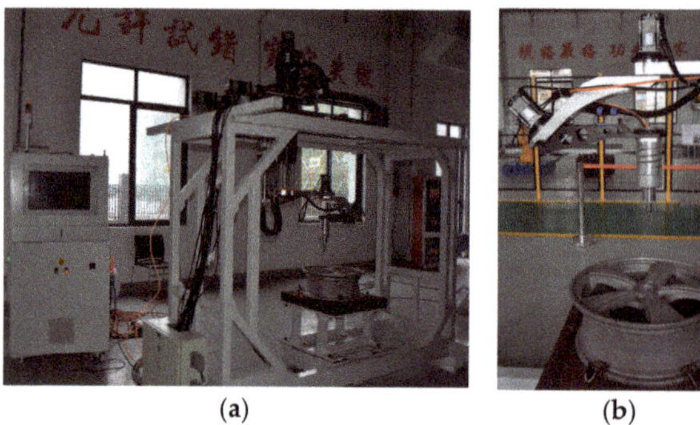

(a) **(b)**

Figure 15. Experimental platform and automobile hub: (**a**) original experimental platform of robot manipulator; (**b**) automobile hub workpiece.

Figure 16. Experimental deburring disc of hub on original experimental platform.

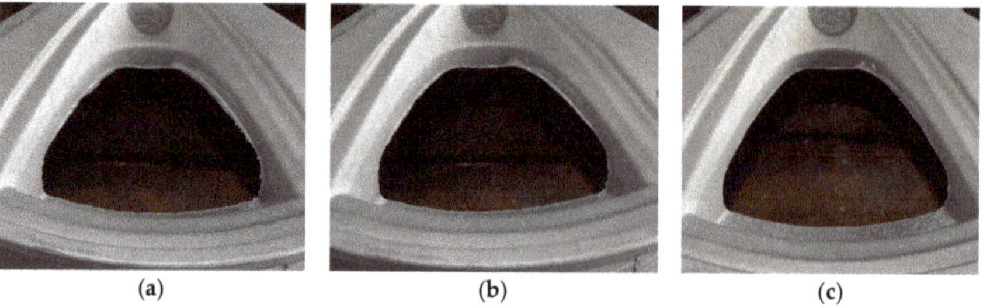

(a) **(b)** **(c)**

Figure 17. Results of layered experimental deburring disc of hub: (**a**) first layered deburring; (**b**) second layered deburring; (**c**) third layered deburring.

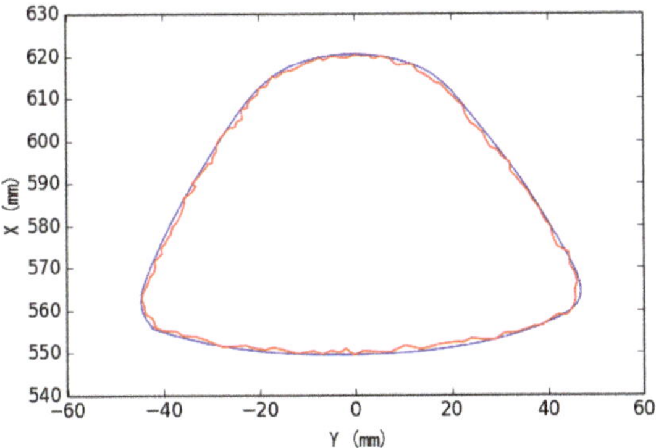

Figure 18. Results of target planning path and actual tool path of experimental deburring.

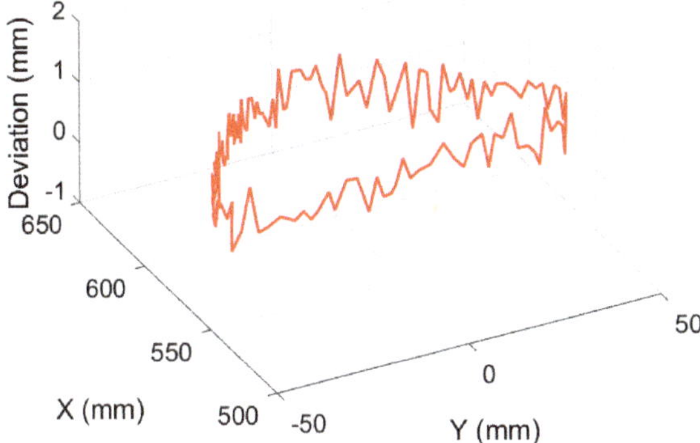

Figure 19. Deviation results of target planning path and actual tool path of experimental deburring for disc of hub.

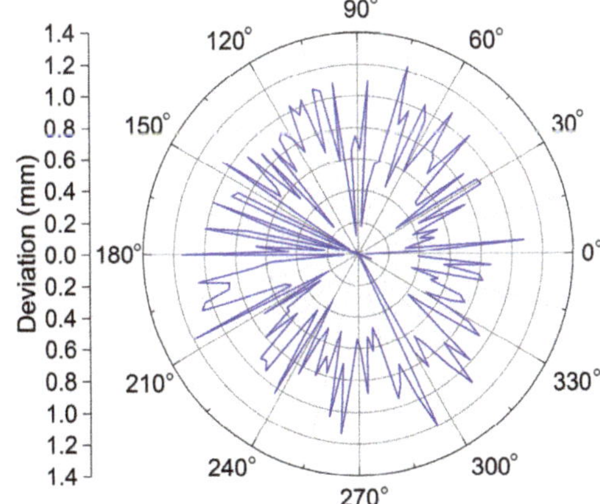

Figure 20. Deviation results of target planning path and actual tool path of experimental deburring for disc of hub (polar representation).

6.2. Multifaceted Edges Robotic Deburring Experiment

In the second robotic deburring experiment, multifaceted edges deburring for an automobile steering booster housing were conducted on the improved experimental platform of the robot manipulator (shown in Figures 2b and 8a). The experimental object was an aluminum alloy casting automobile steering booster housing (shown in Figure 8b), and the experimental deburring object were multifaceted edges of the automobile steering booster housing, as shown in Figure 21, i.e., orifice edges (blue lines are showed in Figure 21) and facet edges (red lines are showed in Figure 20) on top facet, distal side facet and proximal side facet relative to the initial position of the robot manipulator, as shown in Figure 21a,b,c, respectively.

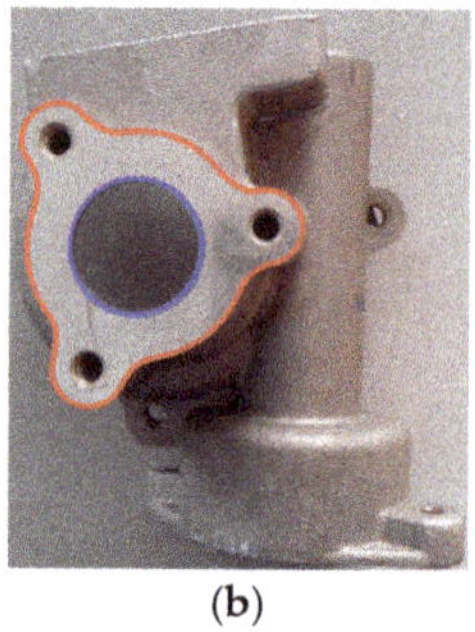

(a)   (b)   (c)

Figure 21. Experimental deburring of multifaceted edges for automobile steering booster housing: (**a**) top facet; (**b**) distal side facet; (**c**) proximal side facet.

A double-cutting arch round-head carbide rotary tool with the side cutting edge (shown in Figure 12c and its type—FX1020M06) was selected to conduct the deburring for multifaceted edges of automobile steering booster housing in the second experiment. In the multifaceted edges deburring experiment, the planned movement directions of the robotic deburring tool were clockwise and counterclockwise for orifice edges deburring and facet edges deburring, respectively.

Since burrs on multifaceted edges are very small, the entire allowance of each facet edge burrs could be deburred completely only once. Similarly, the proposed robotic deburring tool path planning method and the proposed robotic deburring process parameter control method were applied as in the first deburring experiment. Among them, the robotic spindle speeds for orifice edges deburring and facet edges deburring are selected to be 10,000 rpm and 8000 rpm, respectively; and the line speeds of the robotic feed for top facet edges deburring and other edges deburring were set to 30 mm/s and 20 mm/s, respectively. Note that these selected values for robotic spindle speeds and robotic feed speeds are not guaranteed to be very suitable as they are selected only according to limited past experiences. As mentioned above, the most appropriate way of selecting extremely suitable robotic deburring feed speed and tool spindle speed for specific deburring workpiece needs to refer to some related research issues for the technical details and be verified by a series of deburring experiments.

The detailed robotic experimental deburring for orifice edges and facet edges on the top facet, distal side facet and proximal side facet are shown in Figures 22 and 23, Figures 24 and 25, and Figures 26 and 27, respectively. Finally, experimental deburring results for multifaceted edges of the automobile steering booster housing are shown in Figure 28. And robotic experimental deburring results of target planning path and actual tool path of top facet, distal side facet and proximal side facet are shown in Figures 29–31, respectively (here, the blue line and the red line are the target planning path and the actual tool path, respectively). Deviation results of target planning path and actual tool path of robotic experimental deburring for top facet, distal side facet and proximal side facet are shown in Figures 32–34, respectively. In each deviation figure, the vertical axis indicates the magnitude of the deviation, and other two horizontal axes in the horizontal plane indicated the corresponding positions of the target planning path in the deburring experiment (here, the blue line and the red line are the facet edges deburring deviation results and the orifice edges deburring deviation results, respectively). In addition, three figures in the form of plane polar coordinates illustrating deviation results of this experiment deburring for top facet, distal side facet and proximal side facet are shown in Figures 35–37, respectively. Among them, the maximum path deviations of top facet edge, distal side facet edge and proximal side facet edge were 0.97 mm, 1.13 mm and 1.21 mm, respectively. These robotic deburring results can satisfy the experimental deburring requirements.

It can be showed that the effectiveness of the proposed robotic deburring tool path planning method and the proposed robotic deburring process parameter control method were also verified in the second deburring experiment. Furthermore, the highly efficient and dexterous manipulation and deburring capacity of the robot manipulator for multifaceted deburring in one setup was totally demonstrated in the second deburring experiment. In addition, it should be noted that the proposed methods can be now only applied to soft material machining applications and low machining requirements due to the rigidity defect of the robot manipulator and lacking compensation for vibrations and/or chattering, although it had a very high level of dexterous manipulation and orientation reachability. There are still many meaningful research issues need to be conducted in the next step in order to improve the path accuracy of the robot manipulator, such as offline path correction, compensation for vibrations and/or chattering, high frequency oscillation suppression, structural rigidity improvement, calibration of the robot manipulator for dealing with nonnegligible dynamic effects which are caused by backlash of ball screws and robot manipulator structural deformations.

It is necessary to note that, when the end-effector tool of the robot manipulator is changed, like an abrasive belt or a fabric wheel, the robot manipulator can also conduct deburring, grinding and polishing for edges and surfaces of castings or other materials, such as parting line burrs, flash burrs, pouring risers burrs, and so forth.

Figure 22. Experimental deburring of orifice edges on top facet.

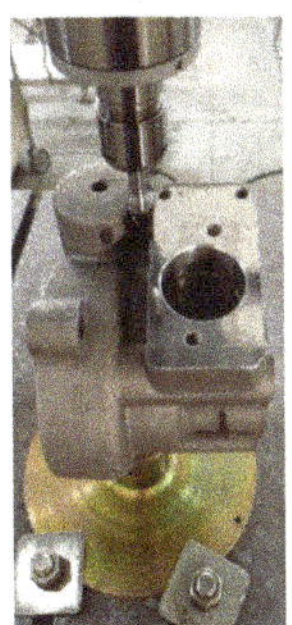

Figure 23. Experimental deburring of facet edges on top facet.

Figure 24. Experimental deburring of orifice edges on distal side facet.

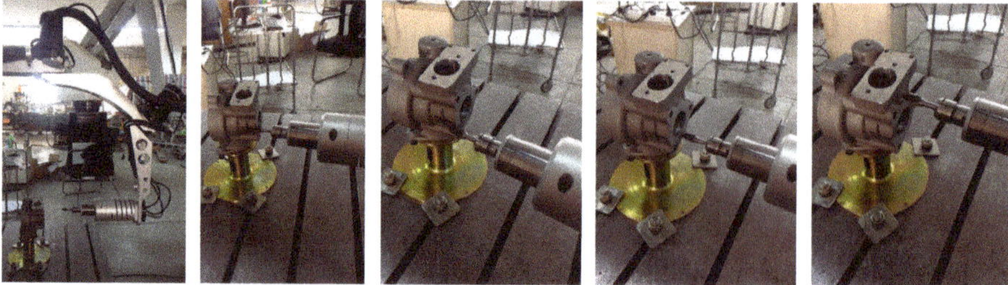

Figure 25. Experimental deburring of facet edges on distal side facet.

Figure 26. Experimental deburring of orifice edges on proximal side facet.

Figure 27. Experimental deburring of facet edges on proximal side facet.

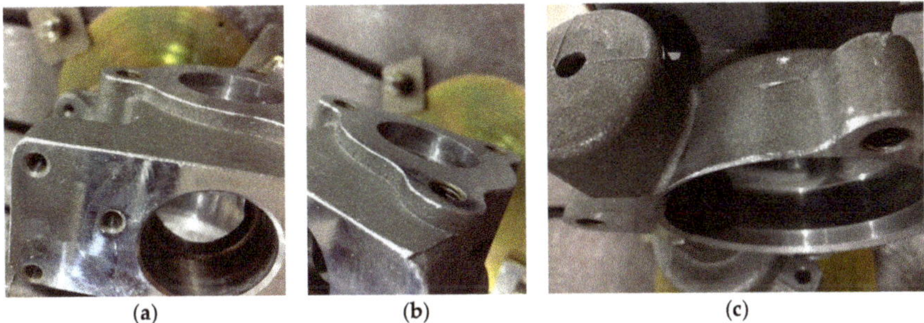

(a)   (b)   (c)

Figure 28. Experimental deburring results for multifaceted edges of automobile steering booster housing: (**a**) top facet; (**b**) distal side facet; (**c**) proximal side facet.

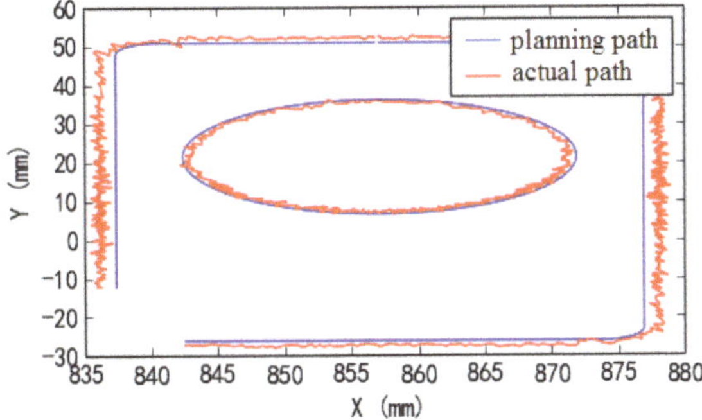

Figure 29. Experimental deburring results of target planning path and actual tool path of top facet.

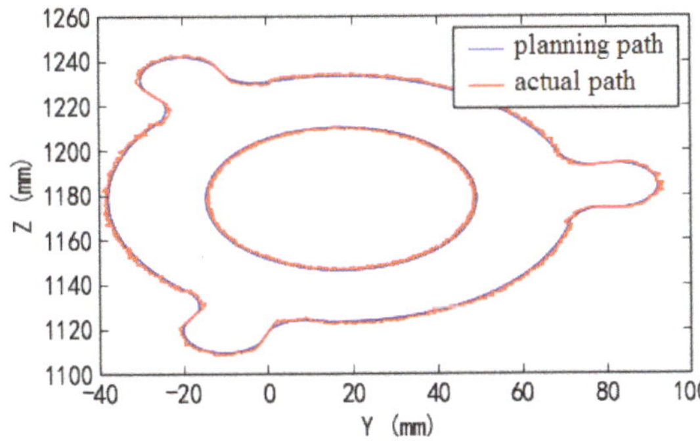

Figure 30. Experimental deburring results of target planning path and actual tool path of distal side facet.

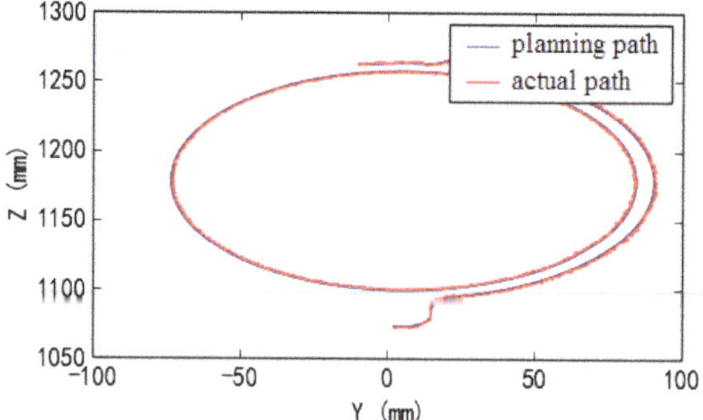

Figure 31. Experimental deburring results of target planning path and actual tool path of proximal side facet.

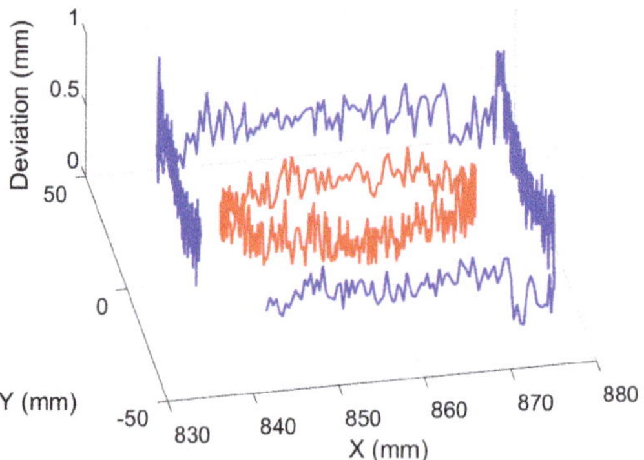

Figure 32. Deviation results of target planning path and actual tool path of experimental deburring for top facet.

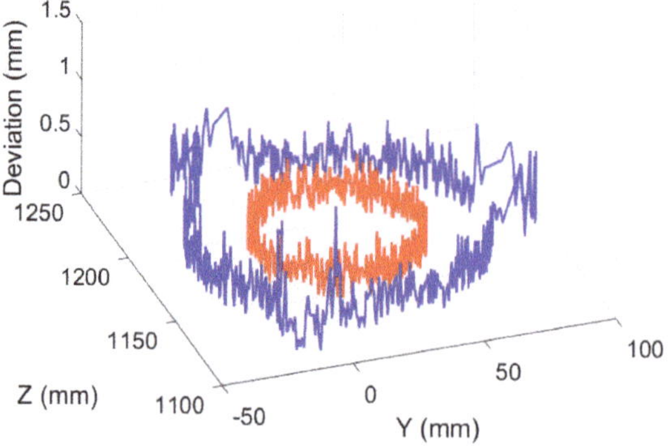

Figure 33. Deviation results of target planning path and actual tool path of experimental deburring for distal side facet.

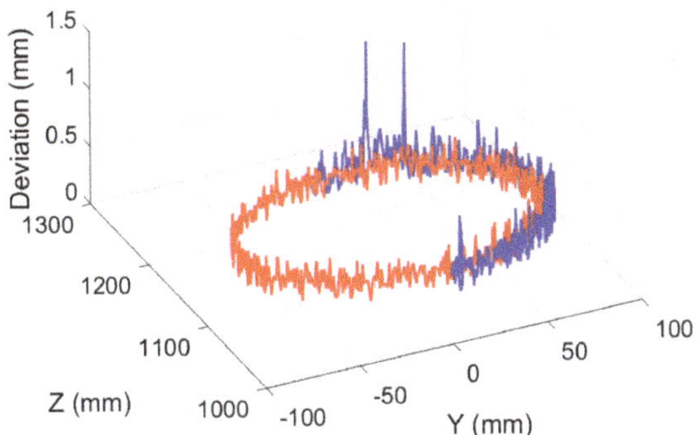

Figure 34. Deviation results of target planning path and actual tool path of experimental deburring for proximal side facet.

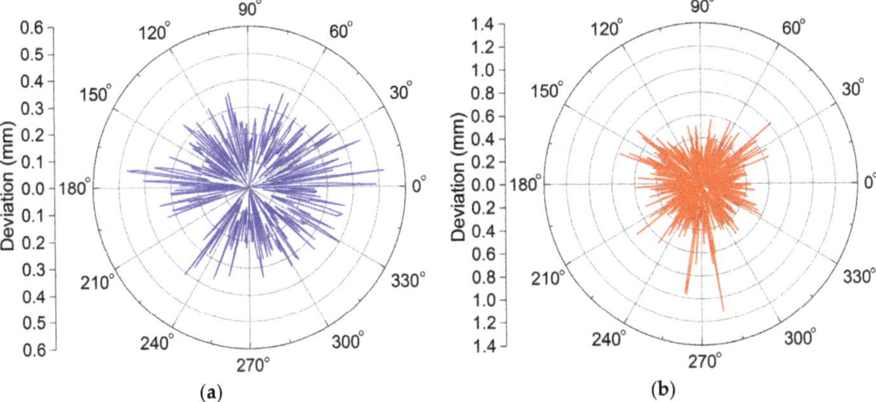

Figure 35. Deviation results of target planning path and actual tool path of experimental deburring for top facet (polar representation): (**a**) facet edges deburring deviation; (**b**) orifice edges deburring deviation.

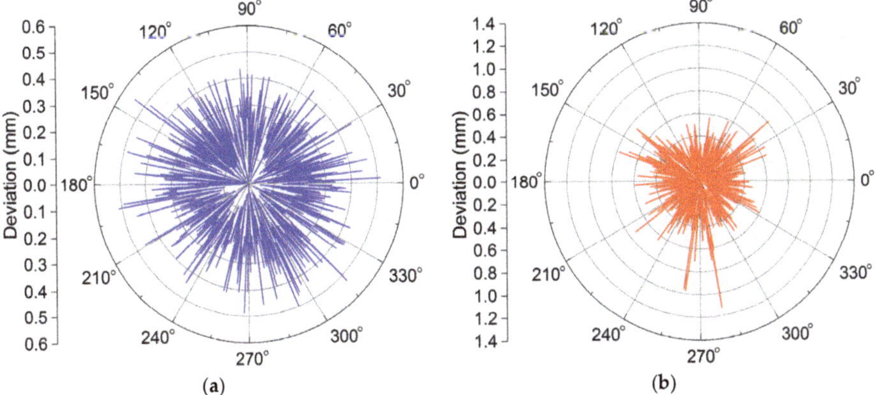

Figure 36. Deviation results of target planning path and actual tool path of experimental deburring for distal side facet (polar representation): (**a**) facet edges deburring deviation; (**b**) orifice edges deburring deviation.

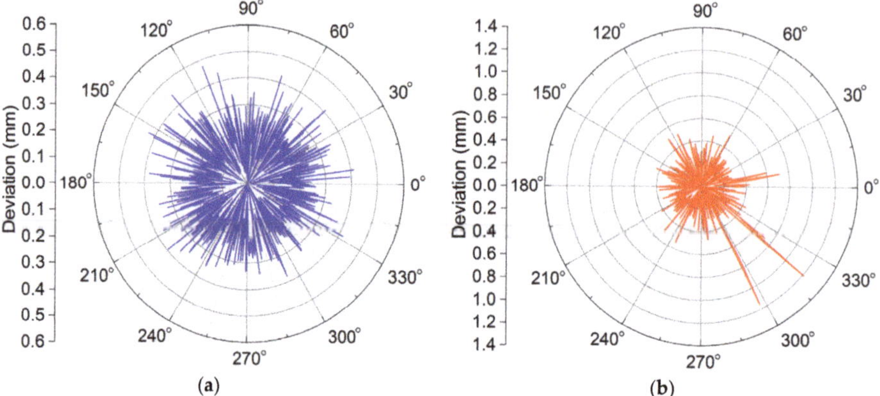

Figure 37. Deviation results of target planning path and actual tool path of experimental deburring for proximal side facet (polar representation): (**a**) facet edges deburring deviation; (**b**) orifice edges deburring deviation.

7. Conclusions

A new five-DOF hybrid robot manipulator and two experimental platforms were developed. For this robot manipulator, a new robotic deburring methodology for robotic deburring tool path planning and robotic deburring process parameter control was presented. The major contributions of this article can be summarized as follows:

1. The design structure and the physical structure of the robot manipulator for the original experimental platform and the improved one were developed, and the property characterizations of the dexterous manipulation reachability and the superior deburring capability in one setup were demonstrated. The robot manipulator can provide a good solution for the selection of the manufacturing equipment, such as deburring of the complex shaped parts.
2. A robotic deburring tool path planning method was proposed for the robotic deburring tool location (position and orientation) planning and the robotic layered deburring planning. Also, a robotic deburring process parameter control method based on fuzzy control for the robotic deburring was proposed, which represents the optimization property, convenience, simplicity, and implementation with automatic-online errors correction. These methods can be extended to handle similar problems for other types of the robot manipulators.
3. A dexterous manipulation experiment was conducted to verify the dexterous manipulation and the orientation reachability of the robot manipulator, especially the dexterous manipulation of a certain processing point. Furthermore, two robotic deburring experiments, i.e., a disc deburring experiment of an automobile hub and a multifaceted edges deburring experiment of an automobile steering booster housing, were conducted on the experimental platforms, and the effectiveness of the two proposed methods was verified, and the highly efficient and dexterous manipulation and deburring capacity of the robot manipulator for multifaceted deburring in one setup was also fully demonstrated.

In future work, several meaningful attempts need to be carried out, such as offline path correction, compensation for vibrations and/or chattering, backlash of ball screws and robot manipulator structural deformations, suppression for high frequency oscillations, and structural rigidity improvement of the robot manipulator. The two proposed methods are expected to provide some insight into the foundational aspects of robotic deburring tool path planning and robotic deburring process parameter control.

Author Contributions: All authors contributed to the research work and have read and approved the final manuscript.

Funding: This research was funded by Natural Science Basic Research Plan in Shaanxi Province of China (No. 2019JQ-426), Fundamental Research Funds for the Central Universities (No. 300102258107, 300102259308, 300102259401, 300102258402, 300102258305, 300103190365, 300102258205), Shaanxi Science and Technology Co-ordinated Innovation Project (No. 2016KTZDGY-02-03), Shaanxi International Science and Technology Cooperation Project (No. 2019KW-015) and Xi'an Science and Technology Project for Talented Personnel Service Enterprise in Colleges and Universities (No. 2017088CG/RC051(CADX001)).

Acknowledgments: Thanks to Mingrui Lv, Yafu Tian, Xiangjin Bu, Lianzheng Ge, Chongyang Wu, Chuqing Cao and Xunwei Tong for their help in the discussion and writing of the paper.

Conflicts of Interest: The authors declare no conflict of interest.

References

1. Iglesias, I.; Ares, J.E.; Gonzalez-Gaya, C.; Morales, F.; Rosales, V.F. Predictive Methodology for Dimensional Path Precision in Robotic Machining Operations. *IEEE Access* **2018**, *6*, 49217–49223. [CrossRef]

2. Roveda, L.; Pedrocchi, N.; Vicentini, F.; Tosatti, L.M. Industrial compliant robot bases in interaction tasks: A force tracking algorithm with coupled dynamics compensation Operations. *Robotica* **2017**, *35*, 1732–1746. [CrossRef]

3. Robla-Gomez, S.; Becerra, V.M.; Llata, J.R.; Gonzalez-Sarabia, E.; Torre-Ferrero, C.; Perez-Oria, J. Working Together A Review on Safe Human-Robot Collaboration in Industrial Environments. *IEEE Access* **2017**, *5*, 26754–26773. [CrossRef]

4. Stipancic, T.; Jerbic, B.; Curkovic, P. A context-aware approach in realization of socially intelligent industrial robots. *IEEE Access* **2016**, *37*, 79–89. [CrossRef]

5. Heydaryan, S.; Bedolla, J.S.; Belingardi, G. Safety Design and Development of a Human-Robot Collaboration Assembly Process in the Automotive Industry. *Appl. Sci.* **2018**, *8*, 344. [CrossRef]

6. Ruppert, T.; Jasko, S.; Holczinger, T.; Abonyi, J. Enabling Technologies for Operator 4.0: A Survey. *Appl. Sci.* **2018**, *8*, 1650. [CrossRef]

7. Chen, Y.; Dong, F. Robot machining: Recent development and future research issues. *Int. J. Adv. Manuf. Technol.* **2013**, *66*, 1489–1497. [CrossRef]

8. Lai, C.Y.; Chavez, D.E.V.; Ding, S. Transformable parallel-serial manipulator for robotic machining. *Int. J. Adv. Manuf. Technol.* **2018**, *97*, 2987–2996. [CrossRef]

9. Barnfather, J.D.; Abram, T. Efficient compensation of dimensional errors in robotic machining using imperfect point cloud part inspection data. *Measurement* **2018**, *117*, 176–185. [CrossRef]

10. Mousavi, S.; Gagnol, V.; Bouzgarrou, B.C.; Ray, P. Dynamic modeling and stability prediction in robotic machining. *Int. J. Adv. Manuf. Technol.* **2017**, *88*, 3053–3065. [CrossRef]

11. Barnfather, J.D.; Goodfellow, M.J.; Abram, T. A performance evaluation methodology for robotic machine tools used in large volume manufacturing. *Robot. Comput.-Integr Manuf.* **2016**, *37*, 49–56. [CrossRef]

12. Sabourin, L.; Subrin, K.; Cousturier, R.; Gogu, G.; Mezouar, Y. Redundancy-based optimization approach to optimize robotic cell behaviour: Application to robotic machining. *Ind. Robot Int. J.* **2015**, *42*, 167–178. [CrossRef]

13. Caesarendra, W.; Pappachan, B.K.; Wijaya, T.; Lee, D.; Tjahjowidodo, T.; Then, D.; Manyar, O.M. An AWS Machine Learning-Based Indirect Monitoring Method for Deburring in Aerospace Industries Towards Industry 4.0. *Appl. Sci.* **2018**, *8*, 2165. [CrossRef]

14. Pandiyan, V.; Caesarendra, W.; Tjahjowidodo, T.; Praveen, G. Predictive Modelling and Analysis of Process Parameters on Material Removal Characteristics in Abrasive Belt Grinding Process. *Appl. Sci.* **2017**, *7*, 363. [CrossRef]

15. Xie, F.; Liu, X.-J.; Wang, C. Design of a novel 3-DoF parallel kinematic mechanism: Type synthesis and kinematic optimization. *Robotica* **2015**, *33*, 622–637. [CrossRef]

16. Cheng, Y.-M.; Peng, W.-X.; Hsu, A.-C. Concentric hole drilling in multiple planes for experimental investigation of five-axis reconfigurable precision hybrid machine. *Int. J. Adv. Manuf. Technol.* **2015**, *76*, 1253–1262.

17. Sangveraphunsiri, V.; Chooprasird, K. Dynamics and control of a 5-DOF manipulator based on an H-4 parallel mechanism. *Int. J. Adv. Manuf. Technol.* **2011**, *52*, 343–364. [CrossRef]

18. Altuzarra, O.; Martín, Y.S.; Amezua, E.; Hernández, A. Motion pattern analysis of parallel kinematic machines: A case study. *Robot. Comput.-Integr. Manuf.* **2009**, *25*, 432–440. [CrossRef]

19. Wang, L.; Wu, J.; Li, T.; Wang, J.; Gao, G. A study on the dynamic characteristics of the 2-dof redundant parallel manipulator of a hybrid machine tool. *Int. J. Robot. Autom.* **2015**, *30*, 184–191. [CrossRef]

20. Wang, L.; Wu, J.; Wang, J.; You, Z. An experimental study of a redundantly actuated parallel manipulator for a 5-DOF hybrid machine tool. *IEEE/ASME Trans. Mech.* **2009**, *14*, 72–81. [CrossRef]

21. Li, Q.; Wu, W.; Li, H.; Wu, C. A hybrid robot for friction stir welding. *Proc. Inst. Mech. Eng. Part C J. Mech. Eng. Sci.* **2015**, *229*, 2639–2650. [CrossRef]

22. Zoppi, M.; Zlatanov, D.; Molfino, R. Kinematics analysis of the Exechon tripod. In Proceedings of the ASME 2010 International Design Engineering Technical Conferences and Computers and Information in Engineering Conference, Montreal, QC, Canada, 15–18 August 2010; pp. 1381–1388.

23. Niknam, S.A.; Davoodi, B.; Paulo Davim, J.; Songmene, V. Mechanical deburring and edge-finishing processes for aluminum parts—A review. *Int. J. Adv. Manuf. Technol.* **2018**, *95*, 1101–1125. [CrossRef]

24. Niknam, S.A.; Songmene, V. Milling burr formation, modeling and control: A review. *Proc. Inst. Mech. Eng. Part B J. Eng. Manuf.* **2015**, *229*, 893–909. [CrossRef]

25. Burghardt, A.; Szybicki, D.; Kurc, K.; Muszynska, M.; Mucha, J. Experimental Study of Inconel 718 Surface Treatment by Edge Robotic Deburring with Force Control. *Strength Mater.* **2017**, *49*, 594–604. [CrossRef]

26. Villagrossi, E.; Cenati, C.; Pedrocchi, N.; Beschi, M.; Tosatti, L.M. Flexible robot-based cast iron deburring cell for small batch production using single-point laser sensor. *Int. J. Adv. Manuf. Technol.* **2017**, *92*, 1425–1438. [CrossRef]

27. Kosler, H.; Pavlovčič, U.; Jezeršek, M.; Možina, J. Adaptive Robotic Deburring of Die-Cast Parts with Position and Orientation Measurements Using a 3D Laser-Triangulation Sensor. *Strojniški Vestnik-J. Mech. Eng.* **2016**, *62*, 207–212. [CrossRef]

28. Caggiano, A.; Marzano, A.; Teti, R. Sustainability Enhancement of a Turbine Vane Manufacturing Cell through Digital Simulation-Based Design. *Energies* **2016**, *9*, 790. [CrossRef]

29. Song, H.-C.; Song, J.-B. Precision robotic deburring based on force control for arbitrarily shaped workpiece using CAD model matching. *Int. J. Precis. Eng. Manuf.* **2013**, *14*, 85–91. [CrossRef]

30. Villagrossi, E.; Pedrocchi, N.; Beschi, M.; Tosatti, L.M. A human mimicking control strategy for robotic deburring of hard materials. *Int. J. Comput. Integr. Manuf.* **2018**, *31*, 869–880. [CrossRef]

31. Gracia, L.; Ernesto Solanes, J.; Munoz-Benavent, P.; Miro, J.V.; Perez-Vidal, C.; Tornero, J. Adaptive Sliding Mode Control for Robotic Surface Treatment Using Force Feedback. *Measurement* **2018**, *52*, 102–118. [CrossRef]

32. Ernesto Solanes, J.; Gracia, L.; Munoz-Benavent, P.; Esparza, A.; Miro, J.V.; Tornero, J. Adaptive robust control and admittance control for contact-driven robotic surface conditioning. *Robot. Comput.-Integr. Manuf.* **2018**, *54*, 115–132. [CrossRef]

33. Pillai, J.U.; Sanghrajka, I.; Shunmugavel, M.; Muthuramalingam, T.; Goldberg, M.; Littlefair, G. Optimisation of multiple response characteristics on end milling of aluminium alloy using Taguchi-Grey relational approach. *Measurement* **2018**, *124*, 291–298. [CrossRef]

34. Tao, Y.; Zheng, J.; Lin, Y. A Sliding Mode Control-based on a RBF Neural Network for Deburring Industry Robotic Systems. *Int. J. Adv. Robot. Syst.* **2016**, *13*, 1–10. [CrossRef]

35. Tao, Y.; Zheng, J.; Lin, Y.; Wang, T.; Xiong, H.; He, G.; Xu, D. Fuzzy PID control method of deburring industrial robots. *J. Intell. Fuzzy Syst.* **2015**, *29*, 2447–2455. [CrossRef]

36. Chen, C.-Y.; Shieh, S.-S.; Cheng, M.-Y.; Su, K.-H. Vision-based Pythagorean hodograph spline command generation and adaptive disturbance compensation for planar contour tracking. *Int. J. Adv. Manuf. Technol.* **2013**, *65*, 1185–1199. [CrossRef]

37. Aggogeri, F.; Borboni, A.; Faglia, R.; Merlo, A.; Pellegrini, N. A kinematic model to compensate the structural deformations in machine tools using fiber Bragg grating (FBG) sensors. *Appl. Sci.* **2017**, *7*, 114. [CrossRef]

38. Borboni, A.; Faglia, R. *Parasitic Phenomena in the Dynamics of Industrial Devices*; CRC Press: Boca Raton, FL, USA, 2011.

39. Aggogeri, F.; Borboni, A.; Merlo, A.; Pellegrini, N.; Ricatto, R. Vibration Damping Analysis of Lightweight Structures in Machine Tools. *Materials* **2017**, *10*, 297. [CrossRef]

40. Wang, M.; Zan, T.; Gao, X.; Li, S. Suppression of the time-varying vibration of ball screws induced from the continuous movement of the nut using multiple tuned mass dampers. *Int. J. Mach. Tools Manuf.* **2016**, *107*, 41–49. [CrossRef]

41. Guo, W.; Li, R.; Cao, C.; Gao, Y. Kinematics, dynamics, and control system of a new 5-degree-of-freedom hybrid robot manipulator. *Adv. Mech. Eng.* **2016**, *8*, 1687814016680309. [CrossRef]
42. Guo, W.; Li, R.; Cao, C.; Gao, Y. Kinematics Analysis of a Novel 5-DOF Hybrid Manipulator. *Int. J. Autom. Technol.* **2015**, *9*, 765–774. [CrossRef]
43. Guo, W.; Li, R.; Cao, C.; Gao, Y. A Novel Method of Dexterity Analysis for a 5-DOF Manipulator. *J. Robot.* **2016**, *2016*, 8901820. [CrossRef]
44. Guo, W.; Li, R.; Cao, C.; Tong, X.; Gao, Y. A New Methodology for Solving Trajectory Planning and Dynamic Load-Carrying Capacity of a Robot Manipulator. *Math. Prob. Eng.* **2016**, *2016*, 1302537. [CrossRef]
45. Kalpakjian, S.; Schmid, S.R. *Manufacturing Engineering and Technology*; Prentice Hall: Upper Saddle River, NJ, USA, 2000.
46. Albertelli, P.; Keshari, A.; Matta, A. Energy oriented multi cutting parameter optimization in face milling. *J. Clean. Prod.* **2016**, *137*, 1602–1618. [CrossRef]
47. Das, S.R.; Dhupal, D.; Kumar, A. Experimental investigation into machinability of hardened AISI 4140 steel using TiN coated ceramic tool. *Measurement* **2015**, *62*, 108–126. [CrossRef]
48. Maher, I.; Eltaib, M.E.H.; Sarhan, A.A.D.; El-Zahry, R.M. Investigation of the effect of machining parameters on the surface quality of machined brass (60/40) in CNC end milling-ANFIS modeling. *Int. J. Adv. Manuf. Technol.* **2014**, *74*, 531–537. [CrossRef]
49. Ko, P.-J.; Tsai, M.-C. H-infinity Control Design of PID-Like Controller for Speed Drive Systems. *IEEE Access* **2018**, *6*, 36711–36722. [CrossRef]
50. Szczecinski, N.S.; Hunt, A.J.; Quinn, R.D. Design process and tools for dynamic neuromechanical models and robot controllers. *Biol. Cybern.* **2017**, *111*, 105–127. [CrossRef]
51. Amezquita-Brooks, L.; Liceaga-Castro, J.; Liceaga-Castro, E. Speed and Position Controllers Using Indirect Field-Oriented Control: A Classical Control Approach. *IEEE Trans. Ind. Electron.* **2014**, *61*, 1928–1943. [CrossRef]
52. Ajwad, S.A.; Iqbal, J.; Ul Islam, R.; Alsheikhy, A.; Almeshal, A.; Mehmood, A. Optimal and Robust Control of Multi DOF Robotic Manipulator: Design and Hardware Realization. *Cybern. Syst.* **2018**, *49*, 77–93. [CrossRef]
53. Reiner, M.J.; Zimmer, D. Object-oriented modelling of wind turbines and its application for control design based on nonlinear dynamic inversion. *Math. Comput. Model. Dyn. Syst.* **2017**, *23*, 319–340. [CrossRef]
54. Prieto-Araujo, E.; Egea-Alvarez, A.; Fekriasl, S.; Gomis-Bellmunt, O. Multi-Shaker Control A Review of the Evolving State-of-the-Art. *IEEE Trans. Power Deliv.* **2016**, *31*, 575–585. [CrossRef]
55. Al-Darraji, I.; Kilic, A.; Kapucu, S. Mechatronic design and genetic-algorithm-based MIMO fuzzy control of adjustable-stillness tendon-driven robot finger. *Mech. Sci.* **2018**, *9*, 277–296. [CrossRef]
56. Chen, Z.; Li, Z.; Chen, C.L.P. Disturbance Observer-Based Fuzzy Control of Uncertain MIMO Mechanical Systems with Input Nonlinearities and its Application to Robotic Exoskeleton. *IEEE Trans. Cybern.* **2017**, *47*, 984–994. [CrossRef]
57. Wang, W.-Y.; Chien, Y.-H.; Leu, Y.-G.; Hsu, C.-C. Mean-Based Fuzzy Control for a Class of MIMO Robotic Systems. *IEEE Trans. Fuzzy Syst.* **2016**, *24*, 966–980. [CrossRef]
58. Mendes, N.; Neto, P. Indirect adaptive fuzzy control for industrial robots: A solution for contact applications. *Expert Syst. Appl.* **2015**, *42*, 8929–8935. [CrossRef]

MDPI

St. Alban-Anlage 66

4052 Basel

Switzerland

Tel. +41 61 683 77 34

Fax +41 61 302 89 18

www.mdpi.com

Applied Sciences Editorial Office

E-mail: applsci@mdpi.com

www.mdpi.com/journal/applsci